Chromosomes to Genome

Chromosomes to Genome

Editor
Rakesh K. Mishra
Centre for Cellular and Molecular Biology
Hyderabad, India

I.K. International Publishing House Pvt. Ltd.
NEW DELHI • BANGALORE

Published by

I.K. International Publishing House Pvt. Ltd.
S-25, Green Park Extension
Uphaar Cinema Market
New Delhi - 110 016 (India)

ISBN: 978-93-80026-21-3

10 9 8 7 6 5 4 3 2 1

Published by Krishan Makhijani for I. K. International Publishing House Pvt. Ltd., S-25, Green Park Extension, Uphaar Cinema market, New Delhi-110 016 and printed by Rekha Printers Pvt. Ltd., Okhla Industrial Area, Phase II, New Delhi-110 020.

Foreword

As a young person in the 1930s and 1940s, I remember my father, a medical doctor, engaged in public health work–telling us often of his then intriguing observations that in villages affected by an epidemic of small pox or plague, there were some individuals who remained uninfected, in spite of close contact with the affected members of their family. This is no longer an enigma today. We recognize that all diseases have a hereditary basis, that is susceptibility or resistance to a particular disease, and the extent of such susceptibility or resistance, is built in our genome.

The journey from the discovery of DNA by Miescher in the 1860s and the naming of chromosomes–the structures that contain the bulk of cellular DNA in living organisms–in 1888, to the sequencing of genomes, and the understanding of the organization and functioning of the genome that makes all living things what they are, has been one of the most exciting journeys in the history of science. This book describes, lucidly and authoritatively, some of the most important manifestations of genomic structure and function in health and disease.

I remember the appearance on the rapidly changing biological scene, of the mysterious satellite band of DNA on ultracentrifugation, in the mid-1960s. Chris Tyler-Smith's narrative of both the history as well as the subsequent development of our understanding (at times, lack of understanding) of satellite DNA from the point of view of its structure and function, makes fascinating reading. He rightly implies that this chapter in the history of biology is still being written.

Similarly, the control of tissue-specific gene expression by fine tuning of chromosomal compacting process (Rakesh Mishra's paper) opens up a new area of investigation. What is the cell-type–specific packing code? A very valid and important question indeed. The three papers of Gunter Reuter, Rachel and Sanjeev Galande (with their colleagues), shed light on a part of the related problem of heterochromatin formation and chromatin organization that eventually leads to the mature chromosome and regulation of gene expression.

The significance of the fact that the total mammalian genome does not consist only of DNA and chromosomes in the nucleus is even today (more than 45 years after the discovery of mitochondrial DNA)–not fully realized. I remember the first mention of mitochondrial DNA that made most of us aware of its existence, was made at a Gordon Conference in 1964. Immediately after this meeting, I was invited for dinner at the residence of M.C. Niu in Philadelphia where I met Margit and Sylvan Nass – the husband and wife team that discovered mitochondrial DNA and published the discovery towards the end of 1963 in the *Journal of Cell Biology* as no more highly cited journal was willing to accept it. It was most appropriate to have included in the present volume the article by Thangaraj on mitochondrial genome in human health.

To understand biological phenomenon, one of the most important tools that biologists have is to look at what goes awry with them and how and why. Thus, to understand fully the implications of the human or animal genome, we need not only look at its normal function but also what goes wrong with even bits of it in disease. From this point of view papers of Peter Vogt, Jamshed Tata, S. Subramanian, Sen Pathak and G. R. Chandak, are most relevant.

Finally, a comment on the first paper of the volume – by Lalji Singh. It is yet another reminder of the role that lowly organisms such as *E.coli*, a nematode or a snake, have played in the development of modern biology. I remember when I was the Director of CCMB–the institute that Lalji directs now–Lalji coming to my room one day and giving me the draft of a paper. He wanted to know if the paper was worthy of publication.

When I went through the paper, I found it absolutely first-rate and, at Lalji's request, communicated it to Ekke Bautz whom I knew and who was one of the editors of *Molecular and General Genetics (MGG)*. In my covering letter, I wrote that implicit in the paper was the possibility of developing a new multilocus probe for human DNA fingerprinting. The paper was, of course accepted for publication in *MGG*, but I forgot all about the letter until six months later when Lalji came to me and said that he had actually not only developed but fully validated the use of his Bkm probe (derived from a deadly snake) for multilocus DNA fingerprinting in human beings. This was the basis of the DNA fingerprinting revolution in India which is one of the most widely known technologies in the country today.

It is, therefore, only appropriate that the present volume is a record of the papers presented at a symposium held at the occasion of Lalji's 60th birthday in which I had the privilege to participate. Even though India claims to have the third largest scientific man-power in the world and thus close to three million trained scientists and technologists, it is unfortunate that the number with high professional ability, integrity, honesty, courage and commitment that Lalji has shown in everything that he has done, is extremely few. One young person recently told me that he can count them on his fingertips. To me, therefore, this volume is not only a celebration of the advances that have taken place in modern biology in one of its most exciting areas, it is also a tribute to one of India's most celebrated scientists, and to the values that practice of the scientific method generates, that are the bedrock of all major scientific discoveries and their application for the largest benefit of the largest number amidst us.

Pushpa M. Bhargava
Member, National Security Advisory Board,
Government of India

Preface

Understanding the information content in chromosomes, that are the physical basis of inheritance, reached a milestone when whole genome sequences became available. This journey from chromosomes to genome passed through a series of discoveries that made what we know today as the field of *molecular biology*. These unprecedented technological advances have completely transformed the study on living systems and with that we are entering the new world of biology and healthcare. While, on the one hand, sophisticated cell biology approach allows us to examine intricate details of chromosome in the living cell, on the other hand, the molecular genetic approach empowers us to study genome to a single nucleotide resolution. Never has the pace, the scope and the social implications of biological research been of such magnitude.

An International Symposium – *Chromosomes to Genome* – was organised by the Centre for Cellular and Molecular Biology (CCMB), Hyderabad, India on July 3-5, 2007. The symposium also coincided with the 60th birthday of Dr. Lalji Singh, Director, CCMB to acknowledge his outstanding contributions to the field of chromosome biology and genomics. Lalji's early work pointed to functional significance of repetitive DNA that constitute much of our genome. Several presentations in this symposium brought forward the fact that repetitive elements are an important and essential component of complex genomes, although much remains to be understood about their function, which makes this field so exciting and active.

The speakers in the symposium represented leaders in their topics of research. It was felt that biology of chromosome and genome is at an extremely important stage. The symposium provided a platform to look back into the history, assess the current state and speculate the future of this fascinating field. The enthusiasm shown by the speakers made it necessary to publish this book.

The book presents complex topics in a simplified manner that is accessible to any student of biology. Each of the chapters provides a historical account of the field and reviews the recent work in this area, puts the subject in appropriate perspective and makes reading enjoyable. The book provides 'genomic account' of the evolutionary history of emergence of complexity from simple living systems. It also gives glimpse of intricate regulatory mechanisms that act on genomic information during differentation of different cell types.

I thank the authors of these articles for their effort in presenting interesting accounts of their research. I also thank G. Giridharan for help in organizing figures and M.W. Pandit for his continuous help. I am thankful to Dr. Lalji Singh for the unhindered cooperation, support and help that I had from him right from the beginning to the completion of this volume, without which it would not have been possible to bring it up to this shape.

Rakesh K. Mishra

Contents

Snake's Eye-view of Adam and Eve: From Chromosomes to Genome

1

Lalji Singh

Centre for Cellular and Molecular Biology, Hyderabad-500 007, India

Introduction

The existence of specialized sex-determining chromosomes poses a range of fascinating and fundamental questions, few of which have been answered satisfactorily. There are a large number of organisms, which do not have specialized morphologically distinct sex chromosomes, yet their sex is determined and sex ratio (1:1) is maintained. In many lowly evolved organisms, sex ratio is determined by environmental factors. This is known to be the case in many species of reptiles in which temperature during the early stages of embryogenesis has a determining effect on the sex ratio; this is so, for example, in alligators, crocodiles and turtles. Normally, in the amphibia, the sex ratio is genetically determined. These different systems embody the evolutionary transition from ancestral conditions, which were dominantly influenced by environmental factors to those where sex determination occurs as a genetic mechanism. Why has the evolutionary process brought about developments, such as those resulting in an entire chromosome, originally containing many other genes unconnected with sex, getting subverted to a singular developmental programme losing most of its genes? Why only some, not all species, exhibit such a characteristic? Why has not the functional monosomy involved in the specialization of the Y or W sex-determining chromosome led to chromosomal imbalance and consequent lethality? Evidently, we do not know the answers to many of these questions.

An important step in the direction of understanding the evolution, structure and function of sex-determining chromosomes is the isolation of sex chromosome-specific repeated DNA. I would particularly like to present how our isolation of highly conserved sex chromosome-specific Bkm (banded krait minor satellite) DNA resolved the puzzle of the evolution of the sex chromosomes and sex reversal in mouse. This has been a logical progression, became possible because we already know that in snakes, the major events in the evolution of chromosomal sex determination (CSD) are exhibited in living species (Singh, 1972). This chapter, therefore, proposes to take you on this fascinating journey, from the beginning to the destination, from chromosomes to genome.

Sex Chromosomes in Snakes

Snake exhibits various states of differentiation of sex chromosomes depending upon the evolutionary status of its various families. In the non-poisonous snakes, Python, Boa, Eryx, etc. of the primitive family Boidae, both sexes are chromosomally alike. It has not been possible to unequivocally identify the female sex-determining W chromosome even by using sex chromosome-specific DNA probes (Fig. 1.1a), the extant method developed so far (Singh, 1972; Singh et al.,

1976; 1980). Those belonging to the second major group of non-poisonous snakes, the rat snakes, water snakes, tree snakes, etc. of the family Colubridae, has morphologically well-differentiated ZW sex chromosomes in the female (Fig. 1.1c) and ZZ in the male. However, there are certain species in which although sex-determining chromosomes are morphologically indistinguishable in both the sexes, it is possible to identify the female sex-determining W chromosome by using W chromosome-specific DNA probe (Fig. 1.1b) (Singh et al., 1976; 1980). In the poisonous snakes, cobras, kraits of the family Elapidae, vipers of the family Viperidae, and sea snakes of the family Hydrophiidae, the sex chromosomes are morphologically highly differentiated. The W chromosome of these species is morphologically distinct (Fig. 1.2), late replicating (Fig. 1.3a), deeply stained by C-banding (Fig. 1.3d), and forms W-chromatin body (Figs. 1.3b and c). The W chromosome of some of the snakes belonging to the highly evolved families, albeit morphologically indistinguishable from the Z chromosome, shows deeply stained C-banding distinct from the Z indicating that it may contain specific repetitive DNA, possibly capable of being isolated by ultra-gradient centrifugation (Singh et al., 1976; 1980). I was obsessed with an idea that W chromosome may have a specific satellite DNA. Since the facilities to carry out the molecular work were not available in India, I made serious attempts to go abroad to carry out this work in the laboratory of Dr. J.G Gall in Yale University, well-known at that time for developing the technique of *in situ* hybridization (Gall and Pardue, 1969). However, when I approached him, he was not convinced with the concept of the presence of W-specific satellite DNA in snake, ideated by me. Then I approached Dr. K.W Jones at the Institute of Animal Genetics, University of Edinburgh, who was internationally famous for his work on satellite DNA in human and chimpanzee and for developing the technique of *in situ* hybridization independently of J.G Gall at the same time (John et al., 1969; Jones, 1970; Jones et al., 1973; Jones et al., 1974). He conditionally agreed for my

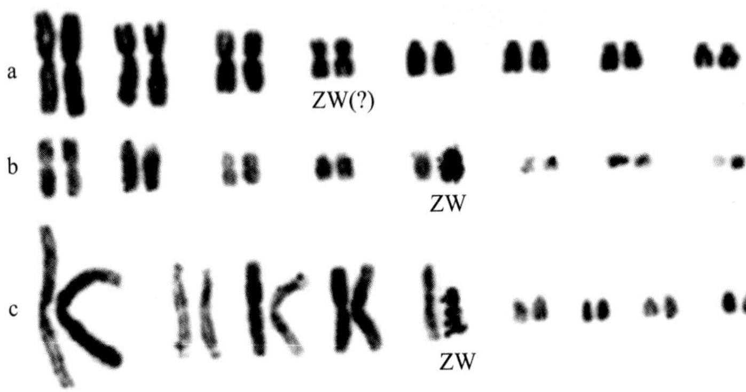

Fig. 1.1 Stages of sex chromosome evolution in snakes (*only macro-chromosomes are shown*). (**a**) Representative female karyotype of the primitive family Boidae in which no specialized sex chromosomes can be distinguished. A pair of chromosomes are presumed in terms of size to be prototypes of the functionally specialized sex chromosomes of more advanced species. (**b**) and (**c**) Female karyotypes of the more evolutionarily advanced snakes illustrating (**b**) homomorphic and (**c**) heteromorphic W chromosomes. These chromosomes have been hybridized *in situ* with tritium-labelled Bkm DNA. The extensive labeling of the W chromosome indicates that Bkm sequence interspersion has occurred throughout the length of both types of W chromosome (Adopted from Jones and Singh, 1985).

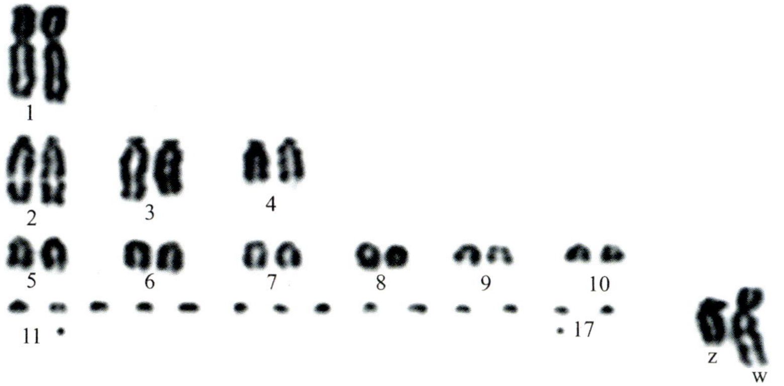

Fig. 1.2 Female karyotype of *Bungarus walliwall* (2n = 36, 22 macro and 14 micro-chromosomes from short-term leukocyte culture showing differentiated sex chromosomes. The W chromosome is unusually large. (Adopted from Singh et al., 1980).

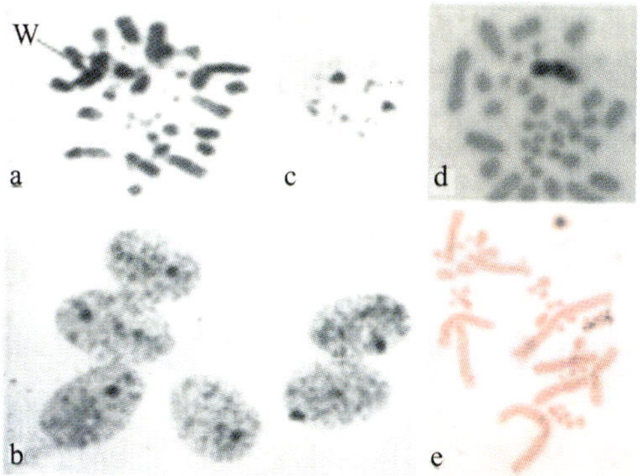

Fig. 1.3 (**a**) Metaphase chromosomes of common Indian Krait *Bungarus caeruleus* after treatment with ^{3}H-TdR showing W chromosome replicating its DNA latest (late replication) in the S phase when all other chromosomes have already completed their DNA replication. (**b**) Interphase nuclei showing W-sex chromatin characteristic of all female snakes with chromosomal sex determination and resembling X chromosome sex chromatin in mammalian females. (**c**) Interphase nuclei of sea snake *Hydrophis fasciatus* having two W chromosomes (W1 and W2) showing two W chromatin bodies. (**d**) Chromosomes of female Elapid snake *Bungarus walliwall* showing the differential heterochromatic nature (C band positive) typical of the functionally specialized W chromosome. (**e**) Chromosomes of female water snake *Natrix piscator* hybridized *in situ* with tritium-labelled Bkm DNA (GATA repeats) showing interspersion of Bkm sequences predominantly throughout the length of W chromosome.

working in his laboratory, provided I bring my own Fellowship. I applied for the Commonwealth Fellowship programme, which was advertised in India and was selected. When I informed Dr. Jones about it, he wrote me saying that he had no knowledge of snakes and that there were no snakes in Edinburgh. Therefore, if I have to go to his laboratory, I will be solely responsible for my proposed work and he will not be able to help me in this regard. I accepted this challenge and joined his laboratory in September, 1974. I carried out my studies using various tissues of numerous species of snakes brought by me from India, preserved in 70% ethanol, but failed to isolate good quality DNA.

The duration of my Fellowship was only 9 months; and I had worked for several months without any success. I wrote about this problem to Professor S. P. Ray-Chaudhuri with whom I did my Ph.D on chromosomes of snakes who supervised over my entire thesis – "Evolution of Karyotypes in Snakes" – that was published in the international journal *Chromosoma* (Singh, 1972). It was because of his untiring effort that within a month I received a consignment, air-lifted by British Airways, of live poisonous snakes, including cobras, common kraits and banded kraits, delivered to me in the laboratory of Dr. Jones in Edinburgh University. Working on these snake samples got me the break I was looking for, namely, successful isolation of good quality DNA. By using analytical density-gradient equilibrium centrifugation of total male and female DNA in $Ag^{+}Cs_2SO_4$ gradient, female-specific satellite DNAs were observed as predicted, which we had also isolated (Singh et al., 1976; 1980).

In situ hybridization of female-specific satellite DNA with female snake chromosomes showed predominant localization on the W chromosome (Fig. 1.3e), providing direct and concluding evidence of the existence of W sex chromosome-specific satellite DNA in snakes. This was the first observation and reporting of this kind of isolation of a chromosome-specific satellite DNA in any species of animal or plant (Singh et al., 1976). Since it was from a poisonous Indian snake banded krait (*Bungarus fasciatus*) that we isolated W chromosome-specific satellite DNA, we designated it as banded krait minor satellite DNA (Bkm) (Singh et al., 1980). When this satellite DNA was used for cross-species hybridization, we could identify W chromosome in all the species of snakes belonging to the highly evolved as well as the intermediate group (Figs. 1.4b and c) possessing C-band positive W chromosome (Fig. 1.4a). However, we failed to identify W chromosome in the primitive family of snakes, such as Python, Sand Boa, etc. by using any of the sex chromosome-specific DNA probes available till date (Singh et al., 1976).

The Hijacking of the W Chromosome

The extant concept that sex chromosomes in primitive species of snakes are homomorphic in both the sexes and that they have not undergone heterochromatinization, whereas, in highly evolved species,

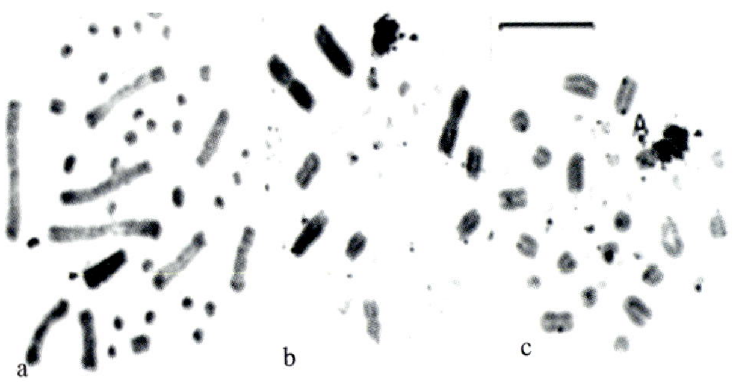

Fig. 1.4 (**a**) Metaphase chromosomes of female banded krait *Bungarus fasciatus* showing C-band positive darkly stained entire W chromosome apart from centromeric C-binding in other chromosomes; (b and c) cross hybridization of female metaphase chromosomes of Indian cobra *Naja naja* (**b**) and common Indian krait *Bungarus caeruleus* (**c**) hybridized *in situ* with tritium-labelled Bkm probe showing identification of W chromosome due to its predominant localization along the length of the W chromosome respectively.

they are heteromorphic in the females with W chromosome undergoing heterochromatinization, raises fundamental questions: How and when was the divergence leading to functional specialization of W chromosome originate? How did it progress? And why did it skip in some of the species? All snakes exhibit genetic control of sex determination and exhibit female heterogamety. It is, therefore, quite reasonable to assume that similar features manifested immediate ancestral forms also, so that all snakes came to possess a functionally definitive W chromosome. The changes at the DNA level, which established a strong genetic control of sex ratio might have involved primarily the W predisposing it to continue to evolve in some snakes so as to attain the status of being specialized as a sex determiner. These changes must have involved some mechanisms, which prevented from recombining the functionally divergent sex alleles on the W and Z chromosomes in order to conserve the novel genetic control of ovaries differentiation on the W. However, it is inconceivable that reduced crossing over alone could have been instrumental in bringing about evolution and differentiation of the W chromosome in preference to the Z. Thus, it is necessary to assume that some sort of mutation in the controlling sequences positioned adjacent to the sex determiner of the W chromosome was involved, which endowed it with novel properties that predisposed the relatively rapid genetic inertness of that chromosome to the functional specialization described above only in some snake species but not in all. This divergence occurring in some of the species as a consequence of evolution in time of the W chromosome seems likely to have been based on a critical structural difference in the W chromosome of the two groups of species narrated above (see Jones, 1983).

This could be explained, if we assume that in species ancestral to those in which the W has continued to specialize, the sex determiner assumed control over an adjacent gene or centre, involved in chromosome condensation. A spreading effect or read-through from the sex gene was such that when it was turned off, the chromosome condensed; the sex determiner thus, fortuitously but effectively, hijacked the entire chromosome. The model postulates that the condensation centre, in some species, which belonged to the primitive family. Boidae, was not located sufficiently close to the sex determiner so as to cause a similar interaction between them. Thus, in the Boidae, there was no inactivation of W chromosome of coincident with the turning off of its sex determiner. This seems plausible as such an event happens to one X chromosome in every mammalian female. The inactivated W chromosome had undergone extensive alterations in its DNA, involving amplifications of GATA repeats (Jones, 1983).

Proposed Future Studies to test the 'Hijacking' Hypothesis

It is possible to test the above hypothesis by making cell lines of female snakes representing the primitive, intermediate and highly evolved states of differentiation of sex chromosomes, sorting out their respective W chromosome by fluorescent activated cell sorter (FACS), making chromosome–specific libraries and by completely sequencing the W chromosome of the respective species. This is expected to unravel the mystery of the molecular basis of the origin, evolution and differentiation of sex chromosomes. We have already established cell lines of male and female embryos of water snake, *Natrix piscator*, grew them for nearly 6 months (Figs. 1.5a–c) and made chromosomes from them (Figs. 1.6a and b). We are in the process of trying to immortalize them and grow them in large quantities. Sorting the W chromosome by FACS is our next step.

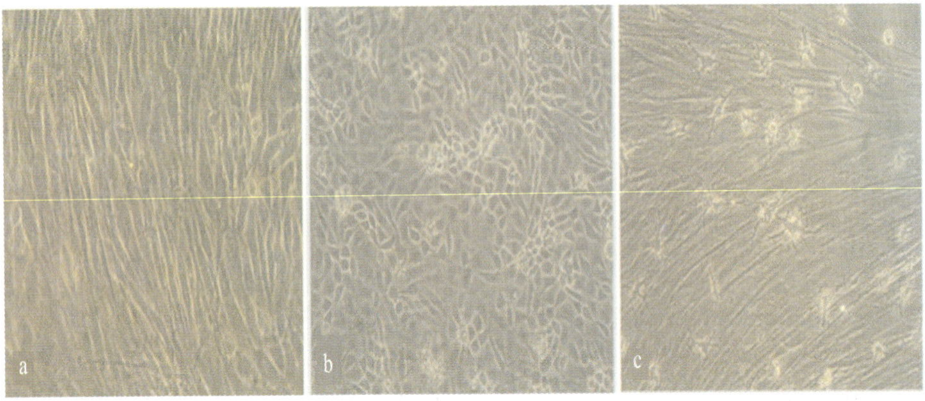

Fig. 1.5 Cells derived from water snake *Natrix piscator* embryos. (**a**) Fibroblast-like cells; (**b**) epithelium-like cells; and (**c**) neuron-like cells.

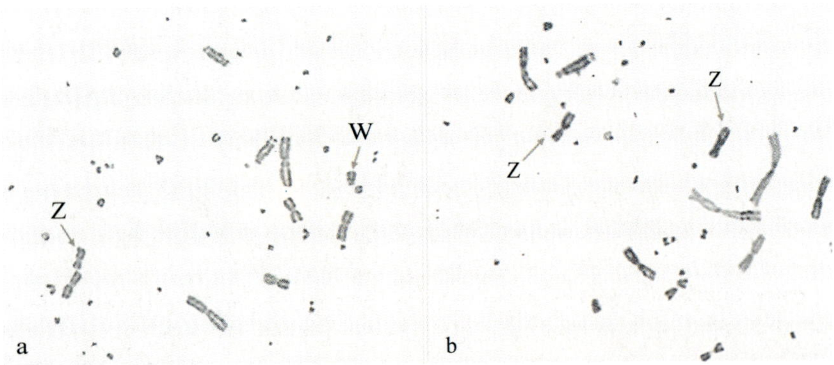

Fig. 1.6 Chromosomes from primary cell lines of water snake *Natrix piscator* male and female embryos cell lines. (**a**) Metaphase plate from female cell line showing two distinct category of macro- and micro-chromosomes with Z and W sex chromosome constitution. (**b**) Male metaphase chromosome with ZZ sex chromosome constitution.

Functional State of the Snake W Chromosome

The W chromosome remains highly condensed, forms a W-chromatin body in interphase nuclei of all somatic tissues of female snake but becomes highly decondensed in developing oocytes (Ray Chaudhuri et al., 1971; Ray Chaudhuri and Singh, 1972; Singh and Ray Chaudhuri, 1975; Singh et al., 1979) implying, according to the present model, derepression of the dominant W sex genes. Interestingly, the inactivated mammalian X chromosome also becomes reactivated in the oocyte (Gortler et al., 1972). This is the only time, when there is a transcriptional activation of the W chromosome. It follows that most of the genes (homologous to those on the Z chromosome) present on the W chromosome at the time when the hijacking occurred, will have been inappropriately expressed subsequently, if indeed they are expressed at all. It is obvious that except those genes that are immediately relevant to early embryonic development, all other genes will have failed to affect the phenotype and, consequently, will have mutated to functionlessness at the maximum possible rate.

Thus, the hijacking of the W chromosome will have ensured its continued evolution to specialization as a sex determiner. This would have led to the spreading of those sequences, which affected the original 'hijacking' throughout the entire W chromosome. This perhaps is the significant cause of the present distribution of Bkm-related DNA (GATA repeats) on this chromosome. This would have effectively limited crossing over with the Z and led to functionlessness of those genes which were homologous to Z but expressed at wrong time and at wrong place because of the hijacking, as this would have been without penalty. Thus, the W would have certainly altered its morphology drastically.

Thus, the origin of Bkm-related (GATA) sequences on the sex-determining W, and colonization of the entire chromosome by this sequence during the course of evolution must have played an important role in the evolution of sex chromosomes. This is substantiated by the fact that quantitative analysis of total counts present on filter-bound DNA of different species of snakes hybridized with 3H-labelled Bkm satellite DNA probe showed the lowest count in the primitive snakes (python) without sex-specific difference and the highest in the poisonous species of snakes showing three- to fourfold increase in the female DNA. This observation was very significant from evolutionary point of view. The fact that satellite DNA, isolated from one species of snake, could identify the female sex-determining W chromosome in all the species found anywhere in the world, excepting in the case of primitive snakes, strongly suggested that this sequence may have been conserved to play an important role in the event of sex determination and differentiation. By using Bkm probe, we tried to study the functional status of W chromosome in different tissues of snakes by virtue of its presence all along the length of the W chromosome (Fig. 1.7a), and observed that W chromosome remained condensed and, therefore probably, was functionally inert in all somatic tissues (Fig. 1.7b), but extensively decondensed and, therefore probably, transcriptionally active in developing oocytes (Figs. 1.8a and b).

We discovered the same phenomenon in the heterogametic sex of the males of human and mouse (Fig. 1.8c) as demonstrated in spermatocytes in the testes. This indicated that there must be a signal present to bring about this conformational change of the W chromatin in sex-and tissue-specific manner. We hypothesized this signal as being a DNA-binding protein. We also predicted that if such a protein exists, it has to be not only sex-specific but also tissue-specific, present only in the developing oocytes of snakes and the testes of mouse and human; and it would be sequence-specific binding only to GATA repeats. We subsequently demonstrated the existence of such a sex- and tissue-specific Bkm (GATA)-binding protein in the ovary of snakes, testis of mouse (Singh et al., 1994) and ovary of *Drosophila* (Sanjay Sachdeva, 1997, Ph.D thesis). Based on these observations, we proposed that Bkm-binding protein specifically binds to GATA repeats interspersed amongst other sequences at intervals of approximately 400kb on the W chromosome of snakes and Y chromosome of mouse and human. We also proposed that this protein brings about coordinated decondensation of the entire chromosome and thereby making all the genes present on these chromosomes transcriptionally accessible. Thus, we proposed that Bkm sequences played a very important role in changing the W-chromatin conformation (Singh et al., 1994).

Bkm Satellite DNA is Evolutionarily Conserved

Consistent with an evolutionarily significant function, it was shown (Singh et al., 1981; Jones and Singh, 1981 a and b) that related sequences were abundant in birds and mammals. The quantitative

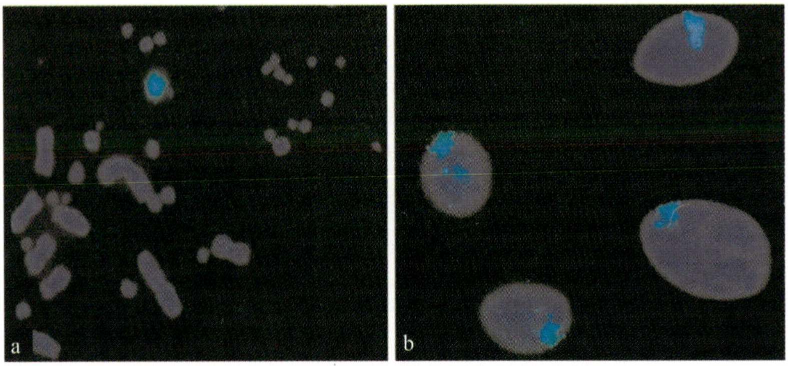

Fig. 1.7 Painting of W chromosome of water snake *Natrix piscator* using $(GATA)_{13}$ probe. (**a**) Female metaphase plate showing W chromosome painted with $(GATA)_{13}$ probe and (**b**) Interphase nuclei from female water snake embryo showing W chromosome body painted with $(GATA)_{13}$ probe.

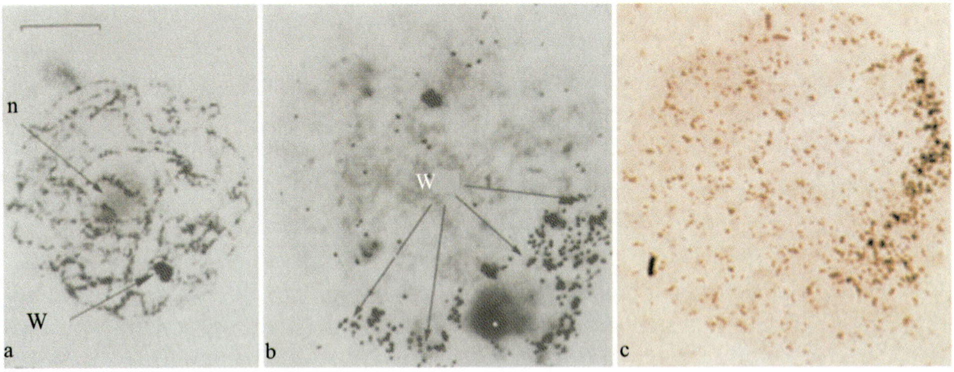

Fig. 1.8 Stained squash preparation of meiotic prophase of female common Indian krait *Bungarus caeruleus* (**a**) showing the persistent condensation of the W chromosome; (**b**) Autoradiograph showing decondensation of the W chromosome (*arrowed*) in the post-meiotic snake oocyte by ^{3}H-Bkm hybridization *in situ*. This is believed to indicate that the W chromosome is involved in germ-cell sex determination; (**c**) Interphase nucleus of primary spermatocyte of mouse hybridized *in situ* with ^{3}H-labelled mouse Y chromosome-specific DNA probe M34 showing extensively decondensed Y chromosome.

excess which characterized female snakes was also found in female birds, indicating that these sequences had become amplified on the avian W chromosome, which was entirely heterochromatic and rich in repetitive DNA (Stefos and Arrighi, 1971). Despite showing a common ancestry in the archosaurian branch of diapsida, birds and snakes evolved CSD independently. Thus, it appears that amplification of Bkm-related DNA is a feature common to all CSD mechanisms.

Bkm (GATA repeats) in Mammals

A mutation was discovered in the laboratory mouse where male mouse when crossed with normal female, produced one-fourth of its progeny consistently, sex reversed. These were chromosomally

(XX) female but developed as a male, although sterile. This mutation was named as Sxr (Cattanach et al., 1971). All over the world, scientists tried to map this mutation, and in the process arrived at the conclusion that probably this was dominant autosomal mutation. This deflected the direction of the concerted efforts of the scientists and they started looking for the sex-determining gene in mouse on autosomes; and the Y chromosome was considered to be genetically inert. We asked the question whether Bkm sequences, which are highly conserved and are predominantly located on the sex-determining chromosome (W) of most of the species of snakes as well as other vertebrates (birds) and also on the Y chromosome of mammals, have any role to play in sex determination. We wanted to answer this question by considering whether the mutation causing sex reversal in mice (XXsxrmale) (Cattanach et al., 1971) would be the ideal starting point. We took a chance and hybridized DNA of many sex-reversed males (XXsxr) along with normal male and female mice on Southern blot with ^{32}P labelled Bkm satellite DNA probe; we were amazed to observe the male-specific hybridization pattern, which lacked the Y chromosome in the DNA of sex reversed individuals (XXsxr), similar to the one found in normal male having a Y chromosome. In females, this pattern was completely absent (Fig. 1.9). *In situ* hybridization on male and female mouse chromosomes demonstrated the presence of Bkm sequences predominantly in the short arm of the Y chromosome (Fig. 1.10a). When we hybridized the sex reversed (XXsxr male) mouse chromosomes with 3H-labelled Bkm probe *in situ*, we observed that it predominantly hybridized at the telomeric region of the long arm of one of the X chromosomes (Fig. 1.10b). The carrier male mouse (XYsxr) revealed duplication of the Bkm positive Y short arm and also its translocation to the distant region of the mouse Y long arm (Fig. 1.10c). The detailed study of the meiosis of carrier mouse by *in situ* hybridization revealed regular crossing over between one chromatid of the X and one chromatid of the Y chromosome, transferring the duplicated Bkm positive (Sxr) region of the Y to one chromatid of the X (Fig. 1.11). This satisfactorily explained how meiotic segregation always produced one-fourth of the progeny with sex-reversal (XXsxr male) (Singh and Jones, 1982; Singh et al., 1984). This observation led to the conclusion that a very small region of Y chromosome was necessary and sufficient to convert a female mouse into a male; and the entire Y chromosome was not necessary for the male sex determination. It also led to the demonstration of the existence of pseudo-autosomal region in the X and Y chromosomes where regular crossing over takes place to insure their proper segregation. This was a sensational discovery (Singh and Jones, 1982; Singh et al., 1984) which led to the detection of a similar phenomenon of sex reversal in human (Gubbay et al., 1990; Koopman et al., 1991; Goodfellow et al., 1993). When we cloned the Bkm satellite DNA, we found that it was predominantly made up of tetra nucleotide repeat, GATA, which did not code for any protein. However, the fact that this sequence was highly conserved and was found in all eukaryotes strongly suggested its functional significance.

GATA Repeats on the *Drosophila* X Chromosome

In the case of *Drosophila melanogaster*, the DNA sequences of heterochromatic region have now become available. The GATA repeats are associated with the heterochromatic region of the X chromosome at 19F and 20AB region of *Drosophila melanogaster* salivary gland chromosomes (Fig. 1.12). In this region, there are two clusters of GATA, one consisting of 40 uninterrupted repeats of GATA and the other having 123 repeats at the terminal end of the X chromosome. This is the largest uninterrupted stretch of GATA repeats in any species reported so far. Embedded in the heterochromatic region are 230–254 protein-coding genes conserved in other *Drosophila* and

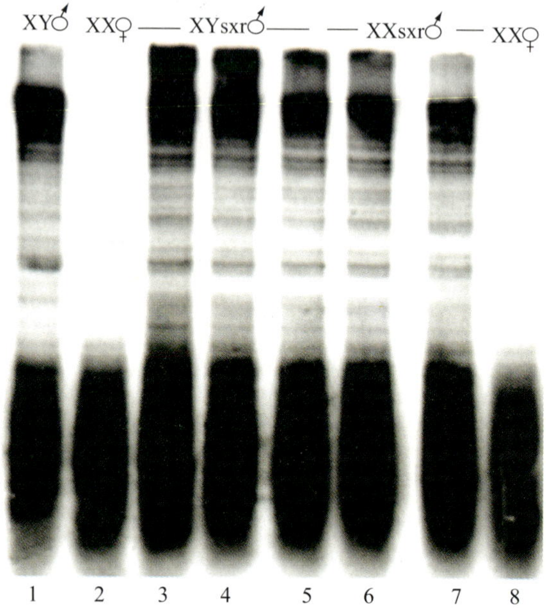

Fig. 1.9 Southern blot of AluI-digested mouse DNA hybridized with a ^{32}P-labelled Bkm probe. The eight tracks are of DNAs obtained from (1) normal male; (2) normal female; (3–5) XYsxr carrier male; (6 and 7) XXsxr male; and (8) normal female. Note the virtual absence of hybridization in the relatively high-molecular-weight regions of normal female DNA and the presence of strong hybridization to this region in the normal male. The sex-reversed male DNA having female chromosome constitution also shows hybridization to this region, indicating the presence of DNA originating from the sex-determining paracentromeric region of the Y chromosome.

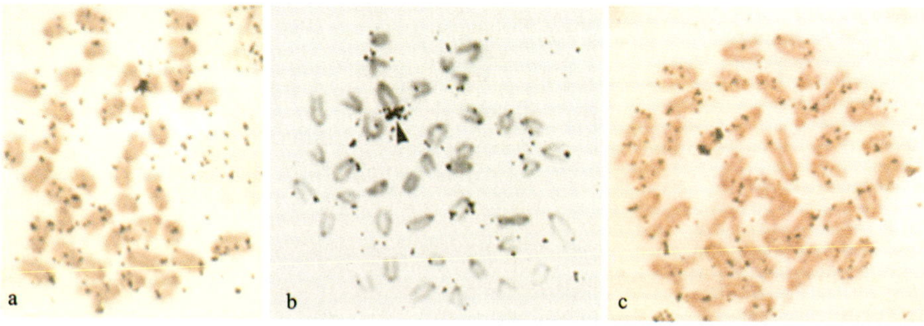

Fig. 1.10 *In situ* hybridization of ^{3}H-labelled Bkm to mouse chromosomes. (**a**) Normal male chromosomes from bone marrow preparation showing the concentration of Bkm-related sequences on the short arm of the Y chromosome. (**b**) Chromosomes from the bone marrow of the sex-reversed XXsxr mouse showing Bkm hybridization to the distal region of the long arm of one of the two X chromosomes. This shows the presence of a cryptic translocation carrying a male sex-determining locus from the Y chromosome. (**c**) Chromosome preparation from the bone marrow of the XYsxr carrier mouse showing the Y chromosome with intensely labelled proximal (normal) and distal (abnormal) locations of Bkm hybridization.

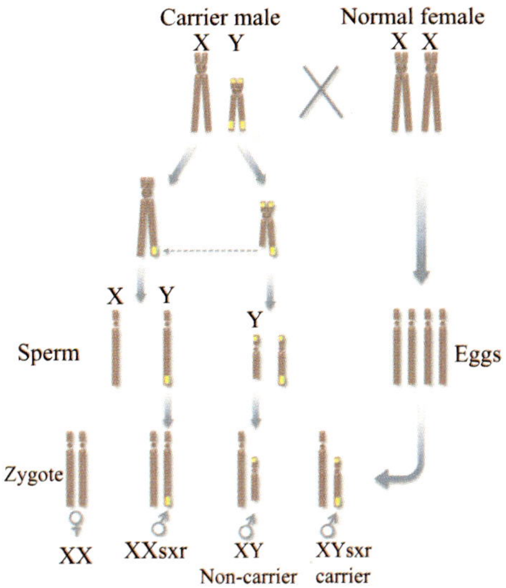

Fig. 1.11 The transmission of the mouse SXR factor. The Y chromosome of the carrier male has two sex-determining locations, one of which lies distal to the cross-over point with the X chromosome. Crossing-over between chromatids of X and Y transfers this sex determiner to one chromatid of the X chromosome. This then determines maleness in one half of the daughters in a mating involving a normal female.

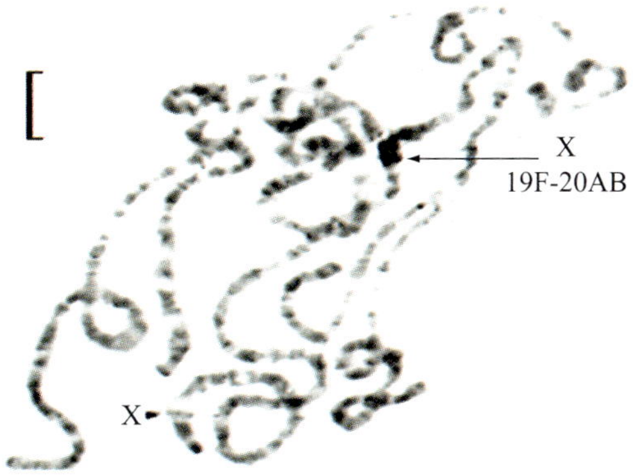

Fig. 1.12 *In situ* hybridization of ^{3}H labelled Bkm probe to *Drosophila melanogaster* salivary gland chromosomes. *Note*: Intense hybridization to the base of the X at location 19F–20AB only. Bar = 10μm.

the more diverged species, 32 pseudogenes and 13 non-coding RNAs. More than 77% of this heterochromatin segment is composed of fragmented and nested transposable elements and other repeated DNAs. Smith et al. (2007) reported that *Drosophila*'s heterochromatin contains islands of highly conserved genes embedded in these "oceans" of complex repeats which may require special expression and splicing mechanisms. It is pertinent to point out that this region of *Drosophila* X chromosome remains highly condensed in somatic cells (Fig. 1.13a) but gets extensively decondensed in oocytes (Fig. 1.13b), suggesting that probably this is where this region becomes functionally active. It would be interesting to see whether all the genes identified recently in this region are

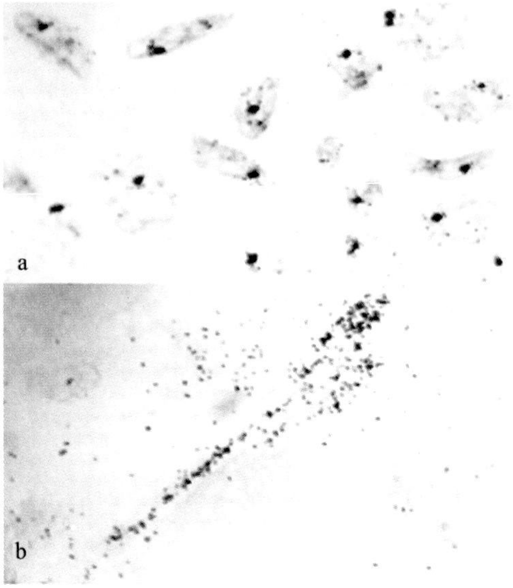

Fig. 1.13 *In situ* hybridization of squash preparation of *Drosophila* ovary with ^{3}H labelled Bkm-2(8) probe. (**a**) Somatic cells showing condensed state of the base of the X chromosome. (**b**) A germ cell showing highly decondensed state of the base of the X chromosome. *Note:* The magnification of both (a) and (b) is the same.

transcribed in developing oocytes. The presence of Bkm (GATA)-binding protein has been shown in *Drosophila* ovary (Sanjay Sachdev, 1997, Ph.D thesis submitted,). It is tempting to speculate that Bkm-binding protein might be responsible in bringing about coordinated decondensation of this region and making the chromatin transcriptionally accessible. This is a transition from higher-order transcriptionally silent chromatin to a less condensed state requisite to accommodate the molecular elements required for transcription. Specificity of such phenomenon to this heterochromatic region may be brought about by the GATA repeat clusters present in this region.

GATA is a Eukaryotic Sequence Family

In our earlier studies, during the survey of the species required for sequences related to Bkm, the absence of Bkm sequences in prokaryotes including Archebacteria was revealed as demonstrable by hybridization. In our present study, we conducted *in silico* analysis for the presence of GATA repeats (Bkm) in the genomic sequences of 2252 species of prokaryotes reported in the database. We observed that while one or two copies of GATA repeats were found in many species, larger than four uninterrupted repeats were absent in most of them except in various strains and species of *Yersinia, Methanosarcina barker, fusaro, Bdellovibrio bacteriovorus,* and *Prochlorococcus marinus* in which single stretch of upto 11 GATA repeats were present. This analysis further demonstrates the absence of more than two GATA repeats in the genome of most of the prokaryotic species sequenced so far (2252 species, March, 2008). There were appreciable quantities of related sequences in the genome of higher eukaryotes as revealed by the direct search for GATA repeats in their genomic sequence database. GATA repeats, therefore, presumably became prominent in organisms exhibiting greater elaboration of form and function than in eubacteria.

Since prokaryotes do not have, or have a very little repetitive DNA and since GATA exists in multiple copies as repetitive DNA, it might be that prokaryotes simply do not have the mechanism necessary to evolve it. This thought has led to the survey of Archebacteria for GATA-related DNA as this group of organisms, which inhabit severe environment, possess considerable amounts of repeated nucleotide sequences (Sapienza and Doolittle, 1982). The survey revealed no trace of GATA repeats in the genome of these organisms. Thus, it appears that the capacity to evolve repeated DNA is not sufficient to generate GATA-related sequences.

No matter how one interprets these results, the one inevitable inference is that these sequences may function in connection with chromatin with control of gene expression. Given that the individual cells in prokaryotes for the most part have their genome under simple operator/repressor type of control, the presence of powerful switching elements capable of long-term inactivation would obviously tend to be strongly selected against. This further suggests that such switching elements and their functional adaptation may have evolved *some compartmentalization* of genome. The switching elements, operating at a distance on covalent chromosomal regions, must have evolved in the non-essential DNA, perhaps on a plasmid. This compartmentalization may have been the origin of the eukaryotic chromosome and the plasmid may have been a sex factor. A plasmid origin of Bkm might explain its evident capacity for spreading on eukaryotic chromosomes. Such switching would have caused an abrupt change in cellular properties which, depending on its precise character, may amount to a lethal event leading to a strong negative selection or a specification or a differentiation, resulting in the cooperation between new varieties of cells in the population. From such cooperation, sex, presumably, first arose. Thus, it is possible to see a thread conceivably connecting the emergence of eukaryotes with the invention of sexual processes. This thread extends to most elaborate life forms where it appears to have played a central role in the evolution of chromosomal sex determination. This concept is similar to that developed on theoretical grounds by Hickey (1982), who suggested that transponders may have caused sexual reproduction to evolve and also spread. According to Hickey, sex and outbreeding may have been the products of parasitic genes (Doolittle and Sapienza, 1980; Orgel and Crick, 1980). Implicit in the present thesis is the idea of existence of DNA sequences whose properties demand adaptive response from the functional organization of the cell. In the absence of such responses, the organism has to avoid acquiring these sequences and/or bear the burden of resulting lethality. Prokaryotes, apparently, have chosen the latter course whereas eukaryotes the former choice. Viewed from the standpoint of this idea, sex chromosomes may be seen as one consequence of trying to control this family of sequences, resulting in the genetic death of the entire linkage group barring the sex-determining function.

Genome-wide Analysis and Functional Significance of GATA Repeats in Vertebrates

After complete DNA sequences of a large number of genomes became available, we carried out comparative genomic study of tetranucleotide repeats and found that GATA was the only repeat in the genome of a large number of species such as human, chimpanzee, mouse, rat, *Drosophila*, etc., which exists as a large stretch of 10–12 copies without any interruption throughout the genome (Fig. 1.14). Other tetranucleotide repeats were found to be only between 1 and 4 uninterrupted copies (Subramanian et al., 2003; present study). This again suggested the evolutionary significance of GATA

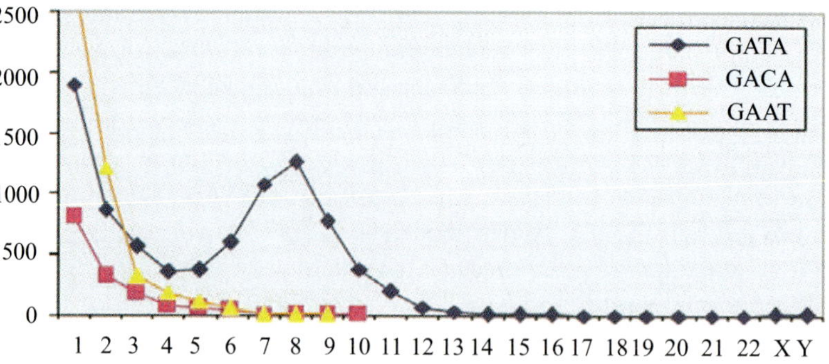

Fig. 1.14 Complete analysis of perfect tandem repeats of GATA in the human genome. The occurrence indicates number of times a repeat length occurs in the genome. Note the peaking of tandem repeats of ($GATA_{11-12}$). This selective accumulation of 11 to 12 GATA repeats is not seen even in the close tetranucleotide variants, suggesting that this trend is unique to GATA repeats (adopted from Subramanian et al., 2003).

repeats (Bkm sequences) in the genome. Our specific analysis of GATA on the Y and X chromosomes of human, chimpanzee and mouse showed highest concentration of the repeats in the genome of these species on the Y chromosome followed by the X compared to other chromosomes. The distribution patterns of GATA repeats on the Y chromosome of human, chimpanzee and mouse (un-published) are superimposable again suggesting its functional significance.

The distribution of (GATA) 10–12 repeats shows unusual features of being flanked by Matrix Associated Regions, MARs. By the virtue of distribution of such repeats at the interval of 400–500 kb on the human Y chromosome, the associated regions may function as boundary elements defining the chromatin domains. The repetitive nature of such boundary elements suggests their role in coordinated regulation of genes associated with these elements. These findings also demonstrate the role of repetitive DNA in genomic packaging that may have a higher order regulatory consequence (Fig. 1.15). GATA-MARs remain associated with the nuclear matrix. The 400–500kb region between the two clustering of GATA and MARs repeats may loop out and become transcriptionally accessible in the tissue where Bkm (GATA) binding protein is present. The GATA and associated MAR sequences, therefore, may define the chromatin boundary and have a role in genome packaging and higher order regulatory mechanisms involving large regions of chromosomes (Kumar et al., 2008). Repetitive elements in higher eukaryotes may function as part of the genomic packaging code with unique regulatory properties. These observations acquire special significance in the light of the fact that the Y chromosome has been demonstrated to modulate expression of hundreds of genes on other chromosomes (Lemos et al., 2008) probably by the mechanism of long range interactions suggested here.

Mapping of a Novel Human Sex-determining Gene linked to the Short arm of the X Chromosome

Although it is a well-established fact that SRY is the major sex-determining gene in humans and other mammals, a large number of cases of sex reversals, such as XX males and XY females and hermaphrodites cannot be satisfactorily explained based on the presence or absence or the mutation

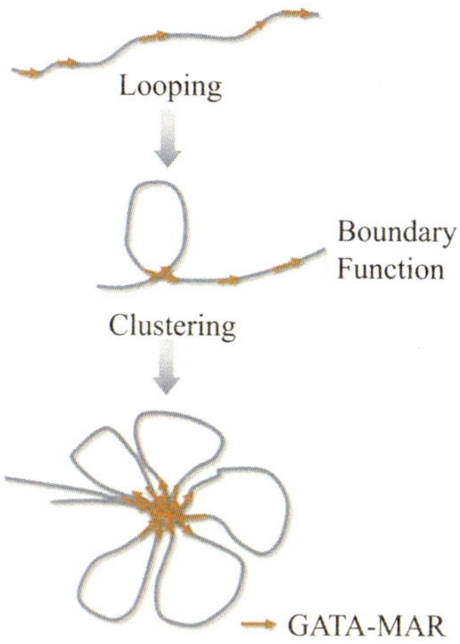

Fig. 1.15 Diagrammatic representation of GATA-MAR elements as packaging code.

in the SRY or in a number of other genes, which have been implicated in sex determination, suggesting that we still have not identified and isolated the major sex-determining gene(s) which may be involved in testis organization in case of male heterogamety or that of the ovary in the case of female heterogamety. In Indian crocodile (*Crocodylus palustris*) when eggs are incubated at high temperature (32.5 °C), they all develop as male while incubation at low temperature (30 °C), develop as female. This fact strongly suggests that the genes responsible for gonadogenesis (testis in male and ovary in female) have to be present in both sexes. We, therefore, set out to identify such genes if they exist. We are studying familial cases of sex reversal in humans; and by genome-wide screening, we have been able to identify a small region at the short arm of the X chromosome which appears to be a male-determining region (Singh et al., 2006). We still have to identify the gene(s). Once this is done, we will be able to explain satisfactorily all the cases, which we are yet unable to explain based on the sex-determining genes that have been isolated so far. Genome-wide screening for identifying a gene is being carried out for the identification of the genes involved in various kinds of diseases. Thus, we have moved from chromosomal abnormalities causing diseases to the detection of changes in the entire genome using genomics and proteomics approaches. It has, thus, been a fascinating journey from chromosomes to genome.

Acknowledgements

I gratefully acknowledge the contribution of those from India and abroad who have been associated with me in this pursuit of unraveling the molecular basis of sex determination and evolution of sex chromosomes. I am particularly indebted to my Ph.D Supervisor and Mentor, Professor S.P. Ray-Chaudhuri, not only for his supervision for my Ph.D programme but also for being a guardian and a friend. Without his help, it would not have been possible for me to achieve what I have so far. I also gratefully acknowledge the encouragement and the excitement I shared with Dr. Jones in whose laboratory I could earlier carry out the molecular aspects. My stay in Edinburgh in his laboratory for 13 years was very exciting and extremely rewarding. What a tremendous journey we had together from chromosome to genome. I am greatly indebted to my students who did their Ph.D with me, to Postdoctoral Fellows and all my colleagues who are closely associated with me in CCMB, Hyderabad particularly Dr. K. Thangaraj, Dr. Rakesh Mishra and their associates. I am thankful to both of them for allowing me to include some of their unpublished data. I gratefully acknowledge the financial assistance rendered by Council of Scientific & Industrial Research (CSIR), Government of India; Department of Science & Technology (DST), Government of India; Department of Biotechnology (DBT), Government of India; Indian Council of Medical Research (ICMR), Government of India. I thank Medical Research Council (MRC) of UK and the Government of UK for the award of Commonwealth Fellowship enabling me to carry out the work whose outlines are presented very briefly in this chapter.

References

1. Cattanach, B.M., Pollard, C.E., and Hawkes, S.G. (1971). Sex reversed mice XX and XO males. Cytogeneties, *10*, 318–337.
2. Doolittle, W.F., and Sapienza, C. (1980). Selfish genes, the phenotype paradigm and genome evolution. Nature *284*, 601–603.
3. Gall, J.G., and Pardue, M.L. (1969). Formation and detection of RNA–DNA hybrid molecules in cytological preparations. Proc. Nat. Acad. Sci. U.S.A. *63*, 378–383.
4. Gartler, S.M. and Andina, R.J. (1976). Mammalian X chromosome inactivation. Adv. Hum. Genet. *7*, 99–140.
5. Goodfellow, P.N., and Lovell-Badge, R. (1993). SRY and sex determination in mammals. Ann. Rev. Genet. *27*, 71–92.
6. Gubbay, J., Collignon, J., Koopman, P., Capel, B., Economou, A., Munsterberg, A., Vivian, N., Goodfellow, P.N., and Lovell-Badge, R. (1990). A gene mapping to the sex-determining region of the mouse Y chromosome is a member of a novel family of embryonically expressed genes. Nature *346*, 245–250.
7. Hickey, D.A. (1982). Selfish DNA: a sexually transmitted nuclear parasite. Genetics *101*, 519–531.
8. John, H.A., Birnstiel, M.L., and Jones, K.W. (1969). RNA–DNA hybrids at the cytological level. Nature *223*, 582–587.
9. Jones, K.W., (1970). Chromosomal and nuclear location of mouse satellite DNA in individual cells. Nature *225*, 912–915.
10. Jones K.W., (1983). Evolution of sex chromosomes. *Development in mammals* (M.H. Johnson, ed.) *5*, 297–320.
11. Jones, K.W., and Singh, L. (1981a). Conserved sex-associated DNA in vertebrates. In *Genome Evolution*. (Drover, G. and Flavell, R., ed.) Academic Press London, UK.
12. Jones, K.W., and Singh, L. (1981b). Conserved repeated DNA sequences in vertebrate sex chromosomes. Hum. Genet. *58*, 46–53.
13. Jones, K.W., Prosser, J., Corneo, G., and Ginelli, E. (1973). The chromosomal location of human satellite DNA III. Chromosoma *42*, 445–451.
14. Jones, K.W., Purdom, I.F., Prosser, J., and Corneo. G. (1974). The chromosomal location of human satellite I DNA. Chromosoma *49*, 161–171.

15. Koopman, P., Gubbay, J., Nigel, V., Goodfellow, P., and Lovell-Badge, R. (1991). Male development of chromosomally female mice transgenic for sry. Nature *351*, 117–121.
16. Kumar R.P., Singh, L., and Mishra, R.K. (2008). Manuscript in preparation.
17. Lemos, B., Araripe, L.O., and Hartl, D.L. (2008). Polymorphic Y chromosomes harbor cryptic variation with manifold functional consequences. Science *319*, 91–93.
18. Orgel, L.E., and Crick, F.H.C. (1980). Selfish DNA: the ultimate parasite. Nature *284*, 604–607.
19. Ray-Chaudhuri, S.P., and Singh, L., (1972). DNA replication pattern in sex chromosomes of snakes. Nucleus *15*, 200–210.
20. Ray-Chaudhuri, S.P., Singh, L., and Sharma, T. (1971). Evolution of sex chromosomes and formation of W chromatin in snakes. Chromosoma *33*, 239–251.
21. Sachdev, S. (1997), Molecular analysis of Bkm-associated DNA sequences in *Drosophila*. (Ph.D. thesis submitted to Jawaharlal Nehru University).
22. Sapienza, C., and Doolittle, W.F. (1982). Unusual physical organization of the *Halobacterium* genome. Nature *295*, 384–389.
23. Singh L., (1972). Evolution of karyotypes in snakes. Chromosoma *38*, 185–236.
24. Singh, L., and Jones, K.W., (1982). Sex reversal in mouse (*Mus musculus*) is caused by a recurrent non-reciprocal crossover involving the X and an aberrant Y chromosome. Cell *28*, 205–216.
25. Singh, L., and Ray-Chaudhuri, S.P. (1975). Localization of C-band in the W sex chromosome of Common Indian krait, *Bungarus caeruleus* Schneider. Nucleus 18(iii), 166–171.
26. Singh, L., Purdom, I.F., and Jones, K.W. (1976). Satellite DNA and evolution of sex chromosomes. Chromosoma *59*, 43–62.
27. Singh, L., Purdom, I.F., and Jones, K.W. (1980). Sex chromosome associated satellite DNA: evolution and conservation. Chromosoma *79*, 137–157.
28. Singh, L., Phillips, C., and Jones, K.W. (1984). The conserved nucleotide sequences of Bkm which define Sxr-in the mouse are transcribed. Cell *36*, 111–120.
29. Singh, L., Purdom, I.F., and Jones, K.W. (1979). Behaviour of sex chromosome- associated satellite DNAs in somatic and germ cells in snakes. Chromosoma *71*, 167–181.
30. Singh, L., Purdom, I.F., and Jones, K.W. (1981). Sex chromosome-associated nucleotide sequences in eukaryotes. Cold Spring Harb. Symp. Quant. Biol. *45*, 805–813.
31. Singh, L., Wadhwa, R., Naidu, S., Nagaraj, R., and Ganesan, M. (1994). Sex- and tissue-specific Bkm (GATA)-binding protein in the germ cells of heterogametic sex. J. Biol. Chem. *269*, 25321–25327.
32. Singh, R., Thangaraj, K., Gupta, N.J., Leelavathy, N., Selvi Rani, D., Nambiar, R.G., Kalavathy, V., Santhiya, S.T., Rajangam, S., Gopinath, P.M., Chakravarty, B., and Singh, L. (2006). A novel human sex-determining gene linked to Xp11.21–11.23. J. Clin. Endocrin. Metab. *91*, 4028–4036.
33. Smith, C.D., Shu, S., Mungall, C.J. and Karpen, G.H. (2007). The release 5.1 annotation of *Drosophila melanogaster* heterochromatin. Science *316*, 1586–1591.
34. Stefos, K. and Arrighi, F.E. (1971). Heterochromatic nature of W chromosome in birds. Exp. Cell Res. *68*: 228-231.
35. Subramanian, S., Mishra, R.K., and Singh, L. (2003). Genome-wide analysis of Bkm sequences (GATA repeats): Predominant association with sex chromosomes and potential role in higher order chromatin organization and function. Bioinformatics. *19*, 681–685.

The Neglected Parts of the Genome: Half a Century of Satellite DNA

2

Chris Tyler-Smith[1,2], Neal Mathias[2,3] and Rebecca J. Oakey[2,4]

[1]*The Wellcome Trust Sanger Institute, Wellcome Trust Genome Campus, Hinxton, Cambs. UK;*
[2]*Department of Biochemistry, University of Oxford, Oxford, UK;*
[3]*Department of Biochemistry and Molecular Biology, Louisiana State University Health Science Center, Shreveport, Louisiana, USA;*
[4]*Department of Medical and Molecular Genetics, King's College London, London, UK*

In one of the triumphs of recent science, the human genome has been sequenced by an international consortium, and the sequence has been finished to a very high standard of accuracy, making it the best-characterized vertebrate genome thus far available (International Human Genome Sequencing Consortium, 2004). So this must mean that we know the complete sequence of the human genome, mustn't it? Well no, not exactly. There is a subtle but important difference between 'finishing' a genomic sequence and 'completing' the sequence. 'Finishing' is a stage in the sequencing procedure in which agreed standards of quality are achieved and a 'finished' sequence is one that has passed successfully through this stage. It may still contain gaps, and these can be megabases (Mb) in size. In the most extreme case in the human genome, that of the Y chromosome, the finished part contains less than half of the total length of chromosome, and the rest consists of gaps. The genome browser Ensembl currently tells us that the chromosome is 57,772,954 base pairs (bp) long, but the finished sequence ends with a C at position 27,228,749 and the remainder consists entirely of Ns, apart from a small region right next to the long arm telomere (http://www.ensembl.org/Homo_sapiens/mapview?chr=y). Even within the first 27,228,749 bp, there are several large gaps filled with Ns. These long runs of Ns consist mainly of a special class of DNA, described below, which is resistant to current sequencing approaches. Yet this missing DNA is not present in our genome to confound sequencers. It represents a class of sequence that has important biological functions and has found multiple uses in areas as diverse as DNA fingerprinting, unravelling the peopling of the Americas, and understanding how chromosomes are transmitted properly when a cell divides. Now, new approaches to resequencing whole human genomes promise to provide massive amounts of data from these DNA sequences whether we want them or not. Here we review aspects of the past, present and future of this neglected and almost forgotten component of the genome. But this chapter should not be seen as a general review of the whole field; it is more a personal view of a few aspects of it.

What is Satellite DNA? – Identification and Definition

In many species, including humans, there are regions of DNA where a short sequence is repeated over and over again in a head-to-tail (tandem) fashion. These unusual DNAs are known as "satellite DNAs". They received considerable attention in the early days of eukaryotic molecular biology when their atypical properties made them the only sequences that could be isolated using the physical methods then available and studied in specific ways, but these same properties then rendered them almost impossible to study by the methods used to sequence and characterize genomes at the end of the twentieth century and in the first years of the twenty-first century. As a result of these technical difficulties and their perceived low biological importance, interest in satellite DNAs declined. This trend is reflected in the number of publications found in PubMed using "satellite DNA" as a query (Fig. 2.1). There was an early surge in publications in the mid-1960s and then a general decline, counteracted by peaks in the late-1970s and mid-1990s reflecting developments in the field that will be described below.

The term "satellite DNA" was introduced in the early 1960s when one of the few techniques available for analysing DNA was to fractionate it by density gradient centrifugation. When a concentrated solution of a salt such as CsCl was centrifuged at very high speed for 2 days, it formed a density gradient and any DNA in the solution banded in the gradient according to its density. Since DNA density is dependent on base composition, crude separation according to base composition was achieved, usually resulting in a roughly symmetrical band. Kit (1961) observed that some DNAs produced additional minor bands and, since earth-orbiting satellites like Sputnik were in the news, called them 'satellites'. Over the next few years, work on these satellite DNAs flourished (Fig. 2.1). It was established that satellite DNA fractions reassociated rapidly after denaturation (Waring and Britten, 1966), suggesting that they were made up of short repeating units. In one of the first applications of the then-novel technique of *in situ* hybridization to metaphase chromosomes, it was shown that they tended to be located in condensed heterochromatic regions such as those found near centromeres (Jones, 1970). And the sequence of the units making up some satellite DNAs was determined as early as 1970, well before the days of Sanger sequencing, when the sequence of guinea pig α-satellite was deduced to consist of $(CCCTAA)_n$ plus variants (Southern, 1970).

Subsequently, any tandemly repeated DNA made up of short units became known as satellite DNA, even if it did not have unusual buoyant density properties. The term was also applied to short- or intermediate-sized arrays of such units in the form of 'microsatellite' and 'minisatellite'.

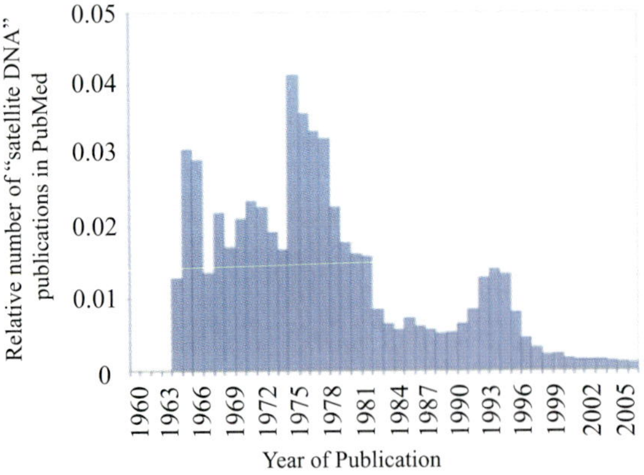

Fig. 2.1 Interest in satellite DNA as reflected by PubMed entries. For each year between 1960 and 2007, the relative proportion of PubMed entries detected using "satellite DNA" as a query is shown, compared with the number detected using "gene" as a query.

We will touch on all of these elements in this chapter, but the long satellite arrays "will be emphasized" because they are the main "neglected sequences".

Satellite DNA on the Y Chromosome – Discovery, Structure and Variation

Satellite DNA is particularly abundant on the Y chromosome, so we will focus on this chromosome, although satellites are found on all the other chromosomes. The high abundance of satellite DNA on the Y may be more than coincidence and could result from the lack of recombination (apart from at the ends of the chromosome, which recombine with the X in male meiosis) or the low gene density, but the relationships between these features and satellite DNA accumulation are not well understood. Whatever the cause of the abundance, the simple experiment of comparing male and female DNA digested with the restriction enzyme *Hae*III and fractionated by agarose gel electrophoresis revealed male-specific bands (Cooke, 1976), an observation that could be explained only if the bands were derived from highly abundant repeated sequences. Cooke (1976) estimated that these sequences must constitute about 50% of the DNA in the Y chromosome, an estimate that is in remarkably good agreement with that from the post-genome-project reference sequence made three decades later and mentioned above (Fig. 2.2). Further work showed that two tandemly repeated sequences were present in Yq: *DYZ1* made up of tandem ~3.5kb units in turn composed of tandemly arranged variants of the 5 bp subunit GGAAT (Cooke and Hindley, 1979; Nakahori et al., 1986) and *DYZ2* made up of tandem ~2.5kb units (Cooke et al., 1982). *DYZ1* and *DYZ2* each form clusters on a 0.1Mb scale, but interspersed with one another on a Mb scale

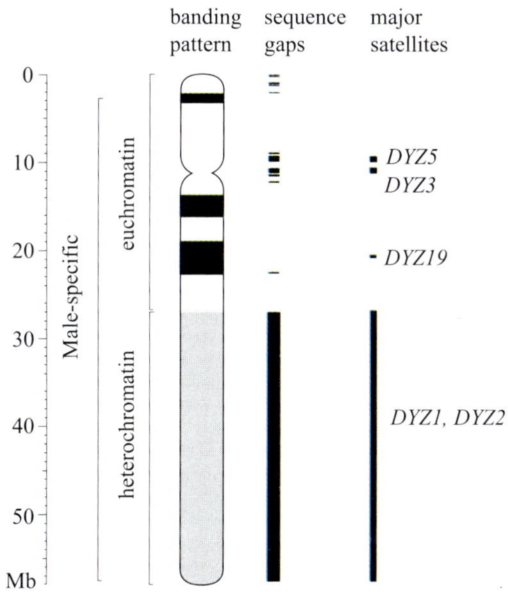

Fig. 2.2 The human Y chromosome reference sequence, showing sequence gaps and the locations of major satellite DNAs. Note the correspondence between the largest gaps and the locations of the satellites, arising because only the edges of the satellite arrays could be included in the reference sequence assembly.

(Babcock et al., 2007; Mathias et al., 1994). These two sequence families are components of the classical human satellites II/III and I, respectively (Prosser et al., 1986), which could be separated by their buoyant density; thus *DYZ1* and *DYZ2* fall well within even the strictest definition of a satellite. The techniques used to characterise these Y-chromosomal sequences, including restriction enzyme digestion, DNA cloning, sequencing, and Southern and *in situ* hybridisation, were of course applied to other satellites as well, and account for the late-1970s peak of publications seen in Fig. 2.1.

Later work identified less abundant Y-chromosomal satellites. Alphoid DNA (*DYZ3*) makes up about 1% of the chromosome and consists of tandem ~5.7kb units containing ~170bp subunits and, as on all human chromosomes, it is located at the centromere (Tyler-Smith and Brown, 1987). It is discussed further in the section on function of satellite DNA. The *DYZ5* sequence also makes up about 1% of the chromosome, but consists of tandem ~20.4kb units (Tyler-Smith et al., 1988) each containing a copy of the *TSPY* gene (Arnemann et al., 1987); the large size of the repeat unit puts it outside our definition of satellite DNA, but it is worth noting here because its tandem structure makes it resistant to current sequencing technology and so it forms one of the large gaps in the Y reference sequence. The last major satellite DNA component to be identified differs from those discussed so far in that it is not resistant to standard sequencing; in fact, it was first described in the paper that presented the reference sequence (Skaletsky et al., 2003). *DYZ19* comprises about 0.7% of the chromosome and consists of tandem ~125bp units; its compatibility with standard sequencing approaches results from the heterogeneity of its units, an important feature discussed later in this section.

Two Y-specific minisatellites are present, *MSY1* (Jobling et al., 1998) and *MSY2* (Bao et al., 2000), and innumerable microsatellites; the count of the latter depends on the criteria used to define them. The useful microsatellites are the polymorphic ones (see below) and a count of those present in the reference sequence using the criteria that the repeat unit size should be ≥3bp and more or less eight copies of the repeat unit should be present yielding 475 (Kayser et al., 2004).

At this point, we will take a digression to consider a satellite sequence which, while not highly abundant on the human Y chromosome, is nevertheless very relevant to this volume, the Bkm satellite. 'Bkm' stands for 'Banded krait minor' and was originally isolated from DNA of a female snake, the banded krait by density gradient centrifugation, where it forms a major satellite (Singh et al., 1980). It was found to be concentrated "by density gradient centrifugation" on the W chromosome of this snake (present in ZW females, not in ZZ males), and might probably create difficulties in sequencing and assembly if the banded krait genome were to be sequenced. Bkm sequences are also abundant on the Y chromosome of some mammals (where the sex determination system is XY males/XX females) such as mouse (Singh and Jones, 1982). On the Y chromosome of human, Bkm sequences are found (Arnemann et al., 1986; Singh and Jones, 1986), as $(GATA)_n$ microsatellites rather than long satellites. Analysis of the human reference sequence shows that the Y does carry the highest density of GATA repeats of any human chromosome, and that these repeats are concentrated in the proximal ~8Mb of Yq; interestingly, while the frequency of $(NNNN)_n$ microsatellites generally falls off with the value of n, $(GATA)_n$ repeats showed an unusually high frequency of n = 11–12 loci (Subramanian et al., 2003). This points towards some features specific to the biology of GATA repeats that remain incompletely understood, and explains why they form a high proportion of the useful microsatellites (Kayser et al., 2004). Bkm sequences also illustrate the links between buoyant density satellites and microsatellites: the same sequence can be a major satellite in one species and a microsatellite in another.

We have alluded above to the importance of the sequence heterogeneity within satellites in facilitating their sequence determination. Homogeneous satellites, in addition to being difficult for sequencers to assemble, have an important biological property: they can readily undergo recombination in a misaligned state, leading to changes in length by 1 unit, or multiples of this

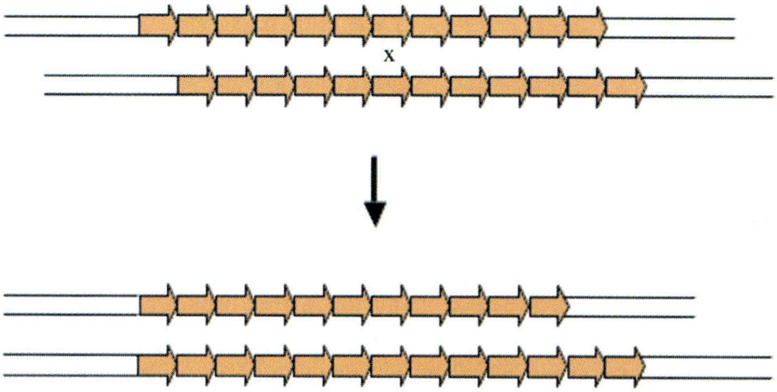

Fig. 2.3 Model for the generation of length variation in a tandem array by homologous but unequal exchange. The misaligned parental molecules are shown at the top, and the products at the bottom. On the Y chromosome, the exchange events would be between sister chromatids; on other chromosomes they could also be between homologues. Each unit of the tandem array is represented by a red arrow, while the flanking sequence is white. 'X' indicates the exchange event.

(Fig. 2.3). As a consequence, their length varies in the population, and for *DYZ1*, *DYZ2*, *DYZ3* and *DYZ5* the lengths given above or in the reference sequence were average values. The length of variation in *DYZ1* and *DYZ2* is large enough to be detected cytogenetically as variation in the size of the Yq12 heterochromatic band (e.g. Bobrow et al., 1971; Laberge and Gagne, 1971), and length variation in all these loci has been detected by pulsed-field gel electrophoresis (Mathias et al., 1994; Oakey and Tyler-Smith, 1990; Tyler-Smith and Brown, 1987; Tyler-Smith et al., 1988). The mutation rates of the different loci, however, have not been previously reported except in D.Phil theses (Mathias, 1993; Oakey, 1989) and we present them here to a wider audience. DNA from families with established paternity was analysed by pulsed-field gel electrophoresis as described in the publications on population variation, and fathers were compared with sons (Fig. 2.4). *DYZ3* and *DYZ5* were each visualized as a single band, or occasionally as two bands when a cleavage site was present within the array, and rare father–son differences were noted (Fig. 2.4 and Table 2.1).

Table 2.1 Measurement of Mutation Rates in Y-chromosomal Satellites

Locus	Number of meioses	Number of mutations	Mutation rate (95% confidence interval*)
DYZ3	121	5	4×10^{-2} ($1.4 \times 10^{-2} - 8.8 \times 10^{-2}$)
DYZ5	121	0	0 ($0 - 2.5 \times 10^{-2}$)

*Confidence intervals were calculated as described by Heyer et al. (1997).

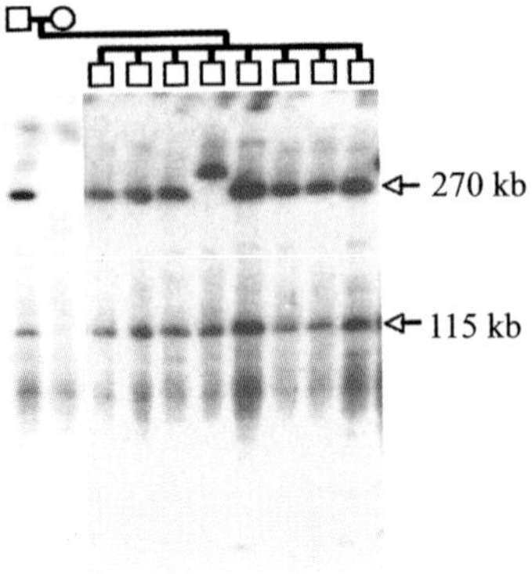

Fig. 2.4 Detection of a mutation in the *DYZ3* array. DNA samples from two parents (left) and eight sons (right) were digested with *Bgl*II, fractionated by pulsed-field gel electrophoresis and probed with a *DYZ3* probe. *Bgl*II cuts once within the *DYZ3* array to produce two fragments. These are transmitted unchanged to seven of the sons, but the larger fragment has increased in size in the fourth son.

Although no mutations were detected in *DYZ5*, the mutation rate for *DYZ3*, at around 4% per meiosis, is among the highest in the human genome, equivalent to the most mutable minisatellites (5%, Jeffreys et al., 1988). Results for *DYZ1* and *DYZ2* were more complex because these loci are so long (~30Mb together) that they could not be visualized as a single band. Instead, they produced complex profiles composed of multiple bands under the conditions used, but father–son differences were readily detected at a rate of about one per 40Mb scanned per meiosis (Mathias, 1993). If there is ~30 Mb *DYZ1/DYZ2* per chromosome, this would correspond to a mutation rate of 75% per meiosis, the highest thus far reported in the human genome and helping to account for the cytogenetically visible variation in Yq. In all cases, the mutations consisted of small changes in the size of a band, similar to that seen in Fig. 2.4, which probably correspond to increase or decrease of one or two units in the length of an array, as predicted by the model (Fig. 2.3).

Functions of Satellite DNA – Centromeres and Enigmatic Transcripts

Satellite DNA is found in centromeric regions of most animals and plants (defined as the primary constriction visible in metaphasic stage of the chromosomes). This association is not universal: human 'neocentromeres' can lack satellite DNA entirely (Amor and Choo, 2002), and for example, the Yq heterochromatin is non-centromeric. Nevertheless, a role in normal centromeric function was a possibility that called out for investigation. An analysis of rearranged human Y chromosomes by deletion and mapping located centromere activity to an interval of ~500kb that contained alphoid DNA and adjacent Yp sequences (Tyler-Smith et al., 1993), and a more precise epigenetic mark of centromere activity, topoisomerase II cleavage, identified a ~50kb region within the alphoid array, approximately 120kb from the alphoid/Yp boundary as the centromere (Floridia et al., 2000). A location of the centromere within the alphoid array was subsequently confirmed on other chromosomes, including the X (Spence et al., 2002) and 11 (Spence et al., 2005). A complementary approach set out to construct *de novo* mammalian artificial chromosomes containing functional centromeres. Here, it was found that alphoid DNA could form *de novo* centromeres after introduction into cultured human cells (Harrington et al., 1997) and further investigation of the DNA sequence requirements revealed that alphoid DNA containing a binding site for the centromeric protein CENP-B (a CENP-B box) was needed: neither alphoid DNA lacking CENP-B boxes nor non-alphoid DNA with or without CENP-B boxes formed centromeres (Ohzeki et al., 2002). Thus, it seems well-established that alphoid DNA has a key function as the normal centromeric sequence.

Minisatellites and microsatellites are scattered widely around the genome and sometimes fall within genes, and even within coding regions (e.g. Swallow et al., 1987; The Huntington's Disease Collaborative Research Group, 1993). As expected from the high variability of satellite sequences in general, they lead to very variable gene structures and protein sequences. Unusually long alleles underlie almost 30 human disorders (Mirkin, 2007), but the details of this important area of medical genetics lie outside the scope of this chapter.

Centromeric satellites have generally been thought to be transcribed only at very low levels, or not at all (Beridze, 1986, section 10.1). However, recent evidence has now identified examples of satellite transcription. Single-stranded RNA transcribed from alphoid DNA from several chromosomes was detected in the nucleolus of HeLa cells during interphase, where it bound centromeric proteins such as CENP-C and INCENP, and all these components formed part of the complex present at the centromere at metaphase (Wong et al., 2007). The functional role of the alphoid RNA (if any) is currently unclear but provides an exciting area for future research. A probe from the Yq heterochromatin was found to detect heterogeneous transcription of *DYZ1* repeats in the testis at 36 weeks of development and in the adult, but not earlier in development or in other tissues (Jehan et al., 2007). Remarkably, this study found that a 67-nucleotide *DYZ1* fragment was *trans*-spliced in the testis onto the *CDC2L2* transcript from chromosome 1. Again the functional significance of this event was unclear: a by-product of cellular machinery that exists for other purposes, or an important mechanism for fine-tuning some aspect of the cell division-related and apoptotic activities of CDC2L2? Whichever is closer to the truth, these studies of satellite DNA transcription show that novel aspects of this venerable field continue to be discovered, and that even the non-centromeric long satellites may have undiscovered functions.

The Diverse Uses of Satellite DNA – Identifying Individuals, Disease Genes, Population Movements and Chromosomes

Irrespective of their functions within the cell, satellites have been used by scientists in many ways. DNA fingerprinting, with its potential to distinguish the DNA of all individuals on earth (except identical twins) became possible when minisatellites were discovered (Jeffreys et al., 1985). Of particular relevance to this chapter, satellites such as Bkm have also been used in this way, for example in crocodilians (Lang et al., 1993). In more recent years, such approaches have been superseded by another kind of satellite, microsatellite that is easier to use and provides greater sensitivity. Forensic geneticists now type a set of autosomal microsatellites that provide a highly discriminating profile. In the UK, for example, the SGM Plus set of 10 autosomal microsatellites (and the *AMELX/Y* locus) is used, and is suggested to have a match probability to a random individual of $<10^{-13}$, and similar, but not identical, tests are used in other countries (Jobling and Gill, 2004). For some specialist applications, a set of 11 Y-chromosomal microsatellites and a corresponding database of haplotype frequencies are used (Willuweit and Roewer, 2007).

Medical geneticists followed a similar historical path of using first minisatellites and later microsatellites to construct linkage maps of the human genome that could be used to localize genes for Mendelian disorders. This work was responsible for the mid-1990s peak in "satellite DNA" publications (Fig. 2.1). The tail-off in the last decade reflects the move to investigating complex disorders, where association studies that type single nucleotide polymorphisms (SNPs) are more appropriate.

Evolutionary geneticists also made use of the various classes of satellite DNA, including the major satellites; indeed these provided some of the first markers capable of discriminating between Y-chromosomal lineages (Mathias et al., 1994; Oakey and Tyler-Smith, 1990). While the labour required for these analyses, involving pulsed-field gel electrophoresis, limited their application, some variants occurring within satellite arrays could be typed easily and have been more widely used. Among these were a system for amplifying variant Y alphoid repeats and analysing them as heteroduplexes, which revealed 13 haplotypes (Santos et al., 1995). Application of this system to populations from the Americas provided the first evidence for a single major male founder lineage during the prehistoric peopling of the Americas some 15–20,000 years ago (Pena et al., 1995), an important finding amply confirmed by subsequent studies (Zegura et al., 2004). Some SNP variants (Hurles et al., 1999; Mathias et al., 1994) and retroposon insertions (Santos et al., 2000) within satellites are readily typed and have been widely used, as they have the highly variable microsatellites (Jobling and Tyler-Smith, 2003).

We thus see that the use of satellites has been important for the development of genetics over the last few decades, but that, apart from the microsatellites that continue to be essential for forensic genetics and to find limited uses in evolutionary and medical genetics, these loci have been superseded by other markers. In one application, however, large satellites remain central: fluorescence *in situ* hybridization (FISH, Pinkel et al., 1986). Here, a probe is hybridized to chromosomal or cellular preparations to visualize the corresponding target sequence, and the large size and repetitive nature of satellites produces bright signals that are particularly suitable for this analysis. Centromeres or pericentric regions can be identified collectively or individually, and chromosomes identified, marked and counted. The use of satellites in cytogenetic research and diagnosis looks likely to continue for the foreseeable future.

The Future of Satellite DNA Studies – The Promise of New Technologies

Our understanding of the biology of satellite DNA will continue to accumulate through specific studies such as those mentioned above, but there is one current development that promises to increase our understanding of satellite DNA structure in a more global fashion: new sequencing technologies. As we write, the current generation of new technologies is provided by 454 (http://www.454.com), Solexa/Illumina (http://www.illumina.com/) and AB SOLiD (http://www.appliedbiosystems.com/). These all share, to greater or lesser degrees, a set of properties that distinguish them from conventional Sanger sequencing by gel or capillary electrophoresis: they provide more sequence data at a lower cost, the sequence is derived from individual molecules, and the reads (the length of contiguous sequence determined from one template) are short. The advantages of more data at lower cost make resequencing multiple whole human genomes a practical proposition. Craig Venter's DNA was resequenced using conventional capillary methods

(Levy et al., 2007) but few other human genomes are likely to be determined in this way. The sequence of James Watson's genome obtained using 454 technologies and that of an as yet unnamed Chinese individual by Solexa/Illumina are expected to be published soon, and plans to use these technologies to sequence as many as 1000 genomes in the next 2 years are being discussed. If satellites are difficult to sequence and assemble using long capillary reads, it might be even more hopeless a task to sequence them using short reads. Two compensating factors should, however, be taken into account. First, the new technologies differ from that used to generate the reference sequence in some important methodological aspects: they do not require cloning in bacteria, so one stage that reduces representation is eliminated. And with the new technologies, it is easier to resequence the entire genome, including the satellite sequences, than to attempt to purify a non-satellite fraction. So satellite sequence data will be generated on a much larger scale than ever before as an integral part of these efforts. Second, sequence can be obtained as read pairs: two reads that are known to originate from a certain distance apart in the same molecule. Read pairs can be assembled or mapped back to a reference sequence much more accurately than single reads because of the additional information they contain about their relative orientation and distance apart. Currently, read pairs are generated from molecules a few hundred or a few long. These would be sufficient to link flanking sequences to a few hundred "base pairs" or a few kilo bases of satellite DNA, and thus assemble the ends of an array, but not the middle part. But if in the future it were possible to generate read pairs separated by hundreds or thousands of bases, assembly of large satellite arrays would become possible. In addition, a further generation of sequencing methods based on much longer reads from single molecules is under development, although currently these are at the stage of press releases rather than scientific papers (e.g. Sequenom to Develop Third-Generation Nanopore-Based Single Molecule Sequencing Technology, Sep 24 2007; http://www.genengnews.com/news/bnitem.aspx?name=23610133). If such approaches became reality, the data could cover the entire genome in a more contiguous fashion. This exciting possibility of high-quality whole-genome sequences from large numbers of individuals provides a fitting vision of the future for a volume dedicated to 'Chromosomes to Genome'.

Acknowledgements

We thank Yali Xue for his help in producing Fig. 2.4. Our early work in Oxford was supported by the Cancer Research Campaign. The current work of CTS, and attendance at the 'Chromosomes to Genome' symposium, were supported by the Wellcome Trust.

References

1. Amor, D.J., and Choo, K.H.A. (2002). Neocentromeres: role in human disease, evolution, and centromere study. Am. J. Hum. Genet. *71*, 695–714.
2. Arnemann, J., Epplen, J.T., Cooke, H.J., Sauermann, U., Engel, W., and Schmidtke, J. (1987). A human Y-chromosomal DNA sequence expressed in testicular tissue. Nucleic Acids Res. *15*, 8713–8724.
3. Arnemann, J., Jakubiczka, S., Schmidtke, J., Schafer, R., and Epplen, J.T. (1986). Clustered GATA repeats (Bkm sequences) on the human Y chromosome. Hum. Genet. *73*, 301–303.

4. Babcock, M., Yatsenko, S., Stankiewicz, P., Lupski, J.R., and Morrow, B.E. (2007). AT-rich repeats associated with chromosome 22q11.2 rearrangement disorders shape human genome architecture on Yq12. Genome Res. *17*, 451–460.
5. Bao, W., Zhu, S., Pandya, A., Zerjal, T., Xu, J., Shu, Q., Du, R., Yang, H., and Tyler-Smith, C. (2000). *MSY2*: a slowly evolving minisatellite on the human Y chromosome which provides a useful polymorphic marker in Chinese populations. Gene *244*, 29–33.
6. Beridze, T. (1986). Satellite DNA (Berlin: Springer-Verlag).
7. Bobrow, M., Pearson, P.L., Pike, M.C., and el-Alfi, O.S. (1971). Length variation in the quinacrine-binding segment of human Y chromosomes of different sizes. Cytogenetics *10*, 190–198.
8. Cooke, H. (1976). Repeated sequence specific to human males. Nature *262*, 182-186.
9. Cooke, H.J., and Hindley, J. (1979). Cloning of human satellite III DNA: different components are on different chromosomes. Nucleic Acids Res. *6*, 3177–3197.
10. Cooke, H.J., Schmidtke, J., and Gosden, J.R. (1982). Characterisation of a human Y chromosome repeated sequence and related sequences in higher primates. Chromosoma *87*, 491–502.
11. Floridia, G., Zatterale, A., Zuffardi, O., and Tyler-Smith, C. (2000). Mapping of a human centromere onto the DNA by topoisomerase II cleavage. EMBO Rep. *1*, 489–493.
12. Harrington, J.J., Van Bokkelen, G., Mays, R.W., Gustashaw, K., and Willard, H.F. (1997). Formation of *de novo* centromeres and construction of first-generation human artificial microchromosomes. Nat. Genet. *15*, 345–355.
13. Heyer, E., Puymirat, J., Dieltjes, P., Bakker, E., and de Knijff, P. (1997). Estimating Y chromosome specific microsatellite mutation frequencies using deep rooting pedigrees. Hum. Mol. Genet. *6*, 799–803.
14. Hurles, M.E., Veitia, R., Arroyo, E., Armenteros, M., Bertranpetit, J., Pérez-Lezaun, A., Bosch, E., Shlumukova, M., Cambon-Thomsen, A., McElreavey, K., et al. (1999). Recent male-mediated gene flow over a linguistic barrier in Iberia, suggested by analysis of a Y-chromosomal DNA polymorphism. Am. J. Hum. Genet. *65*, 1437–1448.
15. International Human Genome Sequencing Consortium (2004). Finishing the euchromatic sequence of the human genome. Nature *431*, 931–945.
16. Jeffreys, A.J., Royle, N.J., Wilson, V., and Wong, Z. (1988). Spontaneous mutation rates to new length alleles at tandem-repetitive hypervariable loci in human DNA. Nature *332*, 278–281.
17. Jeffreys, A.J., Wilson, V., and Thein, S.L. (1985). Individual-specific 'fingerprints' of human DNA. Nature *316*, 76–79.
18. Jehan, Z., Vallinayagam, S., Tiwari, S., Pradhan, S., Singh, L., Suresh, A., Reddy, H.M., Ahuja, Y.R., and Jesudasan, R.A. (2007). Novel noncoding RNA from human Y distal heterochromatic block (Yq12) generates testis-specific chimeric *CDC2L2*. Genome Res. *17*, 433–440.
19. Jobling, M.A., Bouzekri, N., and Taylor, P.G. (1998). Hypervariable digital DNA codes for human paternal lineages: MVR-PCR at the Y-specific minisatellite, MSY1 (*DYF155S1*). Hum. Mol. Genet. *7*, 643–653.
20. Jobling, M.A., and Gill, P. (2004). Encoded evidence: DNA in forensic analysis. Nat. Rev. Genet. *5*, 739–751.
21. Jobling, M.A., and Tyler-Smith, C. (2003). The human Y chromosome: an evolutionary marker comes of age. Nat. Rev. Genet. *4*, 598–612.

22. Jones, K.W. (1970). Chromosomal and nuclear location of mouse satellite DNA in individual cells. Nature *225*, 912–915.
23. Kayser, M., Kittler, R., Erler, A., Hedman, M., Lee, A.C., Mohyuddin, A., Mehdi, S.Q., Rosser, Z., Stoneking, M., Jobling, M.A., et al. (2004). A comprehensive survey of human Y-chromosomal microsatellites. Am. J. Hum. Genet. *74*, 1183–1197.
24. Kit, S. (1961). Equilibrium sedimentation in density gradients of DNA preparations from animal tissues. J. Mol. Biol. *3*, 711–716.
25. Laberge, C., and Gagne, R. (1971). Quinacrine mustard staining solves the length variations of the human Y chromosome. Johns Hopkins Med. J. *128*, 79–83.
26. Lang, J.W., Aggarwal, R.K., Majumdar, K.C., and Singh, L. (1993). Individualization and estimation of relatedness in crocodilians by DNA fingerprinting with a Bkm-derived probe. Mol. Gen. Genet. *238*, 49–58.
27. Levy, S., Sutton, G., Ng, P.C., Feuk, L., Halpern, A.L., Walenz, B.P., Axelrod, N., Huang, J., Kirkness, E.F., Denisov, G., et al. (2007). The diploid genome sequence of an individual human. PLoS Biol. *5*, e254.
28. Mathias, N. (1993) Y chromosome DNA polymorphisms and human evolution, D.Phil thesis, University of Oxford.
29. Mathias, N., Bayes, M., and Tyler-Smith, C. (1994). Highly informative compound haplotypes for the human Y chromosome. Hum. Mol. Genet. *3*, 115–123.
30. Mirkin, S.M. (2007). Expandable DNA repeats and human disease. Nature *447*, 932–940.
31. Nakahori, Y., Mitani, K., Yamada, M., and Nakagome, Y. (1986). A human Y-chromosome specific repeated DNA family (DYZ1) consists of a tandem array of pentanucleotides. Nucleic Acids Res. *14*, 7569–7580.
32. Oakey, R. (1989). The structure of alphoid satellite DNA on normal and abnormal Y chromosomes, D.Phil thesis, University of Oxford.
33. Oakey, R., and Tyler-Smith, C. (1990). Y chromosome DNA haplotyping suggests that most European and Asian men are descended from one of two males. Genomics *7*, 325–330.
34. Ohzeki, J., Nakano, M., Okada, T., and Masumoto, H. (2002). CENP-B box is required for de novo centromere chromatin assembly on human alphoid DNA. J. Cell. Biol. *159*, 765–775.
35. Pena, S.D., Santos, F.R., Bianchi, N.O., Bravi, C.M., Carnese, F.R., Rothhammer, F., Gerelsaikhan, T., Munkhtuja, B., and Oyunsuren, T. (1995). A major founder Y-chromosome haplotype in Amerindians. Nat. Genet. *11*, 15–16.
36. Pinkel, D., Straume, T., and Gray, J.W. (1986). Cytogenetic analysis using quantitative, high-sensitivity, fluorescence hybridization. Proc. Natl. Acad. Sci. U. S. A. *83*, 2934–2938.
37. Prosser, J., Frommer, M., Paul, C., and Vincent, P.C. (1986). Sequence relationships of three human satellite DNAs. J. Mol. Biol. *187*, 145–155.
38. Santos, F.R., Pandya, A., Kayser, M., Mitchell, R.J., Liu, A., Singh, L., Destro-Bisol, G., Novelletto, A., Qamar, R., Mehdi, S.Q., et al. (2000). A polymorphic L1 retroposon insertion in the centromere of the human Y chromosome. Hum. Mol. Genet. *9*, 421–430.
39. Santos, F.R., Pena, S.D., and Tyler-Smith, C. (1995). PCR haplotypes for the human Y chromosome based on alphoid satellite DNA variants and heteroduplex analysis. Gene *165*, 191–198.
40. Singh, L., and Jones, K.W. (1982). Sex reversal in the mouse (*Mus musculus*) is caused by a recurrent nonreciprocal crossover involving the X and an aberrant Y chromosome. Cell *28*, 205–216.

41. Singh, L., and Jones, K.W. (1986). Bkm sequences are polymorphic in humans and are clustered in pericentric regions of various acrocentric chromosomes including the Y. Hum. Genet. *73*, 304–308.
42. Singh, L., Purdom, I.F., and Jones, K. W. (1980). Sex chromosome associated satellite DNA: evolution and conservation. Chromosoma *79*, 137–157.
43. Skaletsky, H., Kuroda-Kawaguchi, T., Minx, P.J., Cordum, H. S., Hillier, L., Brown, L. G., Repping, S., Pyntikova, T., Ali, J., Bieri, T., et al. (2003). The male-specific region of the human Y chromosome is a mosaic of discrete sequence classes. Nature *423*, 825–837.
44. Southern, E.M. (1970). Base sequence and evolution of guinea-pig α-satellite DNA. Nature *227*, 794–798.
45. Spence, J.M., Critcher, R., Ebersole, T.A., Valdivia, M.M., Earnshaw, W.C., Fukagawa, T., and Farr, C.J. (2002). Co-localization of centromere activity, proteins and topoisomerase II within a subdomain of the major human X alpha-satellite array. EMBO J *21*, 5269–5280.
46. Spence, J.M., Fournier, R.E., Oshimura, M., Regnier, V., and Farr, C.J. (2005). Topoisomerase II cleavage activity within the human D11Z1 and DXZ1 alpha-satellite arrays. Chromosome Res. *13*, 637–648.
47. Subramanian, S., Mishra, R.K., and Singh, L. (2003). Genome-wide analysis of Bkm sequences (GATA repeats): predominant association with sex chromosomes and potential role in higher order chromatin organization and function. Bioinformatics *19*, 681–685.
48. Swallow, D.M., Gendler, S., Griffiths, B., Corney, G., Taylor-Papadimitriou, J., and Bramwell, M.E. (1987). The human tumour-associated epithelial mucins are coded by an expressed hypervariable gene locus PUM. Nature *328*, 82–84.
49. The Huntington's Disease Collaborative Research Group (1993). A novel gene containing a trinucleotide repeat that is expanded and unstable on Huntington's disease chromosomes. Cell *72*, 971–983.
50. Tyler-Smith, C., and Brown, W.R. (1987). Structure of the major block of alphoid satellite DNA on the human Y chromosome. J. Mol. Biol. *195*, 457–470.
51. Tyler-Smith, C., Oakey, R.J., Larin, Z., Fisher, R.B., Crocker, M., Affara, N.A., Ferguson-Smith, M.A., Muenke, M., Zuffardi, O., and Jobling, M.A. (1993). Localization of DNA sequences required for human centromere function through an analysis of rearranged Y chromosomes. Nat. Genet. *5*, 368–375.
52. Tyler-Smith, C., Taylor, L., and Muller, U. (1988). Structure of a hypervariable tandemly repeated DNA sequence on the short arm of the human Y chromosome. J. Mol. Biol. *203*, 837–848.
53. Waring, M., and Britten, R.J. (1966). Nucleotide sequence repetition: a rapidly reassociating fraction of mouse DNA. Science *154*, 791–794.
54. Willuweit, S., and Roewer, L. (2007). Y chromosome haplotype reference database (YHRD): update. FSI Genet *1*, 83–87.
55. Wong, L.H., Brettingham-Moore, K.H., Chan, L., Quach, J.M., Anderson, M.A., Northrop, E.L., Hannan, R., Saffery, R., Shaw, M.L., Williams, E., and Choo, K.H.A. (2007). Centromere RNA is a key component for the assembly of nucleoproteins at the nucleolus and centromere. Genome Res. *17*, 1146–1160.
56. Zegura, S.L., Karafet, T.M., Zhivotovsky, L.A., and Hammer, M.F. (2004). High-resolution SNPs and microsatellite haplotypes point to a single, recent entry of Native American Y chromosomes into the Americas. Mol. Biol. Evol. *21*, 164–175.

Genome to Epigenome

3

Rakesh K. Mishra

Centre for Cellular and Molecular Biology, Uppal Road, Hyderabad, India

Flow of Genetic Information

One of the most fascinating events in the biological world is the development of an embryo from a simple looking egg. Witnessing a germinating seed, hatching of young creature from the eggshell and birth of a baby are the most wonderful scene of biological marvels that we ever see. Wonderful because we do not understand how from a single cell a creature develops and how a small seed gives rise to a giant tree. And all this happens by means of the genetic endowment that every creature inherits from its parents. The genetic information that is present is in the form of its genome. Ultimately, all the development and all the evolution funnel through the information system of the genome. Much of the efforts in modern biology is still about how this information system works. While great progress has been made in the fields of genetics and molecular biology, we still do not quite understand how information is encoded and expressed during embryonic development leading to thousands of molecular and cellular events that flawlessly execute the complex regulatory and growth processes and make a complex organism from a simple zygotic cell. When this information system falters we encounter disease conditions.

During the last century, much became known about DNA – it being the genetic material and how sequence of DNA determines the nature of proteins that it codes for. Following these landmark discoveries, past few decades have seen great advances in our ability to analyse cellular events at molecular levels. One of the epoch making events in the history of modern biology that took place at the very end of last millennium was the sequencing of the human genome. Since then large number of genomes of different organisms have been sequenced. This new field of biology – genomics – has changed the way biology is done now. It has also remarkably enhanced the degree of our understanding of genetic basis of life, evolution, development and disease. In spite of these great achievements, we are far away from full understanding of these very basic processes of the life on the earth. What is important is though that there seems to be a way to reach there.

One Genome Multiple Epigenomes

Every organism begins its life as a single cell. By series of events during embryonic development, this single cell gives rise to a large variety of cells. For example, an embryonic stem cell of human gives rise to more than two hundred different kinds of cells that interact and communicate to make our body and carry out its functions (Mellor et al., 2008). Although different cell types of an organisms look and behave in very distinct manner, like a neuron, a muscle cell or an epithelial cell lining the alimentary canal do, the genetic information present in these different kind of cells is identical. What brings about the differences is the regulation of the genetic information that is expressed in different ways in different cell types. A muscle cell will express unique set of genes and so will a nerve cell or a liver cell – the set of gene products that gives the identity to these cells. Many complex cellular events, signalling processes, cell-cell interactions and gene regulatory events contribute to the making of an organism from a single cell. Much remains to be understood about the precise mechanisms behind these events.

Whatever is the mechanism, the end product of embryonic development may be considered the formation of different cell types. Different set of genes being active and rest safely silenced in respective cell types. It becomes immediately obvious that regulation of genes plays the central role in embryonic development. There are many ways of regulating genes and embryonic developmental process, probably uses all of them. Direct repression or activation of a gene by one or more regulatory protein is only of limited use as regulators also need to be regulated and the cascade of regulatory factors can have only limited steps. Among different ways of regulating genes the most important and effective one is to keep the unwanted genes hidden from the transcriptional machinery and to expose only those that need to be transcribed during the life time of the cell type. Therefore, packaging the genome in cell type specific manner is central to the developmental process (Iqbal and Mishra, 2007; Vasanthi and Mishra, 2008).

In higher eukaryotes, long genomic DNA needs to be packaged in a few micron size nucleus, a process that requires several tens of thousands fold compaction. It is intuitive, in retrospect, that the complex process of genomic packaging – a.k.a. chromatinization – may have major regulatory consequences. For example, the genome can be packaged in different ways that eventually leads to different sets of gene accessible or inaccessible for transcription. Large body of evidence is now accumulating that points to this global strategy of regulation of genes. The packaging of one genome into different structural and functional states is accompanied by numerous molecular marks including DNA methylation, histone tail modifications and the sets of non-histone chromatin proteins that associate with specific regions of the genome. These molecular features are collectively referred to as epigenetic marks and different structural and functional states of a genome as epigenome. Identical DNA sequence in a genome is converted in many cell type specific epigenomes during the process of embryonic development, Fig. 3.1. While several techniques are now available to analyse various features of an epigenome – what regions are more accessible, what parts have cytosine methylation, what part of the genome has which histone modification and what chromatin proteins with active or repressive function are associated at which part of the genome – principles or sequence features that control the formation of different epigenomes is largely unknown.

Packaging Code in Genomes

Recent availability of genome sequences has thrown light on the blueprint of the making of different organisms although we do not understand the full language of this genome blueprint. In fact, one of the biggest challenges of the post-genomics era is to understand the function of various features of genomic sequences beyond the protein coding part. While comparing the genome of various organisms, it has become clear that complex organisms have bigger genome but the surprising finding was that complex organisms do not have more genes in that proportion, Fig. 3.2. It is obvious, therefore, that accumulation of genomic sequences other than those coding for proteins

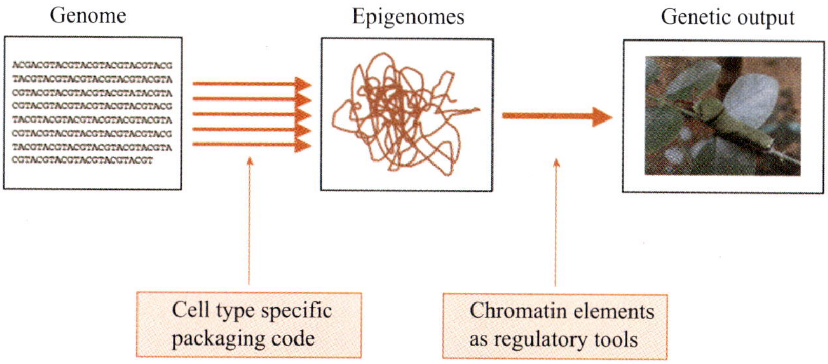

Fig. 3.1 Flow of genetic information. Genome gets packaged into cell type-specific epigenome with the help of chromatin proteins. Although it is likely that the information to package the genome into varying forms lies within the genome, this "packaging code" is not known. Once packaged into epigenome, chromatin-mediated regulatory elements come into existence and lead to the complete functional genetic output of a genome – the organism.

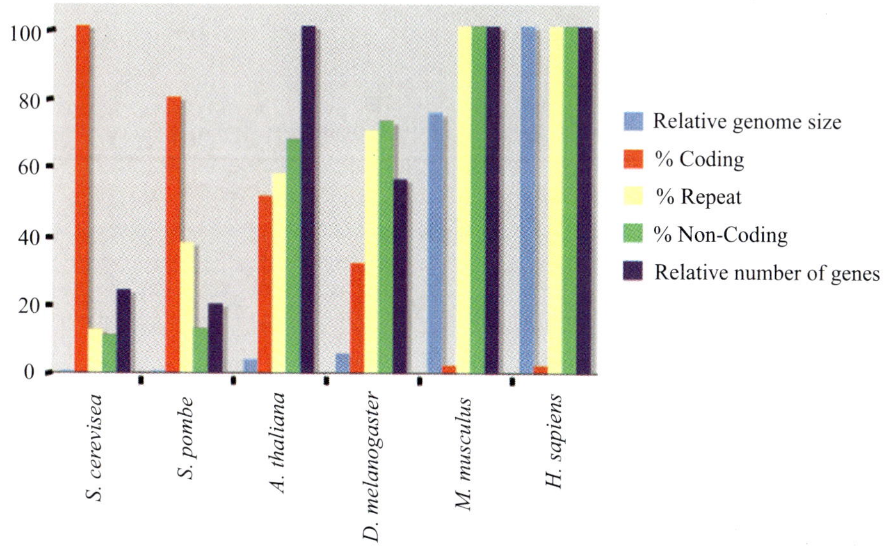

Fig. 3.2 Evolutionary trends in genomes of eukaryotes. Genome size in contrast to number of genes, proportion of coding, non-coding and repetitive DNA in genome of organisms across species of varying degree of complexity is shown. More complex life forms have more genetic material but this increase in the size of the genome does not correspond to more number of genes. In fact, much of the DNA that has accumulated during the evolution of higher eukaryotes is non-coding and repetitive in nature. Less complex organisms have higher proportion of coding genome while more complex organisms have more of non-coding DNA and less of protein coding DNA.

has led to the evolution of complexity (Taft et al., 2007). Since genes code for proteins that are the building blocks for all living forms and function as workhorses for a cell, variety in all life forms reflects the variety in proteome of different cell types. It seems likely, therefore, that the complexity has evolved by not more number of genes but more sophisticated regulation of pre-existing genes.

Major bulk of different classes of sequences that have accumulated during evolution of complex organisms is contributed by repetitive elements, Fig. 3.3. In fact most of the human genome is repetitive in nature, Fig. 3.3. One of the biggest challenges is to find out what is the function of this part of genome that has steadily accumulated during the evolution of complex life forms. Keeping in view the increasing number of cell types or epigenomes in complex life forms, it is tempting to speculate that potential of larger variety of epigenome if a genome is related to both the variety and the quantity of repetitive elements.

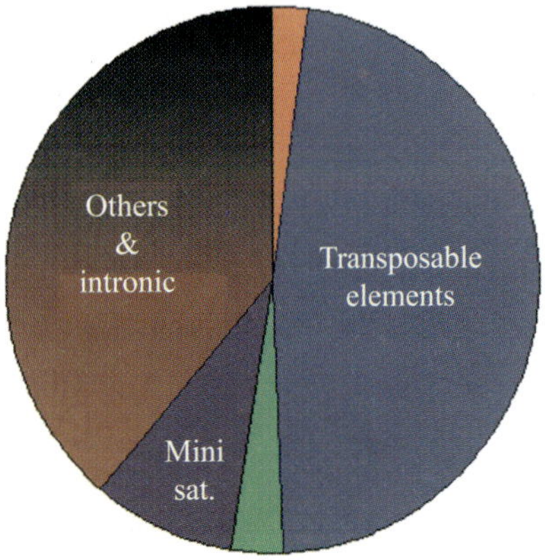

Fig. 3.3 Composition of the human genome. Size of the human genome is 3.2Gb of which 2.95Gb is euchromatic and rest is repetitive. Only ~1.4% of the human is protein coding which is less than the simple sequence repeat component.

Chromatin-Mediated Regulatory Elements

Folding of genome into chromatin not only allows long DNA molecules to fit within the nuclear volume but also by so doing makes only a subset of genes accessible to transcriptional apparatus leaving rest of the genes repressed as they are rendered inaccessible to transcriptional apparatus. Chromatinization also brings into existence the regulatory elements that function by means of higher order chromatin organization, Fig. 3.1. Two such elements are chromatin domain boundaries and the epigenetic cellular memory elements. Chromatin domain boundaries subdivide genome into structural and functional domains. Several studies have shown regulatory role of these boundary elements during embryonic development (Mihaly et al., 1998; Mishra and Karch, 1999; Iqbal and Mishra, 2007). The cellular memory elements help remember the expression state of genome in a cell type and re-establish the same expression state after cell division (Vasanthi and Mishra, 2008). Failure in this cellular memory system often leads to mis-expression of genes resulting to developmental defects and disease conditions.

Functional Elements in the Non-Coding Genome

Chromatin-mediated regulatory elements, viz., boundary and memory elements, have been functionally assayed *in vivo* but we do not know enough about them to identify or predict their occurrence in the genomes. It is, however, expected that such element may constitute as much of the human genome as the protein-coding region. New computational tools need to be developed to annotate the genome for its full functional features. While it is clear that much of the regulatory sequences that help in packaging the genome in cell type-specific manner during development lie within the non-coding part of the genome, this packaging code is yet to be deciphered.

Genomes of complex organisms are full of non-coding DNA with a very small fraction being the protein-coding genes. For example, in human genome only ~1.4% is protein coding. About half of our genome is made of repetitive elements. It has been a major topic of debate for long time whether or not the repetitive DNA has any role for the organisms or it is some kind of 'junk' or 'selfish' DNA (Crick et al., 1979; Orgel and Crick, 1980; Orgel et al., 1980). Several lines of studies have suggested, mostly in recent past, that repetitive DNA is functionally relevant for the cell. Recent studies have shown that variety of repetitive elements indeed have functional significance (Lunyak et al., 2007).

Availability of sequence of number of genomes allows us to compare the genetic material of different organisms and analyse the similarities and differences in the context of the features of those organisms. One of the major surprises of genome projects has been the finding that from simple to complex organisms the size of the genome increased but the number of genes did not increase in that proportion. So much so that in complex eukaryotes like human, most of the genome is non-coding and repetitive in nature with a tiny fraction coding for proteins. Even the genes that have been added in the human genome compared to simpler organisms like *Drosophila* most of them are transcription factors that regulate expression of genes (Adams et al., 2000; Rubin et al., 2000). Why then so much of DNA has been accumulating during evolution of complexity – why humans have 20 times more DNA but only 1.5 times more number of genes? The indications are that it is not the genes but the sophisticated regulation of genes that leads to evolution of complexity and that the regulatory information has been accumulated in the non-coding part of the genome.

There are several trends emerging from the comparative genomics studies that indicate that in the evolution of complex organisms regulation of existing genes and not addition of new genes have been the major factors. Addition of non-coding part of the genome should, therefore, be taken as addition of new regulatory features. The question is how to look for novel regulatory features in the genome.

Non-Random Abundance and Distribution of Simple Sequence Repeats

Several computational methods added with comparative genomic approaches are available to predict protein-coding part of the genome with reasonable accuracy. On the contrary, predicting functional features of non-coding part of genome is still a challenge. There are two simple ways to look for novel regulatory elements in the non-coding part of genome. First one is the comparison of non-coding part of genome between closely related or evolutionarily distant species for specific purposes. A conservation emerging in these analyses can be taken as the indication for functional relevance. Comparison of genome sequence of different species of *Drosophila* has led to identification of enhancer elements that were not known before. Comparison of genomes of distant species, specially vertebrates, have led to identification of ultra-conserved regions – long stretches of hundreds of nucleotides that show close to 100% conservation from fish to human reflecting a time scale of ~450 million years (Sabarinadh et al., 2004). Several thousand such stretches are present in the human genome that constitutes 2–3% of the genome, much more than the coding part of the genome (Bejerano et al., 2004). While it is obvious that these sequences are under tremendous selection pressure, indicative of their functional significance, it is not known what that function is (Drake et al., 2006; Kim and Pritchard, 2007). All vertebrates have these ultra-conserved elements and in most of the cases, these sequences are present near developmentally regulated genes. Furthermore, size of the ultra-conserved elements increases from fish to human suggesting that once recruited to the genome of the common ancestor of vertebrates, these elements have been further strengthened by extending the stretch of conserved region (Sabarinadh et al., 2004). In spite of these exciting features associated with this class of elements, their function remains unknown. It is highly likely that these sequences perform essential and novel function. Large-scale mutational analysis or novel transgene-based reporter assays will be needed to uncover their function.

The second approach to find function of non-coding DNA is for the repetitive part of the genome where emergence of any pattern, for example, unusual abundance and non-random distribution/association with genes, may be taken as indicative of functional relevance. Recent studies have shown that all repeats are not equal in terms of their accumulation in the genome and that distribution of repeats is often non random in the genome, Fig. 3.4 (Subramanian et al., 2003b). One of the very interesting findings has been that among simple sequence repeats, some repeats have higher abundance of longer repeat as compared to smaller ones, a trend that is totally unexpected.

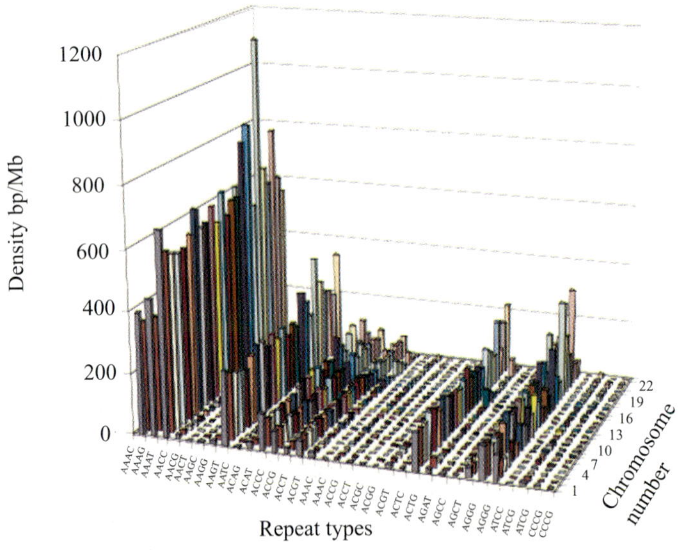

Fig. 3.4 Variable abundance of different repeats in the human genome. All possible tetramer repeats are plotted for their abundance on different chromosomes in the human genome. While few repeats are more abundant than many others, majority of repeats are poorly represented genome wide. This preferential enrichments of selected repeats indicates evolutionary advantage to the genome that may have acted for positive selection of these repeats (taken from Subramanian et al., 2003. Genome Biol. *4*, R13).

For example, in case of GATA tetranucleotide repeat, $GATA_{10-12}$ is far more abundant than $GATA_6$ repeat (Subramanian et al., 2003a). This clearly means that longer repeats have been enriched in the genome that has provided some advantage to the system. We suggest that such longer repeats may function to organize genome and may help in coordinated regulation of large number of genes.

Functional Potential of ncRNA

While a very small fraction of genome codes for proteins in complex organisms, is now known that much of the genome including the non-coding part is transcribed. Why is non-coding DNA transcribed? Many studies have indicated that non-coding RNAs (ncRNA) have variety of regulatory functions including organization of chromatin, packaging of genome, nuclear architecture and post-transcriptional regulations of genes (Mattick, 2004; Mattick, 2005; Kenzelmann et al., 2006; Mattick, 2007; Mattick and Makunin, 2006; Mehler and Mattick, 2006; Pang et al., 2006;

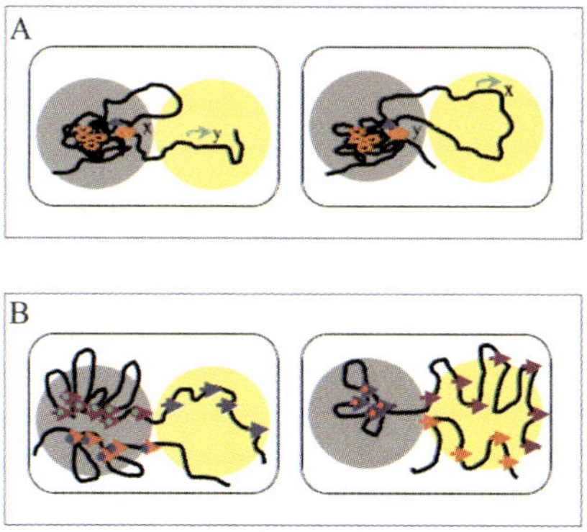

Fig. 3.5 Role of repetitive elements in genomic packaging and gene regulation – a model. Nucleus is highly organized structure that contains various functionally defined compartments, for example, transcriptionally silent heterochromatic compartments shown as gray circle and transcriptionally active compartments shown as yellow circles. We suggest that repetitive elements of the genome contribute to the specific association of genetic loci to the relevant nuclear compartments. (**A**) Switch mediated by association of genes to heterochromatin. Repetitive elements associate with heterochromatin and locate in transcriptionaly silent compartments. Red circles show proteins that interact with repetitive DNA to organize such compartments. Pentagons and triangles are the sequence-specific factors that interact with regulatory regions of the gene on the one hand and heterochromatic foci on the other through protein–protein interactions. When regions of gene 'X', through such specific interactors are brought in the vicinity of heterochromatin the gene is silenced. While a neigbouring gene "Y" stays in the active region and is transcribed, indicated by green arrowhead. In a different cell type where the protein constituents are different, 'Y' is dragged to the repressive compartment and 'X' is allowed to express. (**B**) Global and coordinated regulation of gene mediated by repetitive elements. Genomic regions marked with different repeats (magenta, blue and red arrowheads) are shown. Depending on the presence of the proteins that can specifically interact with these repeats, the corresponding region of the genome is sequestered into transcriptionaly silent or active compartments. For example, in a cell type where proteins interacting with magenta and red repeats (shown as green and blue dots) are expressed the genomic regions associated with these repeats is packaged into the repressive compartment while the regions marked with blue repeat elements is in active compartments and competent for transcription. In a different cell type where repeat DNA interacting proteins scenario changes, blue repeat interacting protein (shown as red dot) is present, the corresponding region assembles in the inactive compartment while other regions where such repeats are absent is retained in the transcriptionaly active compartments.

Dinger et al., 2008). Size of complex genome, therefore, also increased to acquire ncRNA features. Large number of ncRNAs remain to be identified and the subsequent functional analysis of these molecules will be the major task in this area.

Conclusion

The new biology of genome has allowed us to ask questions that were not in the reach before. Now we can hope to understand the full complement of the genetic element of an organism. This goal is quite far though and we need to devise new methods to dissect out the function of non-coding part of the genome. Recent technical advances have enabled us to start this next big task of genome biology. A very powerful approach that has emerged recently to identify functionally relevant parts in the non-coding genome is the high throughput ChIP (chromatin immuno precipitation) that allows identification of sites on the genome associated with specific proteins. It also includes modifications of histone tails that reflect the epigenetic state of the corresponding part of the genome. Such high-throughput methods are now being employed to unravel the epigenetic states of the entire genome in different cell types. In this direction, study of non-coding part of the genome and in particular the repetitive elements is turning out to be important. Repetitive elements can create structural hubs within various compartments of the nucleus, Fig. 3.5. By nature, repetitive elements can be handy to mark large number of loci for coordinated regulation by small number of transacting factors. Non-random distribution, accumulation of longer repeats and the association of repeats with additional regulatory elements of chromatin are strong indicators of this new global regulatory mechanism. With the emergence of novel and high-throughput methods that promise comprehensive analysis of epigenetic states of the genome, next few years are going to be exciting in understanding how genetic information system works.

References

1. Adams, M.D., Celniker, S.E., Holt, R.A., Evans, C.A., Gocayne, J.D., Amanatides, P.G., Scherer, S.E., Li, P.W., Hoskins, R.A., Galle, R.F., et al. (2000). The genome sequence of *Drosophila melanogaster*. Science *287*, 2185–2195.
2. Bejerano, G., Pheasant, M., Makunin, I., Stephen, S., Kent, W.J., Mattick, J.S., and Haussler, D. (2004). Ultraconserved elements in the human genome. Science *304*, 1321–1325.
3. Crick, F.H., Wang, J.C., and Bauer, W.R. (1979). Is DNA really a double helix? J. Mol. Biol. *129*, 449–457.
4. Dinger, M.E., Amaral, P.P., Mercer, T.R., Pang, K.C., Bruce, S.J., Gardiner, B.B., Askarian-Amiri, M.E., Ru, K., Solda, G., Simons, C., et al. (2008). Long noncoding RNAs in mouse embryonic stem cell pluripotency and differentiation. Genome Res. [in press].
5. Drake, J.A., Bird, C., Nemesh, J., Thomas, D.J., Newton-Cheh, C., Reymond, A., Excoffier, L., Attar, H., Antonarakis, S.E., Dermitzakis, E.T., and Hirschhorn, J.N. (2006). Conserved noncoding sequences are selectively constrained and not mutation cold spots. Nat. Genet. *38*, 223–227.
6. Iqbal, H., and Mishra, R.K. (2007). Chromatin domain boundaries: defining the functional domains in genome. Proc. Indian Natn. Sci. Acad. *73*, 239–253.

7. Kenzelmann, M., Rippe, K., and Mattick, J.S. (2006). RNA: networks & imaging. Mol. Syst. Biol. *2*, 44.
8. Kim, S.Y., and Pritchard, J.K. (2007). Adaptive evolution of conserved noncoding elements in mammals. PLoS Genet. *3*, 1572–1586.
9. Lunyak, V.V., Prefontaine, G.G., Nunez, E., Cramer, T., Ju, B.G., Ohgi, K.A., Hutt, K., Roy, R., Garcia-Diaz, A., Zhu, X., et al. (2007). Developmentally regulated activation of a SINE B2 repeat as a domain boundary in organogenesis. Science *317*, 248–251.
10. Mattick, J.S. (2004). RNA regulation: a new genetics? Nat. Rev. Genet. *5*, 316–323.
11. Mattick, J.S. (2005). The functional genomics of noncoding RNA. Science *309*, 1527–1528.
12. Mattick, J.S. (2007). A new paradigm for developmental biology. J. Exp. Biol. 210, 1526-1547.
13. Mattick, J.S., and Makunin, I.V. (2006). Non-coding RNA. Hum. Mol. Genet. 15 *Spec No 1*, R17–29.
14. Mehler, M.F., and Mattick, J.S. (2006). Non-coding RNAs in the nervous system. J. Physiol. *575*, 333–341.
15. Mellor, J., Dudek, P., and Clynes, D. (2008). A glimpse into the epigenetic landscape of gene regulation. Curr. Opin. Genet. Dev. *18*, 116–122.
16. Mihaly, J., Hogga, I., Barges, S., Galloni, M., Mishra, R.K., Hagstrom, K., Muller, M., Schedl, P., Sipos, L., Gausz, J., et al. (1998). Chromatin domain boundaries in the Bithorax complex. Cell. Mol. Life. Sci. *54*, 60–70.
17. Mishra, R.K., and Karch, F. (1999). Boundaries that demarcate structural and functional domains of chromatin. J. Biosci. *24*, 377–399.
18. Orgel, L.E., and Crick, F.H. (1980). Selfish DNA: the ultimate parasite. Nature *284*, 604–607.
19. Orgel, L.E., Crick, F.H., and Sapienza, C. (1980). Selfish DNA. Nature *288*, 645–646.
20. Pang, K.C., Frith, M.C., and Mattick, J.S. (2006). Rapid evolution of noncoding RNAs: lack of conservation does not mean lack of function. Trends Genet *22*, 1–5.
21. Rubin, G.M., Yandell, M.D., Wortman, J.R., Gabor Miklos, G.L., Nelson, C.R., Hariharan, I.K., Fortini, M.E., Li, P.W., Apweiler, R., Fleischmann, W., et al. (2000). Comparative genomics of the eukaryotes. Science *287*, 2204–2215.
22. Sabarinadh, C., Subramanian, S., Tripathi, A., and Mishra, R.K. (2004). Extreme conservation of noncoding DNA near HoxD complex of vertebrates. BMC Genomics 5, 75.
23. Subramanian, S., Mishra, R.K., and Singh, L. (2003a). Genome-wide analysis of Bkm sequences (GATA repeats): predominant association with sex chromosomes and potential role in higher order chromatin organization and function. Bioinformatics *19*, 681–685.
24. Subramanian, S., Mishra, R.K., and Singh, L. (2003b). Genome-wide analysis of microsatellite repeats in humans: their abundance and density in specific genomic regions. Genome Biol. 4, R13.
25. Taft, R.J., Pheasant, M., and Mattick, J.S. (2007). The relationship between non-protein-coding DNA and eukaryotic complexity. Bioessays *29*, 288–299.
26. Vasanthi, D., and Mishra, R.K. (2008). Epigenetic regulation of genes during development: a conserved theme from flies to mammals. J. Genet. Genomics *35*, 413–429.

Coordinated Demethylation and Methylation of Histones Controls Heterochromatin Formation and Gene Silencing in *Drosophila*

4

Gunter Reuter* and Sandro Lein

Institute of Biology/Genetics, Martin Luther University Halle-Wittenberg, Weinbergweg, Germany

Introduction

Epigenetic programming in higher organisms represents a fundamental process in control of development. During differentiation, one genome becomes successively converted into many different epigenomes, a process connected with determination of specific developmental programmes. Differential higher order packaging of DNA defines which genes become activated or remain silent throughout consecutive development. Consequently, epigenetic programming results in selection of the cells specific programme from all possible developmental programmes of an organism depending on a network of spatial and temporal constraints. The selected program is maintained throughout the consecutive development because epigenetic states are stably inherited through mitosis. Reversion of differentiation therefore requires epigenetic reprogramming within the genome. All these processes depend on higher order chromatin structures, which are primarily controlled by differential indexing of histones with covalent modifications.

Molecular dissection of epigenetic processes depends on the identification of phenotypes causally connected with defects in epigenetic programming. Phenotypes caused by epigenetic defects were studied in many genetic model organisms including mating type switching and telomere position effects in *Saccharomyces cerevisiae*, centromere position-effects in *Schizosaccharomyces pombe*, repeat induced point mutations (RIP) in *Neurospora crassa*, X chromosome silencing in the germ line of *Caenorhabditis elegans*, heterochromatic gene silencing in *Drosophila melanogaster*, *Arabidopsis thaliana* and in different mammalian systems (cf. Allis et al., 2006). One of the first experimental systems used in epigenetic research represents the phenomenon of position-effect variegation (PEV) in *Drosophila*.

Already in 1930, Muller (1930) described the first X-ray induced rearrangements showing a white variegated phenotype. Genetic and cytological analysis consecutively revealed that heterochromatization of the *white* gene causes its variegated expression in these rearrangements (Schultz, 1936; Demerec, 1940; Prokofyeva-Belgovskaya, 1947; Rudkin, 1965; Hartmann-Goldstein, 1976). These classical studies demonstrated that in PEV, the variegated expression of euchromatic loci reflects a change in epigenetic programming of euchromatic regions after their placement into a new neighbourhood of constitutive heterochromatin. Heterochromatic regions, which remain condensed and densely stained throughout the cell cycle are regularly found as blocks of pericentromeric heterochromatin, at nucleus organizer regions and at telomeres. These regions contain high levels of histone H3 methylated at lysine 9 (Schotta et al., 2002; Peters et al., 2003; Rice et al., 2003). Using chromatin immunoprecipitation (ChIP) analysis, spreading of histone H3K9 methylation into the euchromatic *white* gene region relocated adjacent to heterochromatin in the *In(1)w^{m4}* rearrangement could recently be shown (Rudolph et al., 2007). In *Drosophila*, dimethylation of H3K9 is enriched in pericentromeric heterochromatin (Schotta et al., 2002; Ebert et al., 2004) and this mark spreads in the *w^{m4}* PEV rearrangement into neighbouring euchromatin (Fig. 4.1) not only showing the molecular basis of spreading effects in PEV rearrangements but also demonstrating that the easily identifiable variegated expression of the *white* genes is caused by a defect in epigenetic programming. Therefore, PEV represents a unique model system for genetic dissection of molecular processes controlling epigenetic programming in *Drosophila* (Schotta et al., 2003b; Ashburner et al., 2005).

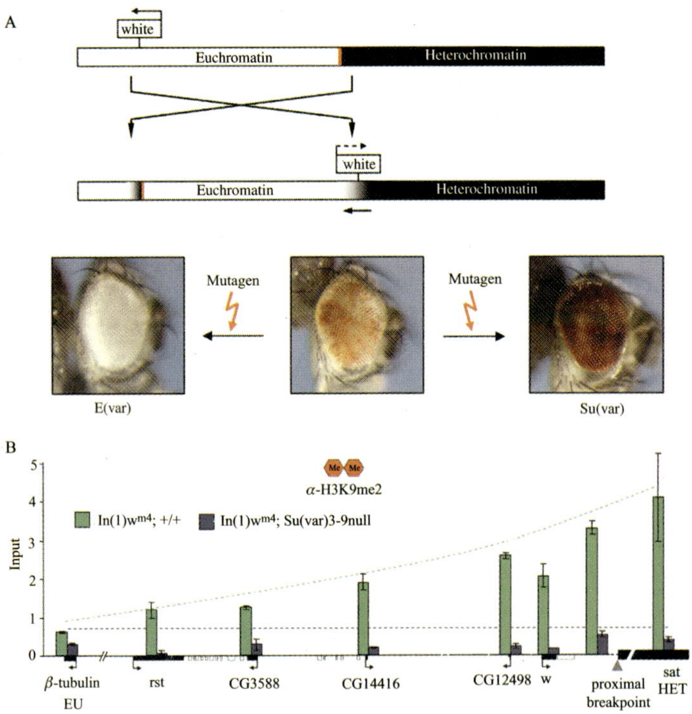

Fig. 4.1 In position-effect variegation (PEV) the variegated expression of euchromatic genes reflects defects in epigenetic programming. (**A**) The *In(1)w^{m4}* PEV rearrangements represent a model system for genetic dissection of epigenetic functions. In *w^{m4}*, the *white* gene is juxtaposed to constitutive heterochromatin of the X chromosome located distally to the nucleolus organizer regions. All *w^{m4}* flies show a salt-and-pepper variegation pattern of *white* allowing easy identification of dominant *Su(var)* and *E(var)* mutations after mutagen treatment. (**B**) Spreading of the heterochromatic H3K9me2 histone methylation mark along the euchromatin flanking the heterochromatic breakpoint can be demonstrated by chromatin immunoprecipitation (ChIP) analysis. In *w^{m4}*, a gradient of H3K9me2 along the euchromatic *white-roughest* region is found which depends on the function of the SU(VAR)3–9 histone methyltransferase.

Modifiers of Position-Effect Variegation and the Genetic Dissection of Epigenetic Processes

Many different PEV rearrangements have been isolated in *Drosophila* (Lindsley and Zimm, 1992; Flybase, 2008) but only very few show genetic characteristics desirable for systematic genetic dissection of epigenetic processes involved in heterochromatization of euchromatic genes. One of these experimentally advantageous rearrangements is the X chromosomal inversion $In(1)w^{m4}$ (Fig. 4.1). The chromosome is homozygous viable and all flies show a clear-cut salt-and-pepper white variegation pattern, which is very sensitive to any of the known modifying factors such as addition or deletion of heterochromatin (Reuter et al., 1983).

In PEV, chromatin packaging in a heterochromatic configuration spreads into neighbouring euchromatin because the buffer zone usually found between pericentromeric heterochromatin and proximal euchromatin is not present. Consequently, mutations, that suppress or enhance heterochromatic spreading in PEV rearrangements should affect epigenetic programming in both euchromatin and heterochromatin. It has been suggested that *Su(var)* mutations affect establishment of a heterochromatic chromatin state whereas *E(var)* mutations indicate genes which function in control of the euchromatic chromatin conformation. Using the $In(1)w^{m4}$ rearrangements, mass isolation screens of dominant PEV *Su(var)* and *E(var)* mutations were performed in order to identify genes controlling high order chromatin conformation in *Drosophila* (Reuter and Wolff, 1981; Sinclair et al., 1983). Altogether more than 500 mutations have been isolated and genetically characterized (cf. Schotta et al., 2003b). Since the first mutations, which suppress or enhance PEV were identified (Schultz, 1950; Spofford, 1967), at present about 150 modifiers of PEV genes have been mapped. This high number of genes suggests complex molecular regulation of epigenetic processes. However, until now only a small set of these genes has been molecularly defined (cf. Elgin and Reuter, 2006).

Genetic analysis of PEV modifier loci allowed differentiation of four groups of genes. All genes as indicated by deficiency analysis show a haplo-insufficient modifier phenotype (Szidonya and Reuter, 1983; Reuter et al., 1987; Wustmann et al., 1989). Only a small number of loci show triplo-dependent phenotypic effects. The three genes *Su(var)2–5*, *3–7* and *3–9*, all characterized by loss-of-function *Su(var)* and triplo-dependent *E(var)* effects have first been studied molecularly. All these genes encode heterochromatin-associated proteins. *Su(var)3–7* encodes a zinc finger protein (Reuter et al., 1990; Delattre et al., 2000; Jaquet et al., 2006), *Su(var)2–5* the heterochromatin protein HP1 (Eissenberg et al., 1990) and *Su(var)3–9* a histone H3K9 methyltransferase (Tschiersch et al., 1994; Rea et al., 2000; Schotta et al., 2002). Other molecularly defined genes do not show triplo-dependent effects and encode (*Su(var)3–6*) the serine/threonine protein phosphatase PP1 (Baksa et al., 1993), the DNA- and RNA-binding protein Modulo (Krejci et al., 1989), the PIAS protein (*Su(var)2–10*) which is a negative regulator of JAK/STAT signalling pathway (Hari et al., 2001) or *Su(var)3–1* encoding the H3S10 kinase JIL1 (Wang et al., 2001; Ebert et al., 2004). Many others of the cloned PEV modifier genes encode enzymes controlling histone modification. Enzymatic activity in histone methylation was first shown for SU(VAR)3–9 proteins (Rea et al., 2000; Nakayama et al., 2001; Schotta et al., 2002). Other *Su(var)* genes encode histone acetylases or deacetylases (Mottus et al., 2000; Grienenberger et al., 2002). Antagonistic interactions in epigenetic programming between histone methylation and demethylation are indicated by the strong suppressor effect of *Su(var)3–3* mutations. The gene encodes the *Drosophila* homolog of the H3K4me1 and H3K4me2 demethylase LSD1 (Rudolph et al., 2007). In summary, molecular studies of *Su(var)* genes contributed substantially to our recent knowledge about molecular factors controlling epigenetic programming in higher organisms.

Su(var) Mutations and the Detection of a Conserved Pathway Controlling Heterochromatic Gene Silencing

With molecular cloning of *Su(var)2–5* and *Su(var)3–9,* two key components controlling heterochromatin formation have been identified. *Su(var)2–5* encodes the heterochromatin-associated protein HP1 and has been cloned using a monoclonal antibody recognizing an antigene preferentially associated with heterochromatin (James and Elgin, 1986; James et al., 1989). Sequence comparison between the HP1 and Polycomb (Pc) proteins resulted in identification of the chromodomain (Paro and Hogness, 1991), which is shown to recognize specific histone methylation marks. The chromodomain of HP1 binds preferentially to H3K9me2 and H3K9me3, whereas the chromodomain of Pc binds strongly to H3K27me3 (Lachner et al., 2001; Bannister et al., 2001; Fischle et al., 2003).

The *Su(var)3–9* gene was first isolated in *Drosophila* by transposon tagging (Tschiersch et al., 1994) and resulted in detection of the SET domain as a conserved C-terminal region in the three chromatin proteins SU(VAR)3–9, E(Z) and TRX. All these proteins represent histone methyltransferases. Beside the SET domain the SU(VAR)3–9 protein also contains a chromodomain whose function is still unknown. Like HP1 SU(VAR)3–9 also preferentially associates with heterochromatin (Schotta et al., 2002). The two proteins are highly conserved and meanwhile HP1 and SU(VAR)3–9 homologues have been identified in more than 40 different species (Krauss et al., 2006). The SU(VAR)3–9 specific combination of a chromodomain and a SET domain is found in most species with the only exception of *Dictyostelium* and *Euascomycetes*-like *Neurospora.* Another interesting finding of these evolutionary studies consists of the observation that 11 of the 19 described deleterious point mutations isolated in *Drosophila* SU(VAR)3–9 (Ebert et al., 2006) appear to be compensable through accompanying substitutions (Krauss et al., 2006).

Functional conservation between *Drosophila* and human SU(VAR)3–9 proteins has been documented by an analysis of transgenic *Drosophila* lines (Aagaard et al., 1999; Schotta et al., 2002). Human SUV39H1-EGFP fusion protein expressed in *Drosophila* associates with heterochromatin, shows *in vivo* interaction with *Drosophila* HP1 and if expressed correctly rescues perfectly the silencing defects caused by a loss of the endogenous *Drosophila Su(var)3-9* gene.

Functional analysis with mutations of the *Su(var)2–5*, *Su(var)3–9* and *Suv4–20* genes revealed a molecular pathway of heterochromatic gene silencing (Fig. 4.2A), which is conserved between *Drosophila* and mammalian systems (Schotta et al., 2002; Schotta et al., 2003a; Ebert et al., 2004 and Schotta et al., 2004). The *Suv4–20* gene encodes a H4K20 methyltransferase (Schotta et al., 2004).

SU(VAR)3–9 is the main H3K9 KMTase in *Drosophila* (Schotta et al., 2002; Schotta et al., 2003a; Ebert et al., 2004) and in loss-of-function mutations most of the H3K9me2 is lost with exception of the fourth chromosome, the telomeric regions and several sites located along the

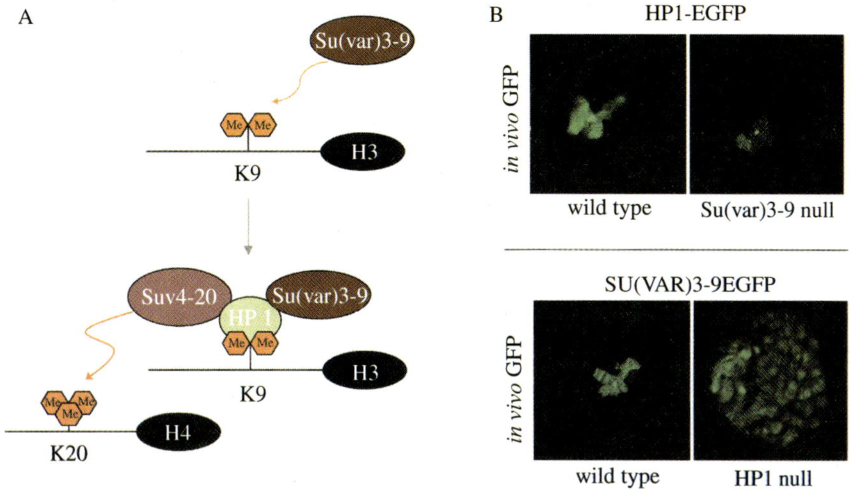

Fig. 4.2 A conserved epigenetic pathway controls heterochromatic gene silencing in *Drosophila* and mammals. (**A**) Heterochromatin association of SU(VAR)3–9 and histone H3K9 methylation establishes binding sites of the HP1 protein which recruits the SUV4–20 methyltransferase catalysing H4K20 trimethylation. (**B**) Heterochromatin association of SU(VAR)3–9 and HP1 are interdependent. In *Su(var)3-9* null cells binding of HP1 to chromocenter heterochromatin is strongly reduced due to a reduction in H3K9 methylation. In HP1 null cells, SU(VAR)3–9 which binds to HP1 is not retained within heterochromatin and binds ectopically along the chromosome.

chromosomes where H3K9me2 is catalysed by the SETDB1 histone methyltransferase (Seum et al., 2007; Tzeng et al., 2007). However, SU(VAR)3–9 is found at all these sites because of its association with HP1 but seems to be enzymatically compromised (Schotta et al., 2002). How this is controlled is still unknown.

SU(VAR)3–9 binding to heterochromatic sequences is controlled by its chromo and SET domains (Schotta et al., 2002). Its enzymatic activity causes dimethylation of H3K9, a mark recognized by the chromodomain of HP1 (Bannister et al., 2001; Lachner et al., 2001). Consequently, in *Su(var)3–9* null cells, binding of HP1 to chromocenter heterochromatin is impaired (Fig. 4.2B). HP1 binds by its chromoshadow domain the N-terminus of SU(VAR)3–9 as shown by yeast two-hybrid tests and immunoprecipitation (Schotta et al., 2002). This is essential for stable heterochromatin association of SU(VAR)3–9. In HP1 null cells, the protein is also found along euchromatin (Fig. 4.2B) and causes ectopic H3K9 methylation, which finally results in lethality (Schotta et al., 2002). The N-terminus of SU(VAR)3–9 furthermore interacts with the C-terminal domain of SU(VAR)3–7, another heterochromatin-associated protein (Schotta et al., 2002). This protein interacts at three different sites with the chromoshadow domain of HP1 (Delattre et al., 2000). SU(VAR)3–7 interactions between the three proteins are essential for stable assembly of the heterochromatin protein complex.

If SU(VAR)3–9 distribution is studied in transgenic flies expressing a chimeric HP1–Pc protein in which the chromodomain of HP1 is replaced with the Polycomb (Pc) chromodomain (Platero et al., 1996), the SU(VAR)3–9 protein is found at the euchromatic Pc binding sites

(Schotta et al., 2002). The Pc chromodomain of the chimeric protein recognizes H3K27me3 (Fischle et al., 2003) and recruits SU(VAR)3–9 through its interaction with the HP1 chromo-shadow domain to Pc binding sites. At these ectopic sites, SU(VAR)3–9 appears to be enzymatically compromised and does not cause H3K9 methylation.

In SU(VAR)3–9 null cells, another heterochromatin-specific methylation mark, H4K20me3 which is catalysed by the SUV4-20 histone methyltransferase, is strongly reduced (Schotta et al., 2004). Recruitment of SUV4-20 to heterochromatin depends on binding of SU(VAR)3–9 and HP1 to heterochromatin (Fig. 4.2A). Among mammalian SUVH39H1, HP1 and SUVH4-20 proteins, an identical functional interrelationship is found suggesting that the molecular pathway controlling heterochromatin formation is evolutionary conserved between *Drosophila* and mammals (Schotta et al., 2004).

In *Drosophila,* H3K27 methylation is found in euchromatin and heterochromatin. Enzymatic activity of the ENHANCER OF ZESTE protein complex in H3K27 methylation has been shown by *in vitro* analysis (Czermin et al., 2002; Müller et al., 2002). In *Drosophila* all methylation levels of H3K27 depend on the E(Z) methyltransferase as shown by immunocytological and mass spectrometric analysis of null mutants (Ebert et al., 2004; 2006). A role of H3K27 methylation in heterochromatin formation and gene silencing in w^{m4} is indicated by the suppressor effect of *E(z)* mutations and the enhancer effect of additional *E(z)* gene copies (Laible et al., 1997).

With the *Su(var)* mutations isolated for different factors of heterochromatin control, new interesting aspects of heterochromatin function during development have been resolved. These studies include control of nuclear architecture and stability of rDNA and satellite DNA repeats by histone H3K9 methylation (Peng and Karpen, 2007) or crossing over suppression by heterochromatin (Westphal and Reuter, 2002). Crossing over is almost completely suppressed in pericentric heterochromatin and strongly reduced within proximal euchromatin. Combining different *Su(var)* mutations, dominant recombinogenic effects increasing crossover within proximal euchromatic region by a factor of ten have been identified, suggesting a role of heterochromatin in control of total crossover frequencies within the genome.

SU(VAR)3–9 Homologues and Heterochromatin Formation in Different Organisms

SU(VAR)3–9, a key factor in heterochromatin formation and heterochromatic gene silencing has been identified first in *Drosophila* (Tschiersch et al., 1994). The gene has been selected for molecular analysis because of its haplo- *Su(var)* and triplo-dependent *E(var)* effects. These dosage-dependent effects of *Su(var)3–9* on PEV suggested a special role in heterochromatic gene silencing. Beside extensive analysis in *Drosophila,* the function of SU(VAR)3-9 homologues in heterochromatin formation has been studied in *Schizosaccharomyces pombe, Neurospora crassa, Arabidopsis thaliana* and the mouse. In *S. pombe,* Clr4 is the homologue of SU(VAR)3–9 (Ivanova et al., 1998). The gene has been identified by genetic dissection of silencing at the $ura4^{+}$ marker gene inserted into the *mat2–mat3* region (Egel et al., 1989; Thon and Klar, 1992; Thon et al., 1994; Ekwall and Ruusula, 1994). Clr4 controls H3K9 methylation at the centromeric outer repeats, the region containing the silent mating type loci and at a region adjacent to the terminal telomeric repeats (Noma et al., 2001; Nimmo et al., 1994; Kanoh et al., 2005). H3K9 methylation by Clr4 results in recruitment of the HP1 homologues Swi6 and Chp2 and of another chromodomain protein Chp1 (Nakayama et al., 2001; Sadaie et al., 2004). In *S. pombe,* the RNAi machinery is involved in heterochromatin formation and controls recruitment of Clr4 to centromeric heterochromatin via a double-stranded RNA produced by transcription of outer centromeric repeats (Volpe et al., 2002; Sadaie et al., 2004; Verdel et al., 2004).

In *Neurospora crassa,* the SU(VAR)3–9 homologue is DIM-5 that controls trimethylation of H3K9 (Tamaru and Selker, 2001; Tamaru et al., 2003). H3K9 trimethylation by DIM-5 creates binding sites for the HP1 homologue HPO, which recruits the DNA methyltransferase DIM-2. HPO binding to heterochromatin depends on DIM-5 (Freitag et al., 2004). Interestingly, mutations for components of the RNAi machinery including all three RNA-dependent RNA polymerase genes and both Dicer genes did not reveal any evidence that RNAi is involved in heterochromatin formation in *Neurospora* (Chicas et al., 2005).

In *Arabidopsis* altogether 10 SU(VAR)3-9 homologues (SUVH) and five SU(VAR)3– 9 related (SUVR) proteins have been identified (Baumbusch et al., 2001). Phylogenetic analysis between different plant species revealed four subgroups of SUVH genes in angiosperms, indicating an early phylogenetic split of SUVH genes during evolution of seed plants. Probably, the intron-less genes of the SUVH1 (SUVH1, 3, 7, 8 and 10), SUVH2 (SUVH2 and 9) and SUVH5 (SUVH5 and 6) groups originated from an intron-containing SUVH4-like gene via retrotransposition (Naumann et al., 2005). Heterochromatin association has been shown for the three SUVH proteins SUVH1, SUVH2 and SUVH4 (Naumann et al., 2005; Fischer et al., 2006). Although *in vitro* enzymatic activity in histone H3K9 methylation could only be shown for SUVH2, SUVH4, SUVH5 and SUVH6 (Jackson et al., 2002; Jackson et al., 2004; Naumann et al., 2005; Ebbs and Bender, 2006), all the 10 SUVH proteins, are suggested to catalyse H3K9 methylation. From the five SUVR proteins the SUVR4 protein shows a specific enzymatic activity and converts H3K9me1 into H3K9me2 (Thorstensen et al., 2006). SUVR4 and a specific isoform of SUVR1 localize to the nucleolus.

From all SUVH proteins tested, only SUVH2 shows strong dosage-dependent effects on heterochromatin formation and gene silencing. After SUVH2 over-expression ectopic heterochromatization is observed accompanied with elevated levels of H3K9me2, H3K27me1 and H4K20me1, which are all heterochromatin-specific histone methylation marks in *Arabidopsis* (Naumann et al., 2005; Fischer et al., 2006).

In human and mouse, two SU(VAR)3-9 homologues (SUV39H1 and SUV39H2) have been identified, which differ in their expression profile (Aagaard et al., 1999; O´Carroll et al., 2000). Disruption of both genes revealed a predominant function of Suv39h in H3K9 trimethylation at pericentric heterochromatin of the mouse (Peters et al., 2001). In double knockouts, viability is impaired, significant chromosomal instabilities are found and the risk to develop tumours is significantly increased. Studies of telomere length in Suv39h double knockout cells further indicated a function of SUVH39H-dependent H3K9 methylation in control of telomere length in mammals (Garcia-Cao et al., 2003).

H3K4 Demethylation, Heterochromatin Formation and the Control of Gene Silencing in Somatic and Germ Line Cells

Histone methylation has been regarded as a robust histone modification especially suited for long-term epigenetic programming. However, with the discovery of histone lysine demethylases (Shi et al., 2004; Tsukada et al., 2006) this view had to be reconsidered. Functional analysis of histone demethylation requires mutants eliminating specific demethylases from the cell in order to understand their function in epigenetic programming and development. Recently, we showed that the *Drosophila Su(var)3–3* gene encodes a homologue of the human H3K4me1 and H3K4me2 demethylase LSD1 (Rudolph et al., 2007). Altogether we isolated 24 *Su(var)3–3* mutations and all show a spectrum of interesting phenotypic effects consisting in female and male sterility, butyrate sensitivity and lethal interaction with additional heterochromatin (Reuter et al., 1982; Szabad et al., 1989). SU(VAR)3–3 *in vitro* and *in vivo* demethylates H3K4me1 and H3K4me2 but does not affect $H3K4me^3$ or any of the H3K9 methylation marks. Over-expression of the enzyme *in vivo* causes significant reduction of H3K4me2 in polytene chromosomes. In polytene chromosomes, this mark as well as the SU(VAR)3–3 protein are found in interbands (Ebert et al., 2006; Rudolph et al., 2007). H3K4 methylation in higher organism is a mark associated with an active chromatin conformation (Zhang and Reinberg, 2001; Bernstein et al., 2002; Kouzarides, 2002), explaining binding of SU(VAR)3–3 to interband regions in polytene chromosomes, which are enriched in H3K4 methylation. However, the dominant PEV suppressor effect of *Su(var) 3–3* mutations in all of the tested PEV rearrangements and especially their lethal interaction with additional heterochromatin cannot be explained with this observation.

The analysis of phenotypic interactions between *Su(var)3–3* mutations and duplications for genes encoding the heterochromatin proteins SU(VAR)3–9, HP1 and SU(VAR)3–7 provided first clues about the function of the SU(VAR)3–3 demethylase in heterochromatin formation and gene silencing. Additional gene copies of SU(VAR)3–9, HP1 and SU(VAR)3–7 result in enhancement of *white* gene silencing in w^{m4} (Schotta et al., 2002). In *Su(var)3–3/+* heterozygotes containing in addition a duplication of the wild type allele of *Su(var)3–9*, *Su(var)2–5* (HP1) or *Su(var)3–7,* the suppressor effect of *Su(var)3–3* mutations strongly dominates the strong enhancer effect induced by additional gene copies of *Su(var)3–9*, *Su(var)2–5* (HP1) and *Su(var)3–7*. This epistatic effect of *Su(var)3–3* mutations on enhancement of heterochromatic gene silencing suggests that the SU(VAR)3–3 demethylase functions upstream of the SU(VAR)3–9 methyltransferase which initiates heterochromatic gene silencing by H3K9 methylation. Furthermore, clonal analysis suggests early initiation of heterochromatic gene silencing by SU(VAR)3–9 during embryonic development of *Drosophila* (Rudolph et al., 2007). Induction of Su^+/Su^+ clones in heterozygous *Su(var)3–9/+* cells using the flipase/FRT mitotic recombination system (Xu and Rubin, 1993) showed only restored *white* gene variegation in large early induced clones. All small late induced clones expressed a *Su(var)* phenotype, although they were Su^+/Su^+ in genotype. The size of the clones with restored *white* gene variegation (large clones) suggests that suppression of heterochromatic gene silencing in w^{m4} is established during embryonic development and thereafter stably maintained. We also tested whether additional cell divisions can affect maintenance of the suppressed silencing phenotype and repeated all experiments with *Su(var)3–9/+* heterozygotes carrying a dominant *Minute* mutation on the *Su(var)3–9* mutant chromosome. The genetic configuration results in Su^+/ Su^+ clones, which are *Minute*$^+$ and overgrow larger territories because the neighbouring M^+/+ cells are significantly delayed in development (Garcia-Bellido et al., 1973). Again the size of Su^+/Su^+ clones showing restored *white* gene silencing corresponds to early induced clones.

Early induction of SU(VAR)3–9 dependent gene silencing and a function of SU(VAR) 3–3, which is downstream of SU(VAR)3–9 suggests that this H3K4 demethylase has an effect on heterochromatin formation and establishment of heterochromatic gene silencing in early embryonic development. Using a SU(VAR)3–3 specific antibody, we found that SU(VAR)3–3 is abundantly associated with cleavage chromatin. At cleavage cycle 14–15 when blastoderm nuclei show an apico-basal polarity (Rabl conformation) and differentiation of euchromatin and heterochromatin occurs, SU(VAR)3–3 preferentially associates with prospective heterochromatin (Fig. 4.3). H3K4me2 is first seen at this time in euchromatin. In embryos produced after a cross of *Su(var)3-3/+* heterozygous females and males, a significantly reduced amount of SU(VAR)3-3 in prospective heterochromatin is found, which coincides with elevated levels of H3K4me2 that spreads toward prospective heterochromatin (Rudolph et al., 2007). These findings suggest a protective function of the SU(VAR)3-3 H3K4 demethylase against spreading of the active H3K4me2 mark into heterochromatin. Biochemical analysis furthermore showed the existence of an early embryonic protein complex, which contains SU(VAR)3–3 as well as the heterochromatin proteins

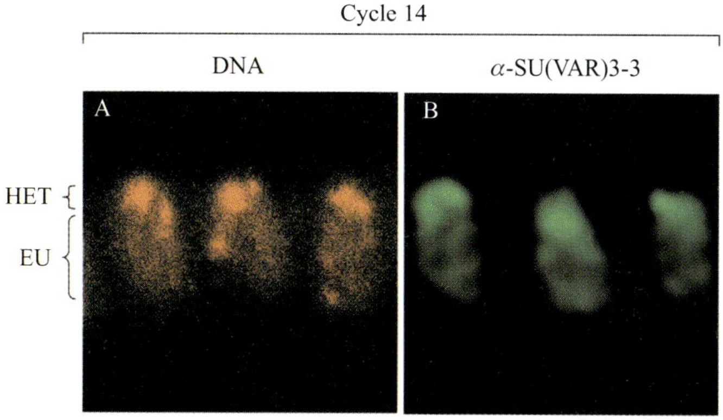

Fig. 4.3 Heterochromatin association of SU(VAR)3-3 in early blastoderm nuclei. (**A**) Early blastoderm nuclei show an apico-basal polarity (Rabl configuration). DAPI intense material reflects prospective heterochromatin and is oriented to the apical site of the embryo. H3K9me2 staining of this material demonstrated establishment of heterochromatin during early embryogenesis in *Drosophila*. (**B**) The SU(VAR)3-3 demethylase associates preferentially with prospective heterochromatin in early blastoderm nuclei. SU(VAR)3–3 has a protective function and inhibits spreading of the active H3K4me2 histone methylation mark into heterochromatic regions.

SU(VAR)3-9 and HP1. In addition, the histone deacetylase RPD3 (HDAC1) is associated with this protein complex (Rudolph et al., 2007). *In vitro* methylation of lysine 4 in histone peptides inhibits methylation of lysine 9 by SU(VAR)3–9. Furthermore, acetylation of H3K9 excludes H3K9 methylation. Assembly of SU(VAR)3–3 and RPD3 with SU(VAR)3–9 into a multimeric heterochromatin protein complex suggests that the protective function of the proteins against H3K4 methylation and H3K9 acetylation, respectively, ensures effective function of SU(VAR)3–9 in H3K9 methylation and establishment of heterochromatin. Integrated function of the different enzymes controls the balance between euchromatin and heterochromatin, which is also supported by the finding that spreading of H3K9me2 into euchromatin in the w^{m4} rearrangement is completely eliminated in *Su(var)3–3* homozygous mutant cells (Rudolph et al., 2007).

The SU(VAR)3–3 demethylase is also essential for maintenance of transcriptional quiescence in primordial germ line cells. In *Drosophila,* primordial germ cells become separated from the somatic lineage at the end of cleavage at cycle 10. These cells are recognized as pole cells at the posterior pole of the embryo. In pole cell nuclei, which are transcriptionally quiescent extensive H3K9 methylation is found but no H3K4 methylation (Schaner et al., 2003; Rudolph et al., 2007). In *Su(var)3-3* mutant embryos, H3K4 methylation appears in pole cell nuclei, whereas H3K9 methylation is significantly reduced. Therefore, it is suggested that precocious genetic activity in primordial germ line cells causes female and male sterility in *Su(var)3-3* mutant homozygotes. Lethal interaction with additional heterochromatin is explained by the protective function of SU(VAR)3–3 against H3K4 methylation during heterochromatin formation. Sodium butyrate is an effective inhibitor of histone deacetylation (Candido et al., 1978; Boffa et al., 1981) and butyrate sensitivity of *Su(var)3–3* mutant flies could be explained by coordinated function of the SU(VAR) 3–3 demethylase and the RPD3 deacetylase during heterochromatin formation in early embryogenesis. We furthermore studied the effect of *Su(var)3–3* over-expression on silencing defects established by *RPD3* mutations. Mutations of *RPD3* show a strong *Su(var)* effect on *white* variegation in w^{m4} (Mottus et al., 2000; Czermin et al., 2001) which is completely rescued by *Su(var)3–3* over-expression (Rudolph et al., 2007), suggesting that increased amounts of SU(VAR)3–3 might efficiently enhance RPD3 recruitment to SU(VAR)3-3/RPD3/SU(VAR)3–9/HP1 chromatin complexes in early embryos.

Functional analysis of the SU(VAR)3–3 demethylase demonstrates the manifold advantages provided by modifier of PEV mutations in the *Drosophila* model system for molecular analysis of epigenetic processes. Further analysis of *Su(var)* and *E(var)* genes in *Drosophila* will increase our understanding of molecular processes involved in epigenetic programming.

Genetic Dissection of Alternative Silencing Processes in *Drosophila*

In *Drosophila,* the heterochromatic silencing processes predominantly control formation of heterochromatin, whereas transcription at many genes during development is repressed by the Polycomb group (PcG) protein silencing system (Pirrotta, 1998; Ringrose and Paro, 2004; Müller and Kassis, 2006; Schwartz et al., 2006). The function of these complexes depends on specific *cis*-regulatory sequences, the Polycomb response elements (PREs) and on histone H3K27me3, H3K9me3 and H4K20me3 methylation (Papp and Müller, 2006). Transgenes with the *bxd* PRE show variegated repression of marker genes, which depends on PcG proteins (Chan et al., 1994).

In PcG gene silencing, the *white* gene represents again a valuable monitoring tool for molecular dissection of the underlying molecular processes.

Recently we started to perform a systematic analysis of *white* variegated P element insertions which were produced within a project to reach complete coverage of the *Drosophila* genome with overlapping deletions (Ryder et al., 2007). Altogether we identified more than 70 *white* variegated inserts at various genomic positions. Several of these inserts have been mapped within terminal repeats of retrotransposons (Phalke et al., unpublished). Genetic analysis revealed that these inserts identify silencing processes independent of heterochromatic and PcG silencing mechanisms. Consequently, alternative molecular mechanisms controlling gene silencing at mobile elements and several other genomic regions could be identified. Genetic dissection of these processes by modifier mutations should help to reveal the underlying molecular processes.

Epigenetic Histone Modification Marks Indexing Heterochromatic Domains in Different Organisms

Identification of conserved molecular pathways in heterochromatic gene silencing between *Drosophila* and mammals implies substantial evolutionary conservation in the functional role of histone modification marks in indexing and epigenetic programming of chromatin domains. Beside *Drosophila* and mammals (mouse and humans) systematic studies of molecular mechanisms of epigenetic programming were performed in the *Arabidopsis* plant. In Fig. 4.4, we summarized our recent knowledge about histone methylation marks, which index heterochromatin in *Drosophila*, mammals and *Arabidopsis*.

Constitutive heterochromatin in *Drosophila* is enriched in H3K9me1, H3K9me2 but only contains little H3K9me3 (Schotta et al., 2002; Ebert et al., 2006). Heterochromatic H3K9me2 and H3K9me3 depends on SU(VAR)3-9 but the methyltransferase controlling H3K9me1 is still unknown. In contrast to *Drosophila* in mammals, H3K9me3 is the abundant marks of pericentromeric heterochromatin and is controlled by SUV39H1 and SUV39H2 (Peters et al., 2003; Rice et al., 2003). In *Drosophila* heterochromatin, all H3K27 methylation states are detected, whereas in mouse H3K27me3 and H3K9me1 are only found at pericentric heterochromatin in *Suv39h* double null cells (Peters et al., 2003). H3K27 methylation however is involved in X inactivation (Plath et al., 2003). Both in *Drosophila* and in mammals H4K20me3 is a heterochromatic mark catalysed by the SUV4-20 histone methyltransferase, which depends on H3K9 methylation by SU(VAR)3-9 (Schotta et al., 2004). Like in *Drosophila* in *Arabidopsis* also H3K9me1 and H3K9m2 index heterochromatin, whereas H3K9me3 appears to be exclusively euchromatic (Naumann et al., 2005). H3K27me1 and H3K27me2 are heterochromatic marks in *Arabidopsis*, whereas in *Drosophila* these marks are found in heterochromatin and in euchromatic regions (bands). In contrast to *Drosophila* H3K27me3 is only euchromatic in *Arabidopsis*. H4K20me1 in *Arabidopsis* is a typical heterochromatic mark but H4K20me2 and H4K20me3 are exclusively euchromatic. Comparing indexing of pericentric heterochromatin in *Drosophila*, mammals and *Arabidopsis* there are significant differences (Fig. 4.4). In plants, proteins recognizing specifically these histone methylation marks are not known. Furthermore, as already discussed epigenetic control of heterochromatic regions in plants appears to be more complex as indicated by the high number of SU(VAR)3–9 homologous proteins.

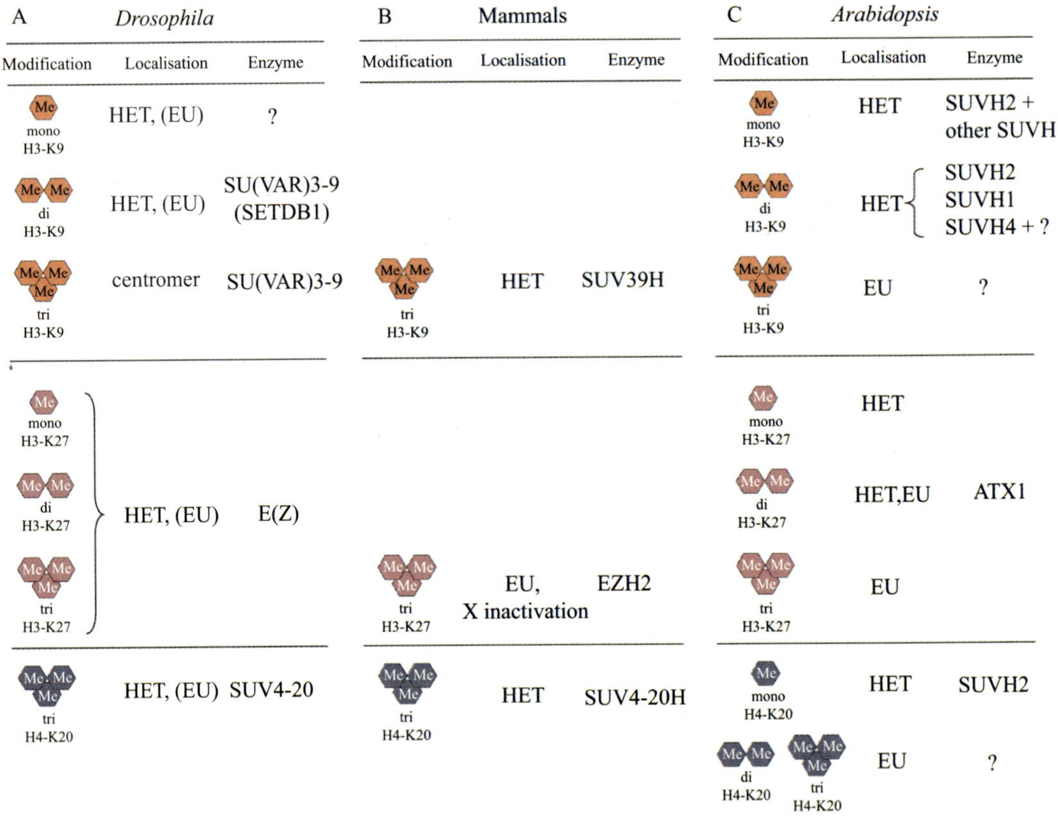

Fig. 4.4 Heterochromatin indexing by histone methylation marks differs between *Drosophila*, mammals and *Arabidopsis*. (**A**) Typical marks indexing heterochromatin in *Drosophila* are H3K9me1 and H3K9m2. H3K9me3 is under-represented in *Drosophila* heterochromatin. H3K9me2 and H3K9me3 is catalysed by SU(VAR)3-9 and SETDB1. The methyltransferase responsible for H3K9me1 is still unknown. In *Drosophila* all H3K27 methylation levels are found in both euchromatin and heterochromatin and depend exclusively on the E(Z) methyltransferase. H4K20me3 is catalysed by SUV4-20. (**B**) In mammals (human and mouse), H3K9me3 catalysed by the SUV39H enzymes represents an abundant mark of pericentric heterochromatin. In *Suv39h1* and *Suv39h2,* double null cells H3K9me1 and H3K27me3 are ectopically recruited to pericentric heterochromatin. H3K27me3 is an abundant mark at the inactivated X chromosome. Like in *Drosophila* H4K20me3 represents a heterochromatic histone methylation mark, which is established by SUV4-20 and depends on the function of SU(VAR)3-9 and HP1. (**C**) Based on immunocytological analysis of interphase nuclei in *Arabidopsis,* H3K9me1 and H3K9me2 are typical heterochromatic marks, whereas H3K9me3 is exclusively restricted to euchromatin. Other heterochromatic marks are H3K27me1, H3K27me2 and H4K20me1. In *Arabidopsis,* H3K27me3, H4K20me2 and H3K20me3 are exclusively found within euchromatic regions. The SUVH2 protein affects all heterochromatic marks but other SUVH proteins contribute in establishment of H3K9 and H3K27 methylation probably in a target sequence specific manner.

In summary, only few histone modification marks appear to have identical indexing properties in *Drosophila*, *Arabidopsis* and mammals. These are all levels of H3K4 and H3K36 methylation, as well as H3K9 acetylation. For all other marks, differences in distribution between euchromatic and heterochromatic domains are detected. These data finally suggest that the histone code is not universal and molecular mechanisms in recognition and interpretation of histone modification marks appear to differ significantly between different species and an interesting field of future epigenetic studies surely will consist in identification and characterization of those proteins which are all involved in recognition and interpretation of epigenetic marks in different organisms.

Acknowledgements

We are grateful to T. Jenuwein for antibodies and to S. Phalke for critical reading the manuscript and discussions. This work was supported by grants (SFB610, Re911/5 and SFB648) from the Deutsche Forschungsgemeinschaft (DFG), the Land Sachsen-Anhalt and the European Union (NoE Epigenome LSHG-CT- 2000-503433).

References

1. Aagaard, L., Laible, G., Selenko, P., Schmid, M., Dorn, R., Schotta, G., Kuhfittig, S., Wolf, A., Lebersorger, A.S., Reuter, G., and Jenuwein, T. (1999). Functional mammalian homologues of the *Drosophila* PEV-modifier *Su-(var)3–9* encode centromere-*associated* proteins which complex with the heterochromatin component M31. EMBO J. *18*, 1923–1938.
2. Allis, C.D., Jenuwein, T., and Reinberg, D. (2006). Epigenetics. Cold Spring Harbor Laboratory Press, Cold Spring Harbor, New York.
3. Ashburner, M., Golic, K.G., and Hawley, R.S. (2005). Position effect variegation. In *Drosophila*: A laboratory handbook, 2nd edition. Cold Spring Harbor Laboratory Press, Cold Spring Harbor, New York, pp 1007–1049.
4. Baksa, K., Morawietz, H., Dombradi, V., Axton, M., Taubert, H., Szabo, G., Török, I., Gyurkovics, H., Szöör, B., Gloover, D., Reuter, G., and Gausz J. (1993). Mutations in the phosphatase 1 gene at 87B can differentially affect suppression of position-effect variegation and mitosis in *Drosophila melanogaster*. Genetics *135*, 117–125.
5. Bannister, A.J., Zegermann, P., Patridge, J.F., Miska, E.A., Thomas, J.O., Allshire, T.C., and Kouzarides, T. (2001). Selective recognition of methylated lysine 9 on histone H3 by the HP1 chromo domain. Nature *410*, 120–124.
6. Baumbusch, L.O., Thorstensen, T., Krauss, V., Fischer, A., Naumann, K., Assalkhou, R., Schulz, I., Reuter, G., and Aalen, R. (2001). The *Arabidopsis thaliana* genome contains at least 29 active genes encoding SET domain Proteins that can be assigned to four evolutionary conserved classes. Nucleic. Acid Res. *29*, 4319–4333.
7. Bernstein, B.E., Humphrey, E.L., Erlich, R.L., Schneider, R., Bouman, P., Liu J.S., Kouzarides, T., and Schreiber, S.L. (2002). Methylation of histone H3 Lys4 in coding regions of active genes. Proc. Natl. Acad. Sci. U.S.A. *99*, 8695–8700.
8. Boffa, L.C., Gruss, R.J., and Allfrey, V.G. (1981). Manifold effects of sodium butyrate on nuclear functions. J. Biol. Chem. *256*, 9612–9621.
9. Candido, E.P.M., Reeves, R., and Davie, J.R. (1978). Sodium butyrate inhibits histone deacetylation in cultured cells. Cell *14*, 105–113.
10. Chan, C.-S., Rastelli, L., and Pirrotta, V. (1994). A Polycomb response element in the Ubx gene that determines an epigenetically inherited state of repression. EMBO J. *13*, 2553–2564.

11. Chicas, A., Forrest, E.C., Sepich, S., Cogoni, C., and Macino, G. (2005). Small interfering RNAs that trigger posttranscriptional gene silencing are not required for the histone H3 Lys9 methylation necessary for transgenic tandem repeat stabilization in *Neurospora crassa*. Mol. Cell. Biol. *25*, 3793–3801.
12. Czermin, B., Melfi, R., McCabe, D., Seitz, V., Imhof, A., and Pirrotta, V. (2002). *Drosophila* enhancer of Zeste/ESC complexes have a histone H3 methyltransferase activity that marks chromosomal Polycomb sites. Cell *111*, 185–196.
13. Czermin, B., Schotta, G., Hülsmann, B.B., Brehm, A., Becker, P.B., Reuter, G., and Imhof, A. (2001). Physical and functional interaction of SU(VAR)3-9 and HDAC1 in *Drosophila*. EMBO Rep. *2*, 915–919.
14. Delattre, M., Spierer, A., Tonka, C.-H., and Spierer, P. (2000). The genomic silencing of position-effect variegation in *Drosophila melanogaster*: interaction between the heterochromatin-associated proteins Su(var)3–7 and HP1. J. Cell Sci. *113*, 4253–4261.
15. Demerec, M. (1940). Genetic behavior of euchromatic segments inserted into heterchromatin. Genetics *25*, 618–627.
16. Ebbs, M.L., and Bender, J. (2006). Locus-specific control of DNA methylation by the Arabidopsis SUVH5 histone methyltransferase. Plant Cell *18*, 1166–1176.
17. Ebert, A., Lein, S., Schotta, G., and Reuter, G. (2006). Histone modification and the control of heterochromatic gene silencing in *Drosophila*. Chrom. Res. *14*, 377–392.
18. Ebert, A., Schotta, G., Lein, S., Kubicek, S., Krauss, V., Jenuwein, T., and Reuter, G. (2004). *Su(var)* genes regulate the balance between euchromatin and heterochromatin in *Drosophila*. Genes Dev. *18*, 2973–2983.
19. Egel, R. Willer, M, Neisen, O. (1989). Unblocking of meiotic crossing-over between the silent mating-type cassettes of fission yeast, conditioned by the recessive, pleiotropic mutant *rik1*. Curr. Genet. *15*, 407–410.
20. Eissenberg, J.C., James, T., Foster-Hartnett, D., Hartnett, T., Ngan, V., and Elgin, S.C.R. (1990). Mutation in a heterochromatin-specific chromosomal protein is associated with suppression of position-effect variegation in *Drosophila melanogaster*. Proc. Natl. Acad. Sci. U.S.A *87*, 9923–9927.
21. Ekwall, K. and Ruusala, T. (1994). Mutations in *rik1*, *clr2*, *clr3*, and *clr4* genes asymmetrically derepress by silent mating-type loci in fission yeast. *Genetics 136*, 53–64.
22. Fischer, A., Hofmann, I., Naumann, K,. and Reuter, G. (2006). Heterochromatin proteins and the control of heterochromatic gene silencing in Arabidopsis. J. Plant Phys. *163*, 358–368.
23. Fischle, W., Wang, Y., Jacobs, S.A., Kim, Y., Allis, C.D., and Khorasanizadeh, S. (2003). Molecular basis for the discrimination of repressive methyl-lysine marks in histone H3 by Polycomb and HP1 chromodomains. Genes Dev. *17*, 1870–1881.
24. Flybase (2008) The *Drosophila* database. Available from World Wide Web at the URLs http://morgan/harvard.edu and http://www.ebi.ac.uk/flybase/.
25. Freitag, M., Hickey, P.C., Khlafallah, T.K., Read, N.D., and Selker, E.U. (2004). HP1 is essential for DNA methylation in *Neurospora*. Mol. Cell *13*, 427–434.
26. Garcia-Bellido, A., Ripoll, P., and Morata, G. (1973). Developmental compartmentalisation of the wing disc of *Drosophila*. Nature New Biol. *245*, 251–253.

27. Garcia-Cao, M., O´Sullivan, R., Peters, A.H.F.M., Jenuwein, T., and Blasco, M.A. (2003). Epigenetic regulation of telomere length in mammalian cells by the Suv39h1 and Suv39h2 histone methyltransferases. Nat. Genet. *36*, 94–99.
28. Grienenberger, A., Miotto, B., Sagnier, T., Cavalli, G., Schramke, V., Geli, V., Mariol, M.C., Berenger, H., Graba, Y., and Pradel, J. (2002). The MYST domain acetyltransferase Chameau functions in epigenetic mechanisms of transcriptional repression. Curr. Biol. *12*, 762–766.
29. Hari, K.L., Cook, K.R., and Karpen, G.H. (2001). The *Drosophila Su(var)2–10* locus regulates chromosome structure and function and encodes a member of the PIAS protein family. Genes Dev. *15*, 1334–1348.
30. Hartmann-Goldstein, I.J. (1976). On the relationship between heterochromatization and variegation in *Drosophila*, with special reference to temperature-sensitive periods. Genet. Res. *10*, 143–159.
31. Ivanova, A.V, Bonaduce, M.J, Ivanov, S.V, Klar, A.J.S., (1998). The chromo and SET domains of the Clr4 protein are essential for silencing in fission yeast. Nature. Genet *19*, 192–195.
32. Jackson, J.P., Johnson, L., Jasencakova, Z., Zhang, X., PerezBurgos, L., Singh, P.B., Cheng, X, Schubert, I., Jenuwein, T., and Jacobsen S.E. (2004). Dimethylation of histone H3K9 is a critical mark for DNA methylation and gene silencing in *Arabidopsis thaliana.* Chromosoma *112*, 308–315.
33. Jackson, J.P., Lindroth, A.M., Cao, X., and Jacobsen, S.E. (2002). Control of CpNpG DNA methylation by the *KRYPONITE* histone H3 methyltransferase. Nature *416*, 556–560.
34. James, T.C., and Elgin, S.C.R. (1986). Identification of a nonhistone chromosomal protein associated with heterochromatin in *Drosophila melanogaster* and its gene. Mol. Cell Biol. *6*, 3862–3872.
35. James, T.C., Eissenberg, J.C., Craig, C., Dietrich, V., Hobson, A., and Elgin, S.C.R. (1989). Distribution patterns of HP1, a heterochromatin-associated nonhistone chromosomal protein of *Drosophila*. Eur. J. Cell Biol. *50*, 170–180.
36. Jaquet, Y., Delattre, M., Spierer, A., and Spierer, P. (2002). Functional dissection of the *Drosophila* modifier of variegation *Su(var)3–7*. Development *129*, 3975–3982.
37. Kanoh, J., Sadaie, M., Urano, T., and Ishikawa, F. (2005). Telomere binding protein Taz1 establishes Swi6 heterochromatin independently of RNAi at telomeres. Curr. Biol. *15*, 1808–1819.
38. Kouzarides, T. (2002). Histone methylation in transcriptional control. Curr. Opin. Genet. Dev. *12*, 198–209.
39. Krauss, V., Fassl, A., Fiebig, P., Patties, I., and Sass, H. (2006). The evolution of the histone methyltransferase gene *Su(var)3–9* in metazoans including a fusion with and re-fission from a functionally unrelated gene. BMC Evol. Biol. *6*, 18–32.
40. Krejci, E., Garzino, V., Mary, C., Bennani, N., and Pradel, J. (1989). *Modulo*, a new maternally expressed *Drosophila* gene encodes a DNA-binding protein with distinct acidic and basic regions. Nucl. Acid Res. *17*, 8101–8115.
41. Lachner, M., O´Carroll, D., Rea, S., Mechtler, K., and Jenuwein, T. (2001). Methylation of histone H3 lysine 9 creates a binding site for HP1 proteins. Nature *410*, 116–120.
42. Laible, G., Wolf, A., Dorn, R., Reuter, G., Nislow, C., Lebesorger, A., Popkin, D., Pillus, L., and Jenuwein, T. (1997). Mammalian homologs of Enhancer of zeste mediate position-effect variegation in *Drosophila* and restore telomeric silencing in *S. cerevisiae*. EMBO J. *16*, 3219–3232.

43. Lindsley, D.L., and Zimm, G.G. (1992). The genome of *Drosophila melanogaster*. Academic Press San Diego and London.
44. Mottus, R., Sobels, R.E., and Grigliatti, T.A. (2000). Mutational analysis of a histone deacetylase in *Drosophila melanogaster*: missence mutations suppress gene silencing associated with position effect variegation. Genetics *154*, 657–668.
45. Muller, H.J. (1930). Types of visible variations induced by X-rays in *Drosophila*. J. Genet. *22*, 299–334.
46. Müller, J., and Kassis, J.A. (2006). Polycomb response elements and targeting of Polycomb group proteins in *Drosophila*. Curr. Opin. Genet. Dev. 16, 476–489.
47. Müller, J., Hart, C.M., Francis, N.J., Vargas, M.L., Sengupta, A., Wild, B., Miller, E.L., O'Connor, M.B., Kingston, R.E., and Simon, J.A. (2002). Histone methyltransferase activity of a *Drosophila* Polycomb group repressor complex. Cell *111*, 197–208.
48. Nakayama, J., Rice, J.D., Strahl, B.D., Allis, C.D., and Grewal, S.I.S. (2001). Role of histone H3 lysine 9 methylation in epigenetic control of heterochromatin assembly. Science *292*, 110–113.
49. Naumann, K., Fischer, A., Hofmann, I., Krauss, V., Phalke, S., Irmler, K., Hause, G., Aurich, A.C., Dorn, R., Jenuwein, T., and Reuter, G. (2005). Pivotal role of at SUVH2 in control of hetero chromatic histone methylation and gene silencing in *Arabidopsis*. EMBO J. *24*, 1418–1429.
50. Nimmo, E.R., Cranston, G., and Allshire, R.C. (1994). Telomere-associated chromosome breakage in fission yeast results in variegated expression of adjacent genes. EMBO J. *13*, 3801–3811.
51. Noma, K., Sugiyama, T., Cam, H., Verdel, A., Zofall, M., Jia, S., Moazed, D., and Grewal, S.I. (2004). RITS acts in *cis* to promote RNA interference-mediated transcriptional and posttranscriptional silencing. Nat. Genet. *36*, 1174–1180.
52. O´Carroll, D., Scherthan, H., Peters, A.H., Opravil, S., Haynes, A.R., Laible, G., Rea, S., Schmid, M., Lebersorger, A., Jerratsch, M., Sattler, L., Mattei, M.G., Denny, P., Brown, S.D., and Jenuwein, T. (2000). Isolation and characterization of Suvh39h2, a second histone H3 methyltransferase gene that displays testis-specific expression. Mol. Cell Biol. *20*, 9423–9433.
53. Papp, B., and Müller J. (2006). Histone tri-methylation and the maintenance of transcriptional on and off states by PcG and trxG proteins. Genes Dev. *20*, 2041–2054.
54. Paro, R., and Hogness, D.S. (1991). The *Polycomb* protein shares a homologous domain with a heterochromatin– associated protein in *Drosophila*. Proc. Natl. Acad. Sci. U.S.A. *88*, 263–267.
55. Peng, J.C., and Karpen, G.H. (2007). H3K9 methylation and RNA interference regulate nucleolar organization and repeated DNA stability. Nat. Cell Biol. *9*, 25–35.
56. Peters, A.H.F.M., Kubicek, S., Mechtler, K., O´Sullivan, J., Derijck, A.A.H.A., Perez-Burgos, L., Kohlmaier, A., Opravil, S., Tachibana, M., Shinkai, Y., Martens, J.H.A., and Jenuwein, T. (2003). Partitioning and plasticity of repressive histone methylation states in mammalian chromatin. Mol. Cell. *12*, 1577–1589.
57. Peters, A.H.F.M., O´Carroll, D., Scherthan, H., Mechtler, K., Sauer, S., Schöfer, C., Weipoltshammer, K., Pagani, M., Lachner, M., Kohlmaier, A., Opravil, S., Doyle, M., Sibilia, M., and Jenuwein, T. (2001). Loss of the Suv39h histone methyltransferases impairs mammalian heterochromatin and genome stability. Cell *107*, 323–337.
58. Pirrotta, V. (1998). Polycombing the genome: PcG, trxG, and chromatin silencing. Cell *93*, 333–336.

59. Platero, J.S., Sharp, E.J., Adler, P.N., and Eissenberg, J.C. (1996). *In vivo* assay for protein–protein interaction using *Drosophila* chromosomes. Chromosoma *104*, 393–404.

60. Plath, K., Fang, J., Mlynarczyk-Evans, S.K., Cao, R., Worringer, K.A., Wang, H., de la Cruz, C.C., Otte, A.P., Panning, B., and Zhang, Y. (2003). Role of histone H3 lysine 27 methylation in X inactivation. Science *300*, 131–135.

61. Prokofyeva-Belgovskaya, A.A. (1947). Heterochromatization as a change of chromosome cycle. J. Genet. *48*, 80–98.

62. Rea, S., Eisenhaber, F., O'Carroll, D., Strahl, B.D., Sun, Z.-W., Schmid, M., Opravil, S., Mechtler, K., Ponting, C.P., Allis, C.D., and Jenuwein, T. (2000). Regulation of chromatin structure by site-specific histone H3 methyltransferases. Nature *406*, 593–599.

63. Reuter, G., and Szidonya, J. (1983). Cytogentic analysis of variegation suppressors and a dominant temperature sensitive lethal in region 23–26 of 2L in *Drosophila melanogaster*. Chromosoma *88*, 277–285.

64. Reuter, G., and Wolff, I. (1981) Isolation of dominant suppressor mutations for position-effect variegation. Mol. Gen. Genet. *182*, 516–519.

65. Reuter, G., Dorn, R,. and Hoffmann, H.-J. (1982). Butyrate sensitive suppressor of position-effect variegation mutations in *Drosophila melanogaster*. Mol. Gen. Genet. *188*, 480–485.

66. Reuter, G., Gausz, J., Gyurkovics, H., Friede, B., Bang, R., Spierer, A., Hall, L.M.C., and Spierer, P. (1987). Modifiers of position-effect variegation in the region from 86C to 88B of the *Drosophila melanogaster* third chromosome. Mol. Gen. Genet. *210*, 429–436.

67. Reuter, G., Giarre, N., Farah, J., Gausz, J., Spierer, A., and Spierer, P. (1990). Dependence of position-effect variegation in *Drosophila* on dose of a gene encoding an unusual zinc-finger protein. Nature *344*, 219–223.

68. Reuter, G., Hoffmann, H.-J., and Wolff, I. (1983). Genetic study of position-effect variegation in *Drosophila melanogaster*: *In(1)w*^{m4h} as a standard rearrangement for the isolation and characterization of suppressor and enhancer mutants. Biol. Zbl. *102*, 281–298.

69. Rice, J.C., Briggs, S.D., Ueberheide, B., Barber, C.M., Shabanowitz, J., Hunt, D.F., Shinkai, Y., and Allis, C.D. (2003). Histone methyltransferases direct different degrees of methylation to define distinct chromatin domains. Mol. Cell *12*, 1591–1598.

70. Rice, J.C., Briggs, S.D., Ueberheide, B., Barber, C.M., Shabanowitz, J., Hunt, D.F., Shinkai, Y., and Allis, C.D. (2003). Histone methyltransferases direct different degrees of methylation to define distinct chromatin domains. Mol. Cell *12*, 1591–1598.

71. Ringrose, L., and Paro, R. (2004). Epigenetic regulation of cellular memory by the Polycomb and Trithorax group proteins. Annu. Rev. Genet. *38*, 413–443.

72. Rudkin, G.T. (1965). The structure and function of heterochromatin. Proceedings of the 11th "International Congress Genetics" The Hague 1963, *2*, 359–374.

73. Rudolph, T., Yonezawa, M., Lein, S., Heidrich, K., Kubicek, S., Schäfer, C., Phalke, S., Walther, M., Schmidt, A., Jenuwein, T., and Reuter, G. (2007). Heterochromatin formation in *Drosophila* is initiated through active removal of H3K4 methylation by the LSD1 homolog SU(VAR)3–3. Mol. Cell *26*, 103–115.

74. Ryder, E., Ashburner, M., Bautista-Llacer R., Drummond, J., Webster, J., Gubb, D., Johnson, G., Morley, T., Sang Chan, Y., Blows, F., Coulson, D., Reuter, G. Baisch, H., Apelt, C., Kauk, A.,

Rudolph, T., Kube, M., Klimm, M., Nickel, C., Szidonya, J., Maróy, P., Pal, M., Rasmuson-Lestander, Å., Ekström, K., Stocker, H., Hugentobler, C., Hafen, E., Pflugfelder, G., Dorner, C., Mechler, B, Schenkel, H., Marhold, J., Serras, F., Corominas, M., Punset, A., Roote, J., and Russell, S. (2007). The DrosDel deletion set: a *Drosophila* genome-wide chromosomal deficiency resource. Genetics *167*, 797–813.

75. Sadaie, M., Iida, T., Urano, T., and Nakayama, J. (2004). A chromodomain protein, Chp1, is required for the establishment of heterochromatin in fission yeast. EMBO J. *23*, 3825–3835.
76. Schaner, C.E., Deshpande, G., Schedl, P.D., and Kelly, W.G. (2003). A conserved chromatin architecture marks and maintains the restricted germ cell lineage in worms and flies. Dev. Cell *5*, 747–757.
77. Schotta, G., Ebert, A., and Reuter, G. (2003a). SU(VAR)3-9 a conserved key function in heterochromatic gene silencing. Genetica *117*, 149–158.
78. Schotta, G., Ebert, A., Dorn, R., and Reuter, G. (2003b). Position-effect variegation and the genetic dissection of chromatin regulation in *Drosophila*. Sem. Cell Dev. Biol. *14*, 67–75.
79. Schotta, G., Ebert, A., Krauss, V., Fischer, A., Hoffmann, J., Rea, S., Jenuwein, T., and Reuter, G. (2002). Central role of *Drosophila* SU(VAR)3-9 in histone H3-K9 methylation and heterochromatic gene silencing. EMBO J. *21*, 1121–1131.
80. Schotta, G., Lachner, M., Sarma, K., Ebert, A., Sengupta, R., Reuter, G., Reinberg, D., and Jenuwein, T. (2004). A silencing pathway to induce H3-K9 and H4-K20 trimethylation at constitutive heterochromatin. Genes Dev. *18*, 1251–1262.
81. Schultz, J. (1936). Variegation in *Drosophila* and the inert chromosome regions. Proc. Natl. Acad. Sci. U.S.A. *22*, 27–33.
82. Schultz, J. (1950). Interrelations of factors affecting heterochromatin-induced variegation in *Drosophila*. Genetics *35*, 134.
83. Schwartz, Y.B., Kahn, T.G., Nix, D.A., Li, X.Y., Bourgon, R., Biggin, M., and Pirrotta, V. (2006). Genome-wide analysis of Polycomb targets in *Drosophila melanogaster*. Nat. Genet. *38*, 700–705.
84. Seum, C., Reo, E., Peng, H., Rauscher, F.J., Spierer, P., and Bontron, S. (2007). *Drosophila* SETDB1 is required for chromosome 4 silencing. PloS Genet. *3*, 709–719.
85. Shi, Y., Lan, F., Matson, C., Mulligan, P., Whetstine, J.R., Cole, P.A., Casero, R.A., and Shi, Y. (2004). Histone demethylation mediated by the nuclear amine oxidase homolog LSD1. Cell *119*, 941–953.
86. Sinclair, D.A.R., Mottus, R.C., and Grigliatti, T.A. (1983). Genes which suppress position effect variegation in *Drosophila melanogaster* are clustered. Mol. Gen. Genet. *191*, 326–333.
87. Spofford, J.B. (1967). Single-locus modification of position-effect variegation in *Drosophila melanogaster*. I. white variegation. Genetics *57*, 751–766.
88. Szabad, J., Reuter, G., and Schröder, M.B. (1988). The effect of two mutations connected with chromatin functions on female germ-line cells of *Drosophila*. Mol. Gen. Genet. *211*, 56–62.
89. Szidonya, J., and Reuter, G. (1988). Cytogenetic analysis of the echinoid(*ed*), dumpy (*dp*) and clot (*cl*) region in *Drosophila melanogaster*. Genet. Res. *51*, 197–208.
90. Tamaru, H., and Selker, E.U. (2001). A histone H3 methyltransferase controls DNA methylation in *Neurospora crassa*. Nature *414*, 277–283.

91. Tamaru, H., Zhang, X., McMillen, D., Singh, P.B., Nakayama, J., Grewal, S.I., Allis, C.D., Cheng, X., and Selker, E.U. (2003). Trimethylated lysine 9 of histone H3 is a mark for DNA methylation in Neurospora crassa. Nat. Genet. *34*, 75–79.

92. Thon, G., Cohen, A., Klar, A.J.S. (1994). Three additional linkage groups that repress transcription and meiotic recombination in the mating-type region of *Schizosaccharomyces pombe. Genetics 138*, 29–38.

93. Thon, G, Klar AJS (1992). The *clr1* locus regulates the expression of the cryptic mating-type loci of fission yeast. Genetics *131*: 287–296.

94. Thorstensen, T., Fischer, A., Sandvik, S.V., Johnson, S.S., Grini, P., Reuter, G., and Aalen, R. (2006). The *Arabidopsis* SUVR4 protein is a nucleolar histone methyltransferase with preference for monomethylated H3K9. Nucl. Acid Res. *34*, 5461–5470.

95. Tschiersch, B., Hofmann, A., Krauss, V., Dorn, R., Korge, G., and Reuter, G. (1994). The protein encoded by the *Drosophila* position-effect variegation suppressor gene *Su(var)3–9* combines domains of antagonistic regulators of homeotic gene complexes. EMBO J. *13*, 3822–3831.

96. Tsukada, Y., Fang, J., Erdjument-Bromage, H., Warren, M.E., Borchers, C.H., Tempst, P., and Zhang, Y. (2006). Histone demethylation by a family of JmjC domain-containing proteins. Nature *439*, 811–816.

97. Tzeng, T.-Y., Lee, C.-H., Chan, L.-W., and Shen, C.-K.J. (2007). Epigenetic regulation of the *Drosophila chromosome* 4 by the histone H3K9 methyltransferase dSETDB1. Proc. Natl. Acad. Sci. U.S.A. 31, 12691–12696.

98. Verdel, A., Jia, S., Gerber, S., Sugiyama, T., Gygi, S., Grewal, S.I., and Moazed, D. (2004). RNAi-mediated targeting of heterochromatin by the RITS complex. Science *303*, 672–676.

99. Volpe, T.A., Kidner, C., Hall, I.M., Teng, G., Grewal, S.I., and Martienssen, R.A. (2002). Regulation of heterochromatic silencing and histone H3 lysine-9 methylation by RNAi. Science *297*, 1833–1837.

100. Wang, Y., Zhang, W., Jin, Y., Johansen, J., and Johansen, K.M. (2001). The JIL-1 tandem kinase mediates histone H3 phosphorylation and is required for maintenance of chromatin structure in *Drosophila*. Cell *105*, 433–443.

101. Westphal, T., and Reuter, G. (2002). Recombinogenic effects of suppressors of position-effect variegation in *Drosophila*. Genetics *160*, 609–621.

102. Wustmann, G., Szidonya, J., Taubert, H., and Reuter, G. (1989). The genetics of position-effect variegation modifying loci in *Drosophila melanogaster*. Mol. Gen. Genet. *217*, 520–527.

103. Xu, T., and Rubin, G. (1993). Analysis of genetic mosaics in developing and adult *Drosophila* tissues. Development *117*, 1223–1237.

104. Zhang,Y., and Reinberg, D. (2001). Transcription regulation by histone methylation: interplay between different covalent modifications of the core histone tails. Genes Dev. *15*, 2343–2368.

Regulation of Higher-order Chromatin Organization and Function by SATB1

5

Prabhat Kumar Purbey, Amita Limaye, Dimple Notani and Sanjeev Galande

National Centre for Cell Science, NCCS Complex. Ganeshkhind, Pune 411007, India

Introduction

Higher-order Chromatin Organization of Genome

This year marks the twelfth anniversary of the discovery of high resolution structure of the nucleosomes, the fundamental unit of chromatin organization (Luger et al., 1997). Most of the living organisms harbor double stranded DNA as genetic material except RNA viruses and some single stranded DNA viruses. The double helical structure of DNA was proposed by Watson and Crick (Watson and Crick, 1953), for which they received Nobel Prize in 1962. Their X ray diffraction analysis revealed that DNA exists in regular helix form, making a complete turn every 34 Å (3.4 nm) with a diameter of ~20 Å (2 nm) (Fig. 5.1A). Genome can be defined as the entire DNA content of an organism. The size of genome varies from 1×10^7 base pair in bacteria to $1\text{-}10 \times 10^8$ in mammals and 1×10^{10} base pairs in certain plants. In eukaryotes, reference to genome typically involves the DNA present in the nuclear compartment. However, certain organelles such as mitochondria also harbor DNA which encodes for few genes.

In eukaryotes, the genomic DNA is packaged at a basal level by winding of the DNA into nucleosomes in a 10 nm beads-on-a-string fiber and at a second level by folding of the nucleoprotein-filament into 30 nm fibers as shown in Figs. 5.1B and C. Each nucleosome contains a core histone octamer composed of histones H2A, H2B, H3, and H4, and often one linker histone either H1 or H5 (Fig. 5.1B). X-ray scattering studies of whole cell and nuclei have shown 30-45 nm reflections that have been interpreted to be arising from the centre to centre compact chromatin fibers (Langmore and Paulson, 1983). A well established fact is that the further compaction of interphase chromatin is effected by formation of loops which operates independently of the basal level packaging (Fig. 5.1D). The looping of genomic DNA is a natural process and is observed from simple prokaryotes to very complex organisms such as humans. The nuclear substructure which provides a framework or platform for the 30 nm chromatin fiber to form loop structures has been termed as the nuclear matrix. The bases of loops which are attached to the nuclear matrix contain specialized DNA sequences called matrix attachment regions (MARs) or scaffold attachment regions (SARs). In 1948, Zbarskii and Debov reported the existence of high salt extracted residual nuclei. But it was in 1974 when this residual structure was redefined as nuclear matrix (Berezney and Coffey, 1974). However the functional implications of nuclear matrix and specialized DNA elements associated with the nuclear matrix or scaffold (Matrix/Scaffold associated region) are relatively new. The formation

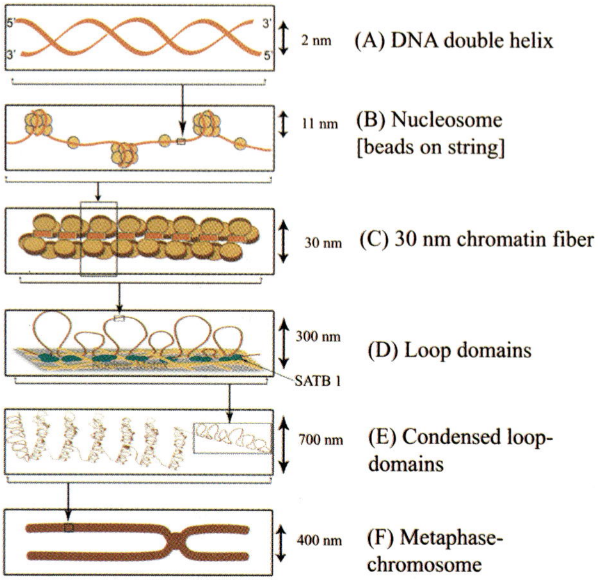

Fig. 5.1 Successive levels of chromatin organization lead to progressive compaction. (**A**) Two anti-parallel strands intertwine to form double helical DNA. (**B**) The DNA duplex is wrapped around core histone octamer to form nucleosomes and histone H1 associates with DNA in linker region. (**C**) The 'beads-on-a-string' assembly of nucleosomes further undergoes coiling induced by binding of histone H1 to linker regions to form the 30 nm fiber comprising 6 nucleosomes per turn. (**D**) The 30 nm fiber gets attached at fixed distances to nuclear matrix by MAR-binding proteins to form 50-100 kb chromatin loops that bring about further compaction and organization of genome into topologically distinct loop domains. (**E**) Chromatin loops are further condensed during metaphase stage to yield coiled and condensed loops. These metaphase condensed loop structures constitute the entire chromatid.

of loop domains involving nuclear matrix and associated MARs have been shown recently to be important for gene transcription (Kumar et al., 2007; Galande et al., 2007).

What is Nuclear Matrix?

The complex higher-order spatial compaction of chromatin is at least in part maintained by the attachment to the nuclear matrix consisting of RNA and protein framework (Berezney et al., 1974). It is operationally defined as the insoluble nucleo-proteinaceous network within the nucleus that resists salt extraction (Berezney, 1991). Biochemical analysis from number of studies revealed presence of more than 200 proteins in the nuclear matrix fraction. Among the most abundant are nuclear lamins, nuclear mitotic apparatus protein (NUMA), nuclear pore complexes, B23, hnRNP proteins and residual nucleoli (Mirkovitch et al., 1984; Gasser et al., 1987; Fey and Penman, 1988; Nelson et al., 1986; Boulikas, 1993; Alvarez et al., 2000; Galande et al., 2000 and Liu et al., 1999). The majority of proteins of the NM (nuclear matrix) are acidic non-histone proteins (Will et al., 1998). Comparison of the protein composition of nuclear matrix by 2-dimensional gel electrophoresis has also revealed the presence of cell type specific and developmental stage specific proteins (Fey and Penman, 1998; Replogle-Schwab et al., 1996; Davido and Getzenberg,

2000). The structure of NM was revealed by immunostaining analysis using polyclonal antibodies against total NM proteins (Fey and Penman, 1988) and transmission electron microscopy of matrix preparations (Nickerson et al., 1997). Evidence for presence of nuclear matrix *in vivo* is indirect, and was provided by comparison of the ultrastructure and composition of nuclear matrix prepared using the extensive crosslinking protocol developed by Wan et al. (Wan et al., 1999), wherein the bulk of chromatin could be effectively removed at physiological ionic strength. With the traditional high-salt extraction nuclear matrix revealed identical internal nuclear matrix structures that are built on a network of branched filaments of about 10 nm diameter. It is estimated that up to 1% of the genomic DNA remains tightly associated with the matrix and a number of studies have been attempted to characterize this fraction. The matrix-bound fraction of DNA is enriched with the flanking sequences from actively transcribed genes (Jackson et al., 1982; Robinson et al., 1982; Ciejek et al., 1983; Hentzen et al., 1984; Jost et al., 1984; Rose et al., 1984; Stratling et al., 1986 and Stratling et al., 1987), repeated DNA sequences (Goldberg et al., 1983; Chimera et al., 1985; Opstelten et al., 1989 and Boulikas T, 1995), sequence with origins of replication (Amati et al., 1990), replicating DNA (Berezney, 1991; Vaughan et al., 1990 and Boulikas T, 1995), and retroviral integration sites (Kumar et al., 2007b). Nuclear matrix association is required for the transcription repression function of SATB1 (Seo et al., 2005). Thus, the MAR-binding proteins may require nuclear matrix association to execute specific functions.

Matrix Attachment Regions/ Scaffold attachment Regions (MARs/ SARs)

Specialized segments of genomic DNA that exhibit high affinity to the nuclear matrix *in vitro* have been designated as matrix/scaffold attachment regions (MARs/SARs). MARs are the DNA sequences which anchor chromatin onto the nuclear matrix, thereby organizing genomic DNA into topologically distinct loop domains that are important not only for the higher-order compaction of chromatin but is also involved in replication, transcription and repair (Gasser et al., 1987 and Galande et al., 2007). In support of this notion, MARs often colocalize or exist in close proximity to regulatory sequences including enhancers, promoters, CNS (conserved non-coding sequences) and DNaseI hypersensitive sites (Kumar et al., 2005; Kumar et al., 2007; Galande et al., 2007 and Cai et al., 2006). The various features of a MAR element and tools used for analyzing them have been reviewed recently (Galande et al., 2002).

MARs are spread all over the chromosomes with irregular spacing and could be identified biochemically by means of their association with nuclear matrix preparation *in vitro* or can be predicted by bioinformatics analysis which can predict the MAR potential of a queried sequence (Singh et al., 1997). The MAR elements or the genomic DNA sequences are tethered to the nuclear matrix via cell type specific or ubiquitously expressed DNA binding proteins. The DNA binding proteins which bind directly to matrix associated regions of DNA are referred to as MAR binding proteins (MBPs). Some of the well known MAR binding proteins are SATB1 (Special AT-Rich Binding Protein 1,) SATB2 (Special AT-Rich Binding Protein 2), BRIGHT, CDP, Lamin A/B/C, SMAR1, nucleolin, Ku, PARP, mutant p53, HMG (High mobility group proteins) etc (Reviewed by Galande, 2002). MARs could be present in non-coding regions, intronic regions, coding regions and promoter regions within the genome. Recent studies have discovered implications of these elements in various cellular processes in addition to the structural organization of the chromatin.

Tools to Analyze the Dynamic Change in Chromatin Organization Involving *cis* Regulatory Elements Including MARs

In several cases, gene activation is accompanied by large-scale changes in chromatin conformation. Activation or repression of genes can involve change in formation of chromatin loops through long-range interactions between regulatory elements (Carter et al., 2002; Tolhuis et al., 2002; Murrell et al., 2004; Spilianakis et al., 2004; Vakoc et al., 2005; Kumar et al., 2006 and Cai et al., 2006). The MARs are one among such elements which act at the bases of loops, where specific MAR binding proteins bind and tether these elements to the nuclear matrix providing a higher-order compaction of chromatin. These loopings are altered during various cellular processes in a similar fashion as observed in case of looping of enhancers. For example, during T_H2 cell activation several segments of DNA within the T_H2 locus get associated with the nuclear matrix, which is required for the coordinated expression of T_H2 locus genes (Cai et al., 2006). The MHC class I locus is dynamically reorganized during IFNγ treatment or upon siRNA-mediated knockdown of SATB1 or PML isoforms (Kumar et al., 2007). Silencing of either SATB1 or various SATB1 interacting PML isoforms lead to changes in the expression of MHC class I associated genes which is directly correlated with change(s) in specific loop organization (Kumar et al., 2007 and Galande et al., 2007). The above mentioned studies were feasible only after formulation of certain methods or tools which can enable to capture such complicated and dynamic structure formed by chromatin in the nucleus. There was a necessity to design certain important tools so as to describe in detail the various conformations of chromatin under different stages of condensed and decondensed states by capturing the conformation of chromatin under a particular physiological condition. For analyzing chromatin dynamics involving MAR elements, amplified FISH, MAR-loop partitioning methods have provided the necessary information to predict changes in the loop organization during cell activation (Cai et al., 1999 and Kumar et al., 2006). These methods could only allow qualitative estimates regarding the presence of a DNA element either in the matrix fraction or loop fraction. More sophisticated methods such as 3C and its derivatives allow detection of the interaction of distant chromatin elements involved in the formation of loop structures (Dekker et al., 2002 and Gheldof et al., 2006). Specific modifications of chromatin conformation capture (3C) have shown significant potential in studying the changes in loop organization, which involves MARs and MAR binding proteins (Cai et al., 2006 and Kumar et al., 2007). Using the combination of chromatin immunoprecipitation (ChIP) and the loop assay (ChIP-loop), the involvement of specific protein factors involved in the formation of such higher-order organization of chromatin can be deciphered (Cai et al., 2006 and Kumar et al., 2007). Various tools employed to study loop organization are detailed in the following sub-section.

Amplified Fluorescence *in situ* Hybridization (Ampli-FISH), A method for Intra-Nuclear Localization of MARs and Other *cis* Regulatory Elements

Employing certain modifications in the prevalent fluorescence *in situ* hybridization (FISH) techniques, Cai and Kohwi-Shigematsu (1999) devised a method which enables detection of the relocalization or delocalization of MARs during various cellular processes. These changes in localization of the MARs (DNA element) either in matrix attachment region or loop region could indicate the change(s) in higher-order loop organization. The method is based on *in situ* hybridization with the tyramide amplification of signal. Using such amplified FISH, one can obtain signal for a relatively small DNA sequence after hybridization with complimentary genomic region (Cai S et al., 1999), whereas in conventional FISH that does not involve any amplification step the size of the probe required to detect the signal is in kilobases. The method requires salt treatment of the nuclei which generate histone depleted nuclei with distended chromatin loops referred to as "Halos". This is then followed by DNase I digestion or restriction digestion by a number of restriction enzymes to digest bulk of the genomic DNA (Cai and Kohwi-Shigematsu, 1999; deBelle et al., 1998 and Kumar et al., 2007). The resulting material contains matrix associated DNA and is then hybridized with specific biotinylated probes. The biotin-labelled probe hybridized to the genomic DNA can be detected by employing tyramide signal amplification (TSA) technology which can enhance the signal more than 1000-fold. Further, the biotin-tyramide is detected using extravidin-linked fluorescein or other chromogenic indicators. This method has been useful in determining the localization of SATB1 binding sequences (SBSs), which in turn are dependent on expression of SATB1 in different cell lines. In SATB1 expressing Jurkat T cells, SBSs are localized in nuclear matrix whereas in SATB1 non-expressing cell line SKBR3 these SBSs are detected in the loops (deBelle et al., 1998).

Chromosome Conformation Capture Methods to Analyze the 3-Dimensional Organization of Chromatin

Chromosome Conformation Capture (3C) Assay

Recent evidence indicates that regulatory elements can act over large genomic distances by engaging in direct physical interactions with their target genes or with other elements (Dekker, 2003; de Laat and Grosveld, 2003; Chambeyron and Bickmore, 2004; West and Fraser, 2005). These observations indicate that the genome may be organized as a complex three-dimensional network that is determined by physical interactions between genes and regulatory elements. These findings were possible only after invention of an efficient 3C technique which can enable capturing of such long range interactions. In 2002, Kleckner's group reported a method to analyze functional relationships between genes and regulatory elements and named it as Chromosome Conformation Capture (3C) method (Dekker et al., 2003). 3C uses formaldehyde cross-linking to covalently trap interacting chromatin segments throughout the genome. Interacting elements are then restriction-enzyme-digested and intramolecularly ligated. The ligated product is used for the semi-quantitative PCR amplification to detect individual chromatin interactions. This method is particularly suited for relatively small-scale studies focused on the analysis of interactions between a set of candidate elements. The frequency with which two restriction fragments become ligated is a measure of the frequency by which they interact in the nucleus (Dekker, 2003). 3C has been employed to study the spatial organization of yeast chromosome III (Dekker, 2003), analysis of several mammalian loci such as the beta-globin locus (Tolhuis et al., 2002; Palstra et al., 2003 and Vakoc et al., 2005), the T-helper type 2 (T_H2) cytokine locus (Spilianakis and Flavell, 2004 and Cai et al., 2006), the immunoglobulin κ locus (Liu and Garrard, 2005), FMR (Fragile X mental retardation 1) promoter conformation (Gheldof et al., 2006) and the *Igf2* imprinted locus (Murrell et al., 2004).

These studies revealed direct interactions between enhancers and promoters of target genes, with the linking DNA looping outward. 3C was also used to detect *trans* interactions between yeast chromosomes (Dekker et al., 2002) and between functionally related elements located on different mouse chromosomes (Spilianakis et al., 2005; Ling et al., 2006 and Xu et al., 2006). A modification of 3C has been developed and employed particularly for capturing the chromatin organization mediated by MARs and MAR binding proteins and has been named as MAR ligation assay (Cai et al., 2006 and Kumar et al., 2006). MAR-ligation assay which is a modification of 3C, does not involve the formaldehyde cross-linking step, instead the matrix associated DNA protein complex fraction is used for the ligation. This method with few modifications has been used to study the organization of loop structure mediated by MARs, which are also SATB1 binding targets in the mouse T_H2 locus and MHC class I locus (Cai et al., 2006 and Kumar et al., 2006). The 3C methodology and its major derivatives are schematically represented in Fig. 5.2.

Circular Chromosome Conformation Capture (4C) Assay

3C method is a powerful tool to capture the *cis*- and *trans*- chromosomal interactions but success of the method entirely depends upon the prior knowledge of the existing interaction between two

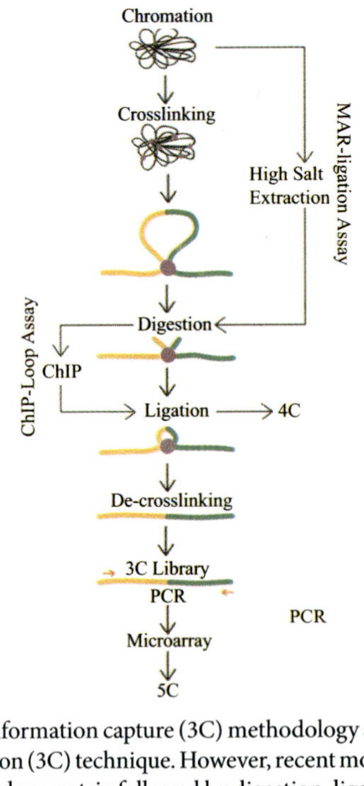

Fig. 5.2 An overview of chromosome conformation capture (3C) methodology and its modifications. Flow diagram depicts the original chromatin capture conformation (3C) technique. However, recent modifications such as omission of crosslinking and introducing high salt extraction of nuclear matrix followed by digestion, ligation and PCR analysis constitute the MAR-ligation assay (Kumar et al., 2007). Before ligation if digested chromatin is subjected to chromatin immunoprecipitation using specific antibodies it is designated as ChIP-loop assay (Horike et al., 2005). Another modification wherein products are ligated such that they are circularized following linker mediated PCR, then it is termed 4C (circular chromosome conformation capture) (Simonis et al., 2006). Similarly, if 3C libraries are amplified, labelled and hybridized onto genomic microarrays then the method is referred to as 5C (chromosome conformation capture carbon copy) (Dostie et al., 2006).

complexes. To overcome this limitation recently Zhao et al., (2006) evolved a technique referred to as circular chromosome conformation capture (4C) which is a modified 3C technique. This approach employs a circularization step that facilitates high-throughput screening of physical interactions between chromosomes without a preconceived idea of the interacting partners. The 4C method is based on the proximity ligation principle, in which DNA-protein/protein-DNA complexes generate circular DNA molecules under high concentrations of ligase and unusually prolonged incubations of upto 1 week. After reverse cross-linking, nested primer PCR is utilized where primers are designed such that they fall within the genomic locus of interest but proximal to the target sequence during ligation which amplifies DNA sequences with physical proximity, without any prior knowledge of their identities. Upon nested primer PCR, only circular plasmids are amplified but linear DNA does not amplify. Further, to identify the *cis*- and *trans*- chromatin interactions, these PCR products are subjected to high-throughput sequencing. To measure the genome-wide most frequent interactions semi-quantitative hybridization to microarrays is employed wherein the 4C target PCR products are labelled and hybridized to an array containing total 4C product spots. Signal ratio between the 4C product and total genomic DNA is compared across the spots which indirectly indicate the frequency of a particular interaction over another interaction. Using this technique, Zhao et al. (2006) addressed the existence of epigenetically controlled networks of chromosomal interactions using the H19 imprinting control region (ICR), a well-known differentially methylated domain 6.

Chromosome Conformation Capture Carbon Copy (5C) Assay

3C uses PCR to detect individual chromatin interactions, which is particularly suited for relatively small-scale studies focused on the analysis of interactions between a set of candidate elements. However, PCR detection is not well applicable for large-scale mapping of chromatin interactions. In 3C method, the ligated products library is extremely complex and the specific ligation products are relatively low in abundance which makes direct large scale analysis difficult. To overcome this problem, an improved 5C (Chromosome conformation capture carbon copy) method was proposed (Dostie et al., 2006). In 5C method, 3C libraries are analyzed using a high-throughput detection method such as microarrays or DNA sequencing. 5C uses highly multiplexed ligation-mediated amplification (LMA) to first "copy" and then amplify parts of the 3C library followed by detection on microarrays or by quantitative DNA sequencing. 5C was developed and validated by analyzing the human β-globin locus and a conserved gene desert region located on human chromosome 16. 5C could be widely applicable to determine the *cis* and *trans* connectivity of regulatory elements throughout large genomic regions. In addition, 5C experiments can be designed so that complete interaction maps can be generated for any large genomic region of interest, which can reveal locations of novel gene regulatory elements and may also provide detailed insights into higher-order chromosome folding.

Matrix-Loop Partitioning Assay

This is modification of technique employed to separate nuclear matrix for the detection of matrix associated proteins (Cai et al., 1999; Seo et al., 2005). In this method DNA is isolated from the nuclear matrix fraction (Kumar et al., 2006; 2007). A classical method is employed to separate nuclear matrix and loop region from the nuclei. The method employs high salt extraction (2M

NaCl) of histones which yields histone depleted halos of chromatin. The extended chromatin in halos is then digested by using DNase I or cocktail of restriction enzymes. The undigested material is referred as nuclear matrix fraction and digested material in the supernatant contains the chromatin in loop regions. The digested and undigested materials are deproteinized by proteinase K treatment followed by phenol chloroform extraction. Finally, the nuclear matrix fraction and loop fraction of DNA are used for PCR amplification using set of primers against a genomic sequence of interest. This method has been employed to detect the presence of HIV integration sequences in nuclear matrix associated region or in the chromatin associated with the loop region (Kumar et al., 2007b).

MAR-Ligation Assay/ 3C MAR-ligation Assay (Chromosome Conformation Capture Involving MARs)

Matrix and loop DNA are prepared essentially as described above. The matrix prepared is then used for ligation. The ligated product is then used for the PCR amplification with specific sets of primers. This method does not require the crosslinking step and has enabled to analyze the chromatin higher-order organization involving MAR elements during cellular stimuli (Kumar et al., 2006). Briefly Jurkat cells are washed with phosphate buffer followed by sequential lysis with CSK-1 (0.5% triton X-100,10 mM PIPES pH 6.8, 100 mM NaCl, 300 mM sucrose, 3 mM $MgCl_2$, 1 mM EGTA, 1 mM PMSF and 1 X protease inhibitor cocktail) and CSK-2 buffers (same as CSK-1 except that triton X-100 is omitted). DNase I digestion or restriction digestion using several enzymes are performed for several hours to digest genomic DNA into approximately 200 bp fragments. Digested supernatant (loop DNA) as well as undigested material (MARs) are independently collected. Isolated matrix DNA is ligated using T4 DNA ligase and the ligated product is then used as a template for PCR amplification with different sets of primers designed against distinct regions of a locus. PCR products are resolved by native polyacrylamide gel electrophoresis, stained and visualized by UV transillumination. The relative PCR intensity is the indirect measure of the relative ligation frequency between two DNA fragments present in the nuclear matrix fraction.

ChIP-Loop Assay to Monitor Dynamics of Proteins Associated with MARs

ChIP-loop assay allows examination of a specific group of chromatin loops that are fastened at their bases with a specific protein or histone modifications, associated either with transcriptionally active or transcriptionally silent chromatin. This method has been used to identify the dynamics of 3-dimensional chromatin organization of T_H2 locus by SATB1 during activation and MHC class I reorganization during IFN-γ activation (Cai et al., 2006; Kumar et al., 2006). In ChIP-loop assay, first *in vivo* crosslinking of cells is performed followed by isolation of crosslinked DNA- protein complex. The crosslinked product is then digested with Sau 3AI before immunoprecipitation with specific antibody. The immunoprecipitated sample is ligated and analyzed by specific sets of primers (Cai et al., 2006 and Kumar et al., 2006). Comparison of the PCR product obtained in ChIP-loop with the PCR product observed in MAR-ligation assay would indicate involvement of a specific DNA binding protein factor in mediating such loop organization.

Significance of Matrix Attachment Regions (MARs)

Matrix attachment regions are specialized DNA sequences present intermittently in the genome and are required for the formation of varying sized loop domains to achieve higher-order compaction of chromatin in the nucleus (Fig. 5.1). The complex organization of eukaryotic genome does not only provide the physical compaction of DNA, but also imposes additional level of complexity in the regulation of DNA associated functions. In eukaryotes, the genome is organized in a very complex manner. Every order of organization of DNA provides additional level of regulation for the nuclear processes involving DNA, such as transcription, replication, DNA repair etc. In recent years MARs, which acts as base of chromatin loops, have been shown to perform several cellular functions in general and gene transcription in particular (reviewed in Galande, 2002).

MARs Regulate Expression of Genes During Variety of Cellular Processes

MARs have been shown to exist both in coding regions, intronic regions as well as 5′ and 3′ flanking regions of genes. Strikingly, many MARs are found in close proximity of the *cis*-acting elements required for transcription and its spatio-temporal regulation (Cockerill et al., 1987; Boulikas, 1993; Cai et al., 2006; Kumar et al., 2007). In recent years, the biological significance of certain MARs in tissue-specific gene expression and three-dimensional chromatin organization has been revealed *in vivo*. In particular, studies on MARs flanking the immunoglobulin heavy chain (*IgH*) enhancer showed that these sequences are essential for the B lymphocyte-specific transcription of a rearranged μ gene (Forrester et al., 1994). These MARs have also been shown to generate long range chromatin accessibility to transcription factors (Jenuwein et al., 1997). This phenomenon correlates with extended demethylation of the gene locus in a transcription-independent manner (Jenuwein et al., 1997). In addition, B cell-specific demethylation at the immunoglobulin kappa (*Igκ*) gene locus requires both the intronic κ enhancer and the nearby MAR (Lichtenstein et al., 1994 and Kirillov et al., 1996). MARs mediate long-range chromatin accessibility presumably by generation of an extended domain of histone acetylation (Fernandez et al., 2001). Transfection studies suggest a specific role for the matrix attachment region (MAR)- and methyl-CpG-binding repressor protein MeCP2/ARBP (von Kries et al., 1991). This MAR-binding protein, through interaction with mSin3A, recruits a corepressor complex containing histone deacetylases that is believed to generate a localized silenced chromatin structure (Stratling et al., 1999). The MAR present in the mouse IL-2 receptor locus, kilobases away from the transcription start site, is associated with SATB1 repressor complex. The knockdown of SATB1 leads to derepression of this gene and histone modifications which is a mark of actively transcribed genes. SATB1 can affect histone modifications over distances of upto several kilobases from its binding site (Yasui et al., 2002). Mouse cMyc, hIL-2 receptor and hIL-2 promoter harbors MARs which are also bound by SATB1 and get de-repressed when SATB1 is knocked down or its association with HDAC1 is abolished (Cai et al., 2003 and Kumar et al., 2005). BCL-2 major breakpoint cluster region (MBR) is another MAR element which is associated with SATB1 *in vivo*. SATB1 negatively regulates BCL-2 expression via associating with this MAR (Ma et al., 2007). Mouse T_H2 locus possesses several of MAR elements that are involved in coordinated regulation of the locus associated genes

(Cai et al., 2006). Similarly, number of MARs have been characterized to be present in the MHC class I locus. These locus associated MARs are bound by SATB1 and PML complex to form a specialized loop structure. The alteration in loop organization either by change in SATB1 level or change in PML level leads to change in the transcription of the locus associated genes (Kumar et al., 2006; Galande et al., 2007).

The Dynamic Associations of MARs are Important for the Regulation of Transcription

The structural elements that organize the nucleus also contribute towards the dynamic changes in chromatin such as enhancer looping, replication, transcription, recombination, chromosome condensation and decondensation. The basal level folding of chromatin in the loops shows dynamic variation based on the activity status of that region, as evidenced by a spatio-temporal interplay of replication and transcription domains (Berezney et al., 1998). MARs/SARs are the only putative *cis* elements that have been identified so far that play an important role in facilitating dynamic changes in higher-order of chromatin organization (Hart et al., 1998; Galande et al., 2002 and 2007). The attachment of certain MARs to the matrix has been found to depend on cell-type or cell cycle and cell-type specific MAR-binding proteins. For example high-resolution fluorescence *in situ* hybridization (FISH) analysis using the human satellite 2 sequence that spans approximately 4 Mbp in the pericentromeric region of human chromosome 1, which was identified as an *in vivo* binding sequence to SATB1 (de belle et al., 1998), revealed that it is firmly anchored onto the nuclear matrix of human Jurkat cells (Cai et al., 1999). However, the same sequence is completely detached from the nuclear matrix of the human breast cancer SK-BR-3 cells that lack SATB1, indicating that this DNA sequence represents a cell-type specific MAR and its association with nuclear matrix is dependent on the MAR-binding protein SATB1 (Cai et al., 1999 and de Belle et al., 1998). The alteration in attachment of these sequences with the nuclear matrix may occur in case of many if not all of the SATB1 binding sequences in T cells that express SATB1 at different level. Interestingly, similar FISH analysis of T cell nuclei using an *in vivo* SATB1-binding sequence TSBS-8 as a probe revealed that its attachment to the nuclear matrix is T cell activation dependent (Cai et al., 1999). During T_H2 activation several MARs associate with the nuclear matrix to form condensed loops that are required for the transcription of T_H2 cytokine locus-associated genes (Cai et al., 2006). In Jurkat T cells, MARs present in the 0.3 Mb MHC class I locus are associated with nuclear matrix via SATB1, undergo rearrangement upon IFN γ treatment and additional new loop is generated due to association of probably non-SATB1-binding MAR element to the nuclear matrix (Kumar et al., 2006). Thus, the dynamic nature of nuclear matrix attachment of MARs may dictate the functional state of cells.

BURs are Specialized MARs

BURs are smaller regions typically identified in MARs that exhibit high affinity to isolated nuclear framework *in vitro*. An important clue to search for the set of proteins that anchor DNA onto the nuclear matrix skeleton is their binding specificity to base unpairing regions (BURs). BURs are represented by unusually high propensity to become continuously base unpaired when subjected to negative superhelical strain. BURs have a unique DNA context, called ATC sequence context, which consists exclusively of well mixed As, Ts, and Cs, but no Gs on one strand (Dickinson et al., 1997). When such ATC sequence stretches of about 16 base pairs or longer are found in clusters, the genomic region of not more than 150 bp in total length unwinds by base unpairing. BURs, as the *in vivo* binding targets of SATB1 and some of the recently identified BUR-binding proteins, are localized at the bases of chromatin loops inside cells. These sites are firmly attached to nuclear framework (nuclear matrix) inside nuclei even after extensive extractions of proteins and DNase I digestion (de Belle et al., 1998). It should be noted that the BUR-mediated attachment to NM is strictly governed by the unwinding potential of a given DNA sequence. Thus, mutations that abolish base unpairing weaken the association with NM (Benham et al., 1997), demonstrating that the unwinding propensity inherent in BURs is biologically significant. The BUR-binding protein SATB1 actively tethers BURs onto the nuclear framework in a cell type-dependent fashion (Cai et al., 1999). Therefore, BURs are specialized MARs that are tethered to the nuclear matrix by specific BUR-binding proteins.

MAR and BUR Associated Proteins: Players Involved in Chromatin Looping and Gene Regulation

The S/MAR and BURs are associated with the nuclear matrix via a number of DNA or chromatin binding proteins. These proteins are expressed either ubiquitously or in cell-type- or developmental stage-specific manner. SATB1 is the first base unpairing region-binding protein identified sixteen years ago (Dickinson et al., 1992). Although it was thought that SATB1 is expressed specifically in T cells but recent studies implied that it also expresses in several other cell types (Kumar et al., 2007). The p114 is a breast carcinoma-specific MAR-binding protein (Yanagisawa et al., 1996; Galande, 2002), subsequently this protein was purified from a breast adenocarcinoma cell line SK-BR-3 using double-stranded BUR-affinity chromatography (Yanagisawa et al., 1996) and identified as PARP1 (Galande et al., 1999). Double-stranded BUR-affinity chromatography is another powerful tool that has been exploited to purify several BUR-binding proteins from a variety of cell lines and tissues (Yanagisawa et al., 1996; Galande et al., 1999, Dickinson et al., 1995). The protocol for this simple, two-step strategy for the purification of BUR-binding proteins is also summarized in a comprehensive review on MARs and MAR-binding proteins (Galande, 2002). PARP1 has been also demonstrated to be involved in the ADP ribosylation, base excision repair, and also the transcription. It facilitates transcription via association with the Pol II and or Yin and Yang 1 (YY1) (Oei et al., 1998). Bright (B cell regulator of IgH transcription) is a B cell specific transcription activator and stimulates transcription driven by immunoglobin heavy chain (IgH) enhancer "Eμ" by binding to an ATC motif within the flanking MAR (Herrscher et al., 1995). Bright binds to the minor groove by the virtue of two identified domains (Webb et al., 1999), a self association tetramerization domain, termed REKLES for a heptapeptide conserved within this region among bright orthologs, and a DNA-binding domain termed ARID (AT-rich interaction domain). A similar domain is also found in SWI 1, a component of the SWI-SNF chromatin remodeling complex (Peterson and Tamkun, 1995). Like SWI1, all ARID containing proteins can bind to AT-rich DNA, but only those family members which contain both ARID and REKLES domains bind specifically to AT-rich MAR motif e.g. Bright, Dri and Dril (Webb et al., 1998). Components of human SWI-SNF are tightly associated with the nuclear matrix (Reyes et al., 1997). This suggests that at least a fraction of the SWI-SNF complex could be involved in the chromatin organization properties associated with matrix attachment region.

SATB1, a Hallmark BUR-binding Protein

SATB1 Possesses Several Characterized Functional Domains and Motifs

Special AT-rich Binding protein 1 (SATB1) was originally identified and cloned by virtue of its ability to bind to the BUR located within the 3′ MAR of the IgH enhancer (Dickinson et al., 1992). SATB1 was the first cell-type-restricted MAR/BUR-binding protein to be identified and is expressed predominantly in thymocytes (Dickinson et al., 1992). Expression of SATB1 is also found in the testis, brain and cell lines of different organ origin. For example HEK 293 cell is from embryonic kidney origin also expresses SATB1 at relatively low level compared to T cell lines such as Jurkat (Kumar et al., 2007). SATB1 is also one of the few gene products that are induced early upon peripheral T cell activation. Mutagenesis experiment demonstrated that SATB1 selectively binds to double stranded DNA sequences possessing high unwinding capability in a specialized AT-rich sequence context where one strand consists of mixed A's, T's, and C's, excluding G's (ATC sequences). Chemical interference assays show that SATB1 binds along the minor groove with very little contact with the bases (Dickinson et al., 1992). Thus it was established that SATB1 exhibits a unique mode of binding to DNA and recognizes a specific DNA sequence context rather than a specific DNA sequence consensus. SATB1 binds to the minor groove of DNA specifically recognizing the specialized AT-rich DNA sequences referred to as BURs. SATB1 probably recognizes ATC sequences indirectly by the altered sugar phosphate backbone determined by the ATC sequence context. Identification of an atypical homeodomain (HD) in the distal region of the protein and presence of two CUT like repeats led to conferring SATB1 with multiple domain function (Dickinson et al., 1997). The presence of an atypical homeodomain and 2 CUT like repeats in a MAR binding protein like SATB1 confers uniqueness as compared with other MAR binding proteins. Report of SATB1 acting as a repressor on MAR-linked reporter genes speculated an important role in T cell development by regulating expression of various target genes by binding to the BURs (Kohwi-Shigematsu et al., 1997). The role played by SATB1 was further identified by creating a SATB1 null mouse model (Alvarez et al., 2000). A variety of genes were shown to be dysregulated due to ablation of SATB1 gene. The 2% genes that were dysregulated due to absence of SATB1 comprised cytokine receptors, cytokines and few other apoptotic genes. The upregulation of certain apoptotic genes predicted a role of SATB1 in T cell apoptosis. During T cell apoptosis SATB1 is cleaved after aspartate at position 254 by caspase 6, as a result of which it was observed that SATB1 can no more bind to its target sequences causing detachment of SATB1 from chromatin (Galande et al., 2001). The cleavage of SATB1 by caspase 6 and its effect on the DNA-binding activity of SATB1 lead to identification of a dimerization domain at N-terminal region. Furthermore, residues 96-204 were narrowed down to be the dimerization domain required for the DNA binding activity of SATB1 *in vitro* (Galande et al., 2001). Interestingly, homology searches predicted that the N-terminal dimerization domain has a PDZ-like domain (Galande et al., 2001). The PDZ-like domain provides the dimerization interface (Galande et al., 2001; Purbey et al., 2008) and ability to interact with multiple protein partners of SATB1 including corepressors and coactivators (Kumar et al., 2005; 2006; 2007). Recent studies on identification of DNA-binding site consensus for SATB1 yielded a homeodomain-like bipartite consensus sequence element for SATB1, shedding light on its mode of DNA binding and contribution of various domains towards DNA binding specificity and affinity of SATB1 (Purbey et al., 2008). SATB1 binds to the inverted consensus palindromic repeats via homeodomain in dimerization-dependent manner via the minor groove, whereas the Cut repeats enhance the binding via hydrogen bonding interactions in the major groove. Furthermore, high affinity binding by SATB1 is dimerization-dependent and its specific high affinity binding is mediated by its homeodomain (HD) in collaboration with the Cut repeat containing domain (CD) (Purbey et al., 2008). The unique activities of various regions of this multi-domain protein evolved a function of its own, the function of a MAR/BUR-binding protein dictated by all these domains in concert. The various functional domains in SATB1 are depicted schematically in Fig. 5.3.

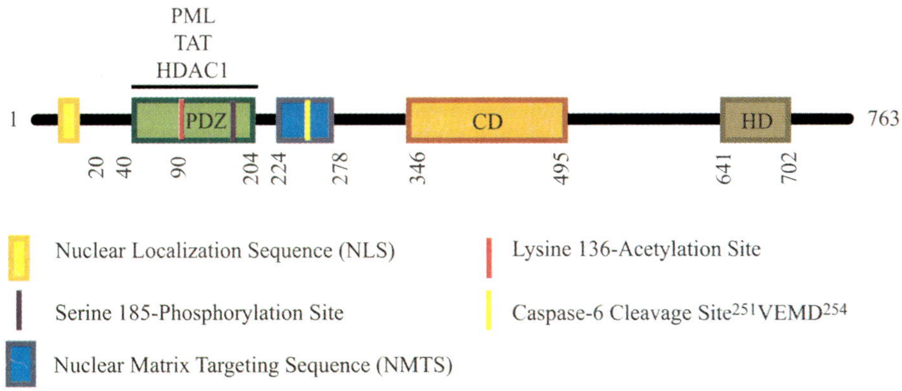

Fig. 5.3 Domain map of SATB1. Schematic representation of various functionally characterized domains of SATB1. Sites of known *in vivo* post-translational modifications of SATB1 are also indicated.

Repertoire of SATB1 Binding Targets Ranges from Virus to Human Genome

It was the IgH MAR base unpairing region-binding property which led to the identification of SATB1 in 1992 (Dickinson et al., 1992). Since then a vast number of SATB1 binding genomic regions have been identified and characterized by several independent researchers. Using *in vivo* crosslinking strategy in Jurkat T cell deBelle et al. (1998) cloned and characterized a set of genomic sequences to which SATB1 is bound *in vivo*. Fluorescent *in situ* hybridization (FISH) analysis with each of these sequences revealed that they are attached to the nuclear matrix indicating that SATB1 binds to MARs *in vivo*. This was the first report showing the direct attachment of binding sites for a MAR-binding protein onto the nuclear matrix *in vivo*. Apart from the ATC consensus sequences, L1 element and CpG islands were also identified as the *in vivo* target sequences of SATB1. The hypersensitive site 1 of T cell receptor β, also a nuclear matrix associated region has been shown to be bound by SATB1. This region represses the enhancer-dependent gene expression but not the endogenous gene expression indicating that nuclear matrix function could be changed if not in its natural position in the genome (Chattopadhyay et al., 1998).

Mouse mammary tumor virus (MMTV) is a retrovirus which is transmitted through milk to the offspring. A variant of MMTV which causes leukemia instead of mammary cancer, is expressed at much higher levels in T cells than the mammary cells. This abnormal expression of virus in T cells is due to a deletion of negative regulatory elements (NREs), which includes a matrix associated region within the promoter proximal to the NREs in the MMTV long terminal repeats (LTR). This MAR region of the MTTV promoter forms two distinct complexes with cellular proteins, and SATB1 was found to be present in both complexes. SATB1 binds these sequences and mutation in the MMTV LTR causes destabilization of the SATB1 complex *in vitro*. SATB1 represses the expression of genes mediated by MMTV LTR. Since there is a mutation in promoter, proximal to the SATB1 binding site MMTV LTR reporter gene expression increases in lymphoid tissue of transgenic mice (Liu et al., 1997).

A DNA (L2a) element which is involved in the regulation of the mouse CD8 alpha gene was shown to be a MAR. Two proteins bound to this MAR were named L2a-P1 and L2a-P2. The L2a-P1 protein was found to be present in the all CD8 positive T cells and was identified as SATB1 (Banan et al., 1997) and the L2a-P2 protein was identified as CDP/Cux (Banan et al., 1997). In fact SATB1 competes out CDP/Cux to favor the transcription of CD8a possibly through alteration of chromatin conformation. HOXB2, a homeodomain protein important in neural development which is also expressed during erythropoiesis binds to the Human γ globin gene. SATB1 and HoxB2 proteins compete with each other for the binding to the same DNA stretch in the γ globin gene. SATB1 positively regulates the gamma globin whereas HOXB2 regulates negatively (Case et al., 1999).

The major breakpoint region (MBR) which resides in untranslated region of exon 3 of BCL2, is involved in the chromosomal translocation that generally occurs during human follicular lymphoma leading to constitutive activation of BCL2 gene. A 33 bp highly AT-rich sequence, also characterized to be a MAR and highly conserved in human and mouse BCL2 gene (37MBR), is bound by SATB1 and thus regulates the expression of BCL2 (Ma et al., 2007). The multi-functionality of SATB1 is also highlighted through the fact that SATB1 interacts with RP3 and recruits it for regulation of its other target gene BCL2 (Yu-Yeh et al., 2005). Thus it is quite evident from the multiple interactions exhibited by SATB1 with various kinds of proteins and its MAR-binding potential highlights the moonlighting property of SATB1. Thus, SATB1, by virtue of its ability to bind to multiple sites where chromatin is fastened to form loop domains, may dictate the organization and structure of chromatin domains in a cell type-specific manner, thereby affecting the transcriptional status of several genes in T cells.

Recently, Purbey et al. (2008) used the approach of systematic evolution of ligands by exponential enrichment (SELEX) to isolate a pool of synthetic DNA sequences that were bound with high affinity by SATB1. They reported a conserved pattern of 10-12 nucleotides in all enriched sequences consisting of two inverted AT-rich (4 to 6 bases) repeats resembling the homeodomain binding site separated by 1-2 non-AT nucleotides. Substitution by cytosine (C) at any position in the conserved homeodomain binding region "TATTAG" abolishes the DNA binding activity of SATB1 indicating that it is mediated primarily by the homeodomain. The consensus SATB1-binding site (CSBS)-driven *in vivo* reporter analysis indicated that SATB1 acts as a repressor upon binding to the CSBS and most of its derivatives and the extent of repression is proportional to SATB1's binding affinity to these sequences. These studies provided evidence for sequence specific binding of SATB1 to target DNA and not only to the ATC context as thought before (Purbey et al., 2008). Availability of CSBS will certainly help the identification of genome-wide targets of SATB1.

SATB1 Complexes with Several Transcriptional Regulators to Regulate the Transcription of Target Genes

SATB1 is a global gene regulator and regulates gene expression by binding to upstream promoter regions, at times kilo bases away from the transcription start site and in the intronic region of individual genes and gene loci. SATB1 also seems to mediate changes in histone tail modifications and nucleosome remodelling. SATB1 makes complex with the components of nucleosome remodeling and histone modifying enzymes as well as the components of ATP dependent nucleosome mobilizing enzymes (ACF/ISWI) (Yasui et al., 2002). The HDAC1 corepressor bound to SATB1 is displaced by Tat protein during HIV-1 infection leading to derepression of IL-2 and its receptor (Kumar et al., 2005). Association of SATB1 with the HDAC1 corepressor complex is determined by its phosphorylation (Kumar et al., 2006). During T cell activation, SATB1 is dephosphorylated by an unknown mechanism and subsequently interacts with the PCAF histone acetyltransferase (HAT). PCAF mediated acetylation leads to dissociation of SATB1 from its genomic targets. SATB1 also interacts with p300 HAT but is not acetylated by this enzyme (Kumar et al., 2006). All of these interactions with the corepressor and coactivator complexes are mediated by the N-terminal PDZ-like domain of SATB1. Additionally, SATB1 makes complex with several isoforms of the promyelocytic leukemia (PML) protein that is a defining component of PML nuclear bodies (Kumar et al., 2007). PML bodies are known to be reservoirs of multiple components of transcriptional machinery, however, PML does not possess DNA-binding activity. Thus, SATB1 and PML bodies are at the crossroads of chromatin architecture and gene regulation.

Post-translational Modifications of SATB1 and Their Impact on the Ability of SATB1 to Act as a Global Gene Regulator

Post-translational modifications such as phosphorylation and acetylation have been shown to modulate interaction of many transcription factors with their interacting proteins as well as their DNA-binding affinities. Number of kinases and acetyltrasferases (HATs) are involved in the phosphorylation and acetylation processes respectively. Activity of many transcription factors is regulated through their interaction with coactivators such as CBP (CREB-binding protein)/p300 (Chan et al., 2001), PCAF/GCN5 (Poux et al., 2003). Phosphorylation also influences the association of different transcription factors with HDACs and HATs. Phosphorylation of NK-κB and p53 enhances their interaction with CBP/p300 (Zhong et al., 2002 and Lambert et al., 1998). Unphosphorylated MyoD showed higher interaction with PCAF (Tintignac et al., 2004), whereas dephosphorylation of pRb promoted the formation of pRb-HDAC1 complex in differentiated myotubes (Puri et al., 2001). Thus, the reversible and coordinated modifications of various transcription factors seem to dictate their functional outcome.

SATB1 recruits various chromatin remodeling proteins via direct physical association thereby suppressing gene expression through histone deacetylation and nucleosome remodeling at SATB1-bound MARs (Yasui et al., 2002). It also regulates gene expression by directly influencing the promoter activity by making complex with p300 (Fujii et al., 2003), CBP (Wen et al., 2005), and HDAC1 (Kumar et al., 2005). MAR-linked luciferase reporter assays using cells treated with kinase and deacetylase inhibitors suggested role of post-translational modifications towards modulating the transcriptional activity of SATB1 (Kumar et al., 2006). Subsequently, Kumar et al. (2006) demonstrated for the first time that SATB1 is both phosphorylated and acetylated *in vivo*. Specifically, the residues K136 and S185 in the PDZ-like interaction domain are modified by PCAF HAT and protein kinase C (PKC) respectively (Kumar et al., 2006). Strikingly, phosphorylation status of SATB1 governs its mutually exclusive association with HDAC1 and PCAF (Fig. 5.4). Furthermore, acetylation abolishes the DNA-binding activity of SATB1. Using point mutants of SATB1 defective in phosphorylation or acetylation, Kumar et al. (2006) unequivocally demonstrated the importance of these modifications towards the ability of SATB1 to function as a global gene regulator.

Post-translational modifications provide a signal-responsive and reversible mode of regulation of the activity of variety of transcription factors, leading to stringent control of gene expression. Many transcriptional factors are phosphorylated by specific kinases that act as mediators in diverse signal transduction pathways. The role of phosphorylation in regulating SATB1 activity provides a unique clue to understand its role as a global regulator in response to various signals.

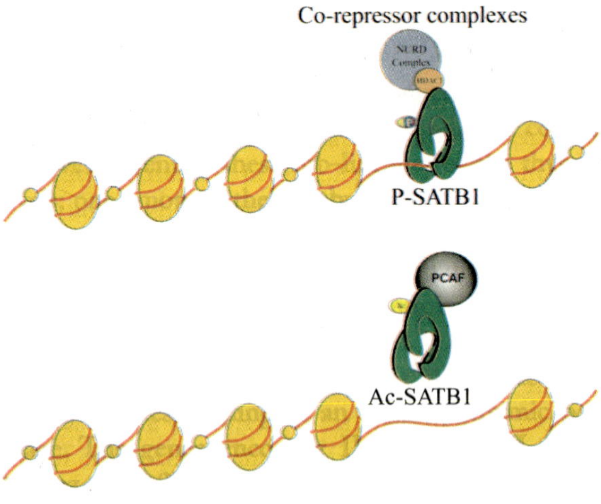

Fig. 5.4 Post-translational modifications of SATB1 act as a molecular switch governing its role as a global transcription regulator. Phosphorylated SATB1 tethers the regulatory element on the nuclear matrix and also recruits the NURD complex containing HDAC1. Thus, the chromatin in the vicinity of SATB1-NURD complex is deacetylated resulting in repression of transcription (lower schematic diagram). Upon activation, the signal amplification cascade ultimately leads to acetylation of SATB1 by the PCAF acetyltransferase. Acetylation abrogates the DNA-binding activity of SATB1. Loss of SATB1 occupancy from its genomic targets results in derepression or upregulation of transcription (upper schematic diagram).

Phosphorylation does not seem to be obligatory for the DNA-binding activity of SATB1; however, it is extremely essential for its association with HDAC1 (Kumar et al., 2006). The DNA binding activity of SATB1 is stimulated upon phosphorylation. The PDZ-like domain of SATB1 acts as a preferred site for recruitment and also as a target site for the kinases and acetyltransferases. Each of these modifications has independent effects on SATB1 function. Acetylation directly and negatively influences the SATB1 DNA-binding activity and phosphorylation indirectly influences the SATB1 transcriptional activity (Kumar et al., 2006).

The contrasting effect of these two post-translational modifications on DNA-binding suggested that the protein recruitments by SATB1 could also be affected. Phosphorylated SATB1 binds DNA with higher affinity and recruits HDAC1 suggesting role of phosphorylation in association with HDAC1. Intriguingly, HDAC1 interacts preferentially with phosphorylated SATB1 as compared to unphosphorylated SATB1. Conversely, phosphorylated SATB1 does not interact with PCAF whereas unphosphorylated SATB1 interacts maximally with PCAF. Thus, phosphorylation directly influences interactions of SATB1 whereas acetylation directly affects the DNA-binding ability of SATB1. *In vivo* reporter assays confirmed this mechanism at the level of transcription, i.e. HDAC1-SATB1 interaction led to downregulation whereas PCAF-SATB1 interaction resulted in upregulation (Kumar et al., 2006). The effect of post-translational modifications on the transcriptional activity of SATB1 thus seems to be predominantly exercised via the choice of interaction partners. This mechanism is partly similar to that reported in case of the inducible factor NF-kB wherein the phosphorylated p65 subunit binds to CBP and dephosphorylated p65 associated with HDAC1 (Zhong et al., 2002).

Kumar et al. (2006) tested SATB1 mediated global gene regulation by microarray analysis using 19k cDNA arrays. The results of such analyses revealed that SATB1 dysregulates more than 10% of genes. Surprisingly, about half of these were repressed by SATB1 while the other half were upregulated. Thus, SATB1 also acts as an activator of gene expression at a global level. Clustering of WT SATB1 and K136A mutant together indicates that K136A mutant which cannot be acetylated by PCAF acts as WT SATB1 and represses many genes. The strikingly dissimilar expression profile upon over-expression of the S185A mutant provided functional validation of the role of phosphorylation of SATB1 in global gene regulation. Non-phosphorylatable form of SATB1 interacts more efficiently with PCAF and is then acetylated. The resulting loss of DNA-binding activity leads to upregulation of the same genes which were downregulated by WT and K136A mutant, providing an equivalent of a dominant negative effect. Thus, a global regulatory mechanism involving phosphorylation operates to determine association of SATB1 with HDAC1 or PCAF. At the level of individual SATB1 targeted genes such as *IL-2* and *IL-2R*, these are repressed in control cells and upregulated upon T-cell activation. SATB1 binds to *IL-2* promoter and downregulates its gene expression (Kumar et al., 2005). *IL-2* is known to be upregulated during T cell activation. Prior to activation, phosphorylated SATB1 occupies the *IL-2* promoter and recruits HDAC1 presumably as part of the NURD complex. During activation, downregulation of PKC presumably results in decrease in steady state level of phosphorylated SATB1, and therefore loss of associated HDAC1. This further promotes interaction with PCAF and acetylation of SATB1 resulting in derepression accompanied by change in the chromatin loop structure of this locus (schematically depicted in Fig. 5.5).

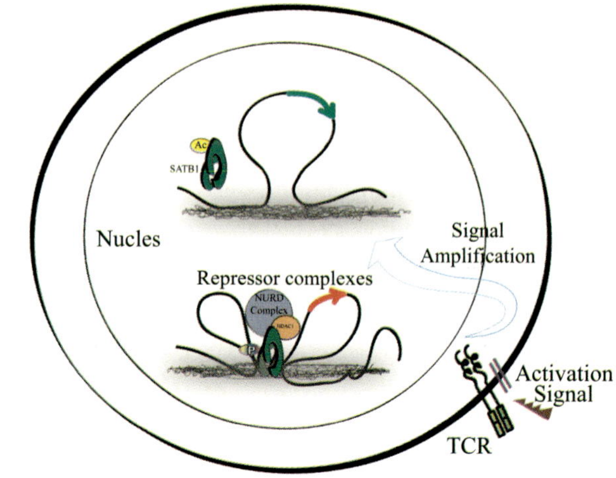

Fig. 5.5 Regulation of *IL-2* transcription by SATB1 in response to T cell activation stimuli. Post-translational modifications of SATB1 determine its association with co-repressor or co-activator complex and DNA binding activity and thereby regulate transcriptional activity of its target genes exemplified by *IL-2*. (**A**) In unactivated T cells SATB1 (green) is phosphorylated (red stars) by PKC and interacts with HDAC1 (blue), a component of NURD complex (brown), at the SBS (SATB1 binding site) within the *IL-2* promoter (red DNA wrapping orange nucleosomes) leading to repression of *IL- 2*. (**B**) In activated T-cells SATB1 (green) is acetylated (yellow stars) by PCAF as a result SATB1 loses its occupancy at the SBS within the *IL-2* promoter (DNA in red wrapped around nucleosomes in orange color) leading to derepression of *IL-2*.

Regulation of Genes by SATB1 Over Large Distance

SATB1 Recruits Chromatin Remodeling Machinery

During the development of thymocytes they undergo induction and repression of large number of genes, ultimately maturing into SP (single positive) population comprising $CD4^+$ and $CD8^+$ T cells. To ensure proper development of these sub-lineages of the thymic population, regulation is imparted at various levels; one amongst them is by the expression of the surface receptors and therefore responsiveness towards the cytokine stimuli, in their presence. The IL-2Rα and IL-7Rα receptor subunits have been implicated to play a role in T cell maturation and during development. Absence or deficiency of either of these receptors leads to autoimmunity and profound lymphopenia (Peschon et al., 1994 and Willerford et al., 1995). SATB1 is a MAR/ BUR binding protein that is found in abundance in thymocytes (Dickinson et al., 1992). Ablation of SATB1 results in impaired development of thymocytes, leading to an arrest of thymic population in $CD4^+CD8^+$ double positive (DP) stage. The genes encoding IL-2Rα and IL-7Rα receptor subunits are regulated by SATB1 (Alvarez et al., 2000). IL-2Rα harbors SATB1 binding site (SBS) upstream of its promoter and within the intronic region, thus regulation of the IL-2R expression is evident from the knockout mouse studies. Association of SBS at the base of the loops in the *in vivo* studies further strengthened the notion that SATB1 acts as an architectural protein on nuclear matrix, and can regulate genes over long distances by acting as a docking site for various components of the chromatin remodeling machinery. Association of various chromatin remodeling or modifying enzymes such as HDAC1, HDAC2, mSin3A, ISWI with SATB1 indicates role of these enzymes in regulating gene expression mediated by SATB1. All these re-modeling enzymes require a DNA binding protein to mediate their effect on gene regulation. Recruitment of these factors at the IL-2Rα locus by SATB1 repressed the gene as is required during the thymic development. However, in absence of SATB1 these factors are not associated at this locus thus leading to its derepression. Thus, SATB1 acts as a docking port for these modifiers to act, bringing about repression of the target gene(s) by not only recruiting repressor complexes but also by affecting the nucleosomal positioning so as to make the chromatin more compact, leaving very little access to the transcription machinery. This is highlighted by the fact that altered nucleosomal positioning is observed over a distance of few kilobases away from the SBS.

The nuclear distribution of SATB1 in thymocyte nuclei revealed a unique cage-like structure circumscribing the heterochromatic region. The presence of various SBSs lying at the base of loops and the intricate structure imparted by SATB1 possibly through multimerization of protein implicated its role in the loop organization and thus inducing a higher level of complexity in gene regulation effected by SATB1 (Galande et al., 2007). The c-Myc gene is repressed in the DP population of T cells but is stimulated only in response to the mitogenic stimuli and is under the control of SATB1. SBS of c-myc co-localizes with the cage-like structure imparted by SATB1 indicative of the occupancy *in vivo* (Cai et al., 2003). The SBSs which exhibited occupancy at the bases of various loops were thrown into loops in SATB1 null mouse. This suggested the fact that SATB1 acts as a loop organizer within the thymocyte nuclei by tethering specific genomic sequences onto the matrix and contributing to the higher-order packaging of chromatin. Though SATB1 regulates the transcription of c-Myc by not acting as a transcription regulator *per se* but as a chromatin organizer, even in presence of SATB1 considerable amounts of c-Myc transcripts are detected as it is also under the control of FBP which probably does not share its occupancy with that of SATB1 (He et al., 2000).

The positioning of a gene within the genome and its chromatin organization further marks it as transcriptionally active or inactive (Francastel et al., 2000 and Schubeler et al., 2000). Studies using SATB1 null cells revealed that SATB1 regulates genes which are in proximity of the SBS i.e. within 70-100 kb distance. Thus, SATB1 regulates genes over long distances. The active and the inactive chromatin regions are associated with different histone modification marks that ensure openness and closeness of chromatin with respect to accessibility to transcription machinery (reviewed by Jenuwein and Allis, 2001). Interaction of SATB1 with different modifiers would further control the gene expression by inducing concurrent histone modifications, thus deciding its fate to act as a repressor or an activator as few SBSs have been localized to the loop region in naïve conditions. SBS T4 which is positively regulated by SATB1, revealed H3K4/K9 acetylation and H3K4 methylation that are an indicative of transcriptionally active chromatin (Cai et al., 2003). However in absence of SATB1, SBS T4 exhibited H3K9 methylation and can be speculated that it is recruited through HP1 onto the matrix, and is thus transcriptionally silenced. However, SBS418 within a neuron specific gene showed an exact reciprocal modification from H3K9 methylation to acetylation as a result of SATB1 ablation (Cai et al., 2003) suggesting that SATB1 differentially regulates the histone modification states of multiple loci. This further suggests that the SBSs might act as hotspots for imparting histone code mediated by SATB1 and for determining the epigenetic modifications of the genes regulated within the specific loci.

Changes in Chromatin Organization and Gene Regulation by SATB1

SATB1 has been shown to regulate genes not only by mediating histones modifications but also by reorganizing the loop structure formed by MARs and SATB1 at least at two gene loci, namely, the T_H2 cytokine locus and the MHC class I locus.

T_H2 Locus Packaging and Cytokine Gene Regulation by SATB1: A Mechanism for the Regulation of Coordinated Expression of Locus-associated Genes by SATB1

In mouse, a 140 kb region of chromosome 11 harbors a cluster of T_H2 specific cytokine genes (IL-5, IL-1 and IL-4), Rad 50 and Kif 3a genes, where the expression of the T_H2 specific genes are coordinately regulated but not that of Rad 50 (Lee et al., 2006 and Cai et al., 2006) (Schematically illustrated in Fig. 5.6A). Several independent investigations have identified generation of T_H2 specific hypersensitive sites and conserved non-coding sequences (CNS1 and CNS2), critical regulatory elements involved in cell type specific expression of these locus associated genes. Hypersensitive sites are the genomic regions that have a relatively open chromatin structure due to distorted nucleosomal arrays and are more susceptible to DNase I digestion. The identification of conserved non coding sequences (CNS) is based on the concept that important regulatory sequences are likely to be evolutionarily conserved. Additionally, epigenetic modifications such as acetylation of histones and methylation of DNA at the regulatory region of the T_H2 cytokine locus are well characterized marks accompanying T_H1/T_H2 differentiation (Fields et al., 2002; Avni et al., 2002; Yamashita et al., 2002; Messi et al., 2003; Yano et al., 2003; Winders et al., 2004 and reviewed in Lee et al., 2006). The involvement of STAT proteins has been shown to mediate polarized acetylation of histones and demethylation of DNA at the locus (Avni et al., 2002 and Fields et al., 2002). GATA3 partly maintains the hyperacetylation of IL-5 promoter gene but the maintenance of acetylation of IL-4 and IL-13 genes are controlled likely by other mechanism. The locus control region (LCR) as a unique *cis* regulatory element lying 6 to 22 kb upstream to the β globin gene was first time identified in $\gamma\,\beta$ thalasemia (Kioussis et al., 1983). This deletion results into a closed chromatin conformation of this locus, which leads to suppression of gene expression (Kioussis et al., 1983 and Forrester et al., 1990). LCRs have been functionally defined in transgenic mice through characterization of human β globin and CD2 LCR (Dillon and Grossveld, 1987 and Greaves et al., 1989). LCRs are distinguished by their unique property that these sequences confer tissue-specific expression of linked genes at physiological level in a position-independent manner. Since then several of LCRs have been defined for more than 10 genes (Li et al., 2002). Recently, advancement of technology which enables capturing long range interactions have led to identification of long range interacting elements and the mechanism involved in transcription regulatory function of LCRs (Tucker et al., 2002; de Laat and Grosveld, 2003; Spilianakis et al., 2004). The developmental regulation of globin gene expression has been elucidated to be outcome of differential interaction of globin gene promoter with the active chromatin hub. The active chromatin hub is formed in the erythroid cells which involves five hypersensitive sites of locus control region (LCR) and three additional 5′ and 3′ hypersensitive sites (Patrinos et al., 2004) which interact with different genes to form an active transcription factory throughout development Tolhuis et al., 2002 and Palstra et al., 2003). The involvement of T_H2 LCR has been shown to be critical in the T_H2 cytokine

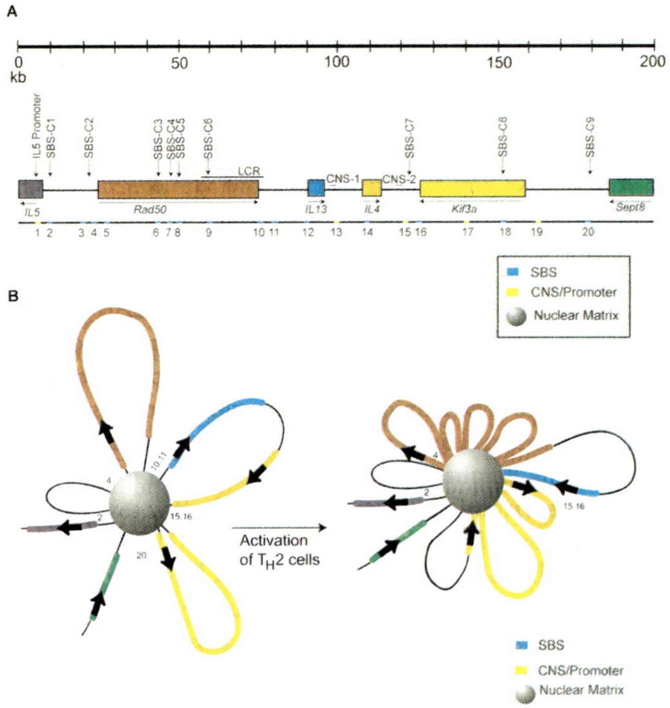

Fig. 5.6 SATB1 mediates condensed loop formation at T_H2 cytokine locus to regulate co-ordinated expression of T_H2 cytokines during T_H2 cell activation. (**A**) The T_H2 cytokine genes are clustered in a 200 kb region of chromosome 6 in mice. (**B**) The genes within this locus are simultaneously regulated by SATB1 via induction of several smaller loops involving multiple cis regulatory elements such as SBSs, promoters, HS sites and CNS within the locus (Adapted from Cai et al., 2006).

expression (Li et al., 2002). Using the 3C (Tolhuis et al., 2002) method, the organization of T_H2 locus has been mapped. In non-lymphoid cell lineages (B cells and fibroblasts) a core complex is observed which is termed pre-poised conformation, whereas in lymphoid cells (T_H2 cytokine non-expressing or expressing cells) a poised conformation is observed which involves T_H2 genes (IL-4, IL-5 and IL-13), CNS1 and the T_H2 LCR. This poised state has been proposed to be ready for transcription of T_H2 cytokines and specific polarizing signals mediated burst of cytokine gene transcription occurs, which is mediated by specific group of transcription factors (Spilianakis and Flavell, 2004; Lee et al., 2006). Although this model depicts that the coregulation of T_H2 cytokine genes is mediated by the conformational change in the chromatin organization mediated by T_H2 LCR and other elements, but does not explain the coregulation of IL-5 since the LCR can only regulate expression of IL-4 and IL-5 (Lee et al., 2003) in accordance with the observation that T_H2 LCR makes strong interaction with IL-4 and IL-13 but not with IL-5 (Spilianakis and Flavell, 2004). These findings explained the coregulation of IL-4 and IL-13 but not that of IL-5, raising the possibility of involvement of other *cis* elements and tissue-specific factors in coregulation of all T_H2 specific genes. The question that how upon T_H2 cell differentiation cell specific transcription factors are recruited to the preformed poised chromatin which is ready to transcribe of all three T_H2 cytokines but not that of Rad50. None of the *trans*-acting factors have been conclusively shown to be involved in this process. However, it is proposed that STAT6 and GATA3 which can recruit various chromatin remodellers and histone modifiers may be involved (Lee et al., 2006). Specific factors like T-bet (in T_H1) and GATA3 (in T_H2) may also be involved in establishing this interaction via protein-protein interactions, since T-bet and GATA3 have been shown to interact in tyrosine phosphorylation of T-bet-dependent manner (reviewed in Lee et al., 2006 and Hwang et al., 2005). All these studies indicate involvement of chromatin organization in the coordinated expression of T_H2 specific cytokine genes. The differential organization of chromatin by SATB1 involves several SBS and CNS at the T_H2 locus, and therefore the study by Cai et al. showed for the first time the direct involvement of any tissue-specific protein factor in the chromatin loop organization-dependent regulation of expression of all the three T_H2 cytokines but not that of Rad50 (Cai et al., 2006). In this study authors extensively examined the organization of a 200 kb region of the mouse T_H2 cytokine locus by SATB1 and the induced compact loop organization during T_H2 activation using the T_H2 specific D10.G4.1 cells and correlated it with the T_H2 cytokine expression. Interestingly, SATB1 itself is upregulated in T_H2 cells (Lund et al., 2005), however, the precise role of SATB1 in the process of T_H cell differentiation is not yet defined.

By using two BAC clones (BAC M01 and BAC M02), encompassing a 200 kb region which contains IL-5, Rad 50, IL-13, IL-4 and Kif3a genes, nine SATB1 binding sites were identified *in vitro*. By urea-ChIP assay of activated D10.G4.1 cells, it was found that SATB1 is also associated with two of the T_H2 specific conserved nucleotide sequences (CNS1 and CNS2) indirectly via some other unknown factors in addition to the nine direct binding sequences (Cai et al., 2006). These two CNS are known to harbor T_H2 specific DNaseI hypersensitive sites. The ChIp-loop assay scores for specific chromatin loops tethered to the bases by specific proteins or histone modifications (Horike et al., 2005). The resting D10.G4.1 cells revealed three-dimensional loops that involve only two distal SBS (SBS C1 and SBS C9), CNS2, SBS C7, IL-5 promoter, SBS C2 and 3′ end of LCR. This result indicated that in the 200 kb T_H2 locus there are several elements involved in the formation of chromatin loops but only at two places the binding by SATB1 (at SBS1 and SBS 9) may mediate the anchorage to the matrix. Using 3C assay it was found that in activated D10.G4.1 cells large number of smaller sized loops are formed whereas results of the ChIP-loop assay revealed that all matrix associated elements involved in the formation of loops are also associated with SATB1 (Fig. 5.6B). Apart from the elements involved in the formation of loop structure in unactivated D10.G4.1 cells, additional elements such as SBS C3, SBS C4, SBS C5, SBS C6, IL-13 promoter and CNS 1 and IL-4 promoter are the newly matrix associated elements due to the SATB1-mediated anchorage in the activated cells. This shows that SATB1 mediates the compartmentalization of a *cis* regulatory element, where under normal condition elements such as SBS and promoters are present in the loop eventually get associated with the matrix upon activation. SBS 1 to SBS 6 that are present upstream of the LCR region are the new elements other than LCR involved in the long range interactions when transcription is to be carried out. Some of these SBSs overlap with the DNaseI hypersensitive sites too, which suggest their involvement in T_H2 specific hypersensitive site formation. This study also supports that the matrix association of the MAR is dynamic at T_H2 locus and its association with the matrix is determined and mediated by the proteins which can directly binds to MARs or/and non-MAR SATB1-binding sequences. Upon activation the loop formation in between IL-5 and IL-13 promoters is induced whereas there is no involvement of Rad 50 promoter in any loop formation (Fig. 5.6). The pattern of GATA3 and STAT6 binding do not seem to be matching with the 3D organization of this locus and SATB1 association. However RNA pol II, cMaf, Brg 1 (2- to 10-fold increased binding upon activation otherwise bound at basal level) and GATA3 colocalized extensively with SATB1 in activated cells. The histone acetylation (H3K9/14) was increased at most of the sites in the 200 kb T_H2 locus region. All these alterations in the T_H2 cytokine locus is mediated by SATB1 induction upon D10.g4.1 cell activation leading to the induction of IL-5, IL-4 and IL-13 have been confirmed by shRNA-mediated knockdown of SATB1 in D10.G4.1 cells, wherein the activation-mediated change in chromatin conformation and cytokine gene expression was not observed. Knockdown of SATB1 also identified a direct SATB1 mediated regulator of IL-4 expression, viz. cMaf. cMaf expression is positively regulated by SATB1 during the activation of D10.G4.1 cells. Taken together, all of these findings suggest that increased expression of SATB1 leads to altered association of various *cis* regulatory elements in this locus and affect its chromatin organization. These SATB1 binding elements can be compared with the LCRs which are also known to regulate expression of genes by altered 3D organization of genome (Lee et al., 2006). These elements mediate long range and higher-order chromatin loop organization in T_H cells. Any alteration in loop structure by increased binding to the bases of loops may attract RNA pol II and other transcription factors which may modify the local chromatin by altering the histone modifications all of which finally culminates into coordinated expression of the T_H2 locus specific cytokine genes.

Reorganization of Loop Structure of MHC Class I Locus: A Mechanism of Gene Regulation by SATB1 in Collaboration with PML Bodies

Given the known function of SATB1 as a chromatin organizer, Kumar et al. (2007) hypothesized that PML might assist SATB1 in regulating chromatin conformation *in vivo*. Towards this, they sought to determine whether PML and SATB1 cooperate in the chromatin loop organization of the gene-rich major histocompatibility complex I (MHC-I) locus. This locus encompasses a region of approximately 1.7 Mb and maps to human chromosome 6p21.3 which was shown to associate specifically with PML NBs. Kumar et al focused on a 0.3 Mb region of the MHC-I locus that is flanked by the *HCG-9* and *HLA-F* genes (Fig. 5.7) and studied its chromatin organization, gene potentiation and PML association.

Interaction between PML and SATB1 is shown to be mediated via their coiled-coil and PDZ domains respectively and to be dependent on the proper sumoylation of PML. Furthermore, SATB1 is actively recruited to PML NBs in different cell types. Studies in PML-deficient fibroblasts have indicated that SUMO-modification of PML is required for proper formation of PML NBs, for the PML-dependent recruitment of PML NB residents to these nuclear domains and for nuclear matrix attachment. Given that both PML NB affiliated PML as well as SATB1 are known nuclear matrix components, it is likely that the physiologically relevant interaction between PML and SATB1 takes place in PML NBs.

Remarkably, Kumar et al. (2007) found that upon IFNγ treatment the chromatin 'loopscape' of the MHC-I locus undergoes a substantial change, which is reflected in the altered expression of a subset of MHC-I genes. Interestingly, specific knock-down of SATB1 not only closely mirrored the expression pattern for this subset of genes, but also led to the genesis of the IFNγ-specific loop reorganization at MAR position 256 kb. The HCG-4 locus, which is about 30 kb away from the nearest site of matrix attachment, gets attached to the MAR at 256 kb region in an IFNγ-inducible manner, suggesting that the positioning of this gene with respect to the nuclear matrix influences its expression. Thus, SATB1-mediated alteration in chromatin loop architecture at the MHC-I locus after IFNγ treatment is translated into differential gene expression for a set of MHC-I genes (Fig. 5.7). SATB1 specifically seems to be involved in the activation of HCG-9, while it represses most other genes within the MHC-I locus. Interestingly, the activation status of HCG-9 is enhanced further if it is pushed into being part of the giant loop formed by dynamic reorganization in that region. Given that loss of SATB1 does not affect all loops equally, yet it affects the transcription of most of the MHC genes under investigation, argues that SATB1 is not only necessary to establish a defined chromatin structure, but maybe even more importantly, also for maintaining a conducive transcriptional environment tailored for coordinated regulation of the MHC-I locus, including its inducibility by IFNγ. Our observations corroborate the notion that SATB1 serves as a landing platform to assemble chromatin remodeling and transcription factors to control transcription. Given the wealth of proteins SATB1 and PML each interact with, they provide an excellent platform to assemble and control chromatin remodeling complexes, modulate the activity status of transcription factors, and determine the fate of the epigenetic modifications at specific gene loci.

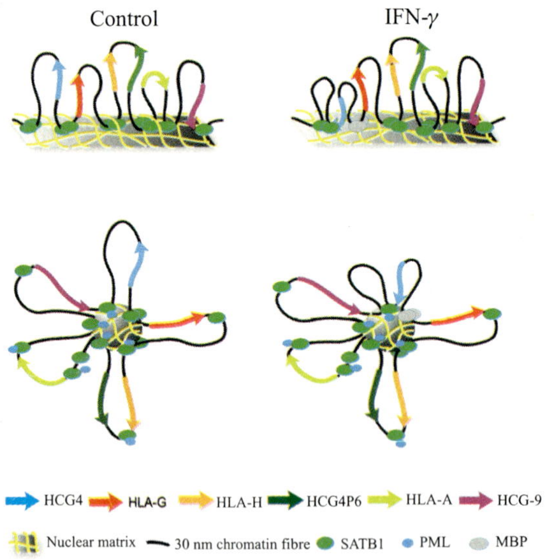

Fig. 5.7 SATB1 coordinates MHC class I loopscape to regulate the transcription of genes within the locus. The MHC class I locus is organized into 5 loops in unactivated T cells, upon IFNγ treatment the number of loops increases to 6 and thereby changes the localization of HCG-4 with respect to nuclear matrix. A specific alteration in genome organization is contributed by various isoforms of PML and SATB1 complexes. The alteration in genome organization affects transcription of locus associated genes (Adapted from Galande et al., 2007).

Acknowledgements

We thank members of our laboratory, especially Pavan Kumar P. for discussion. Work in Galande laboratory is supported by grants from the Department of Biotechnology, Government of India, and the Wellcome Trust, UK. Prabhat Kumar Purbey and Dimple Notani are supported by fellowships from the Council of Scientific and Industrial Research, and Amita Limaye is supported by fellowship from the University Grants Commission, India. Sanjeev Galande is an international senior research fellow of the Wellcome Trust, UK.

References

1. Alvarez, J.D., Yasui, D.H., Niida, H., Joh, T., Loh, D.Y., and Kohwi-Shigematsu, T. (2000). The MAR-binding protein SATB1 orchestrates temporal and spatial expression of multiple genes during T-cell development. Genes Dev. *14*, 521-535.
2. Amati, B., Pick, L., Laroche, T., and Grasser, S.M. (1990). Nuclear scaffold attachment stimulates, but is not essential for ARS activity in Saccharomyces cerevisiae: analysis of the Drosophila ftz SAR. EMBO J. *9*, 4007-4016.
3. Avni, O., Lee, D., Macian, F., Szabo, S.J., Glimcher, L.H., and Rao, A. (2002). T(H) cell differentiation is accompanied by dynamic changes in histone acetylation of cytokine genes. Nat. Immunol. *3*, 643-651.
4. Banan, M., Rojas, I.C., Lee, W.H., King, H.L., Harriss, J.V., Kobayashi, R., Webb, C.F., and Gottlieb, P. D. (1997). Interaction of the Nuclear Matrix-associated Region (MAR)-Binding Proteins, SATB1 and CDP/Cux, with a MAR Element (L2a) in an Upstream Regulatory Region of the Mouse *CD8a* Gene. J. Biol. Chem. *272*, 18440-18452.
5. Berezney, R. (1991). The nuclear matrix: a heuristic model for investigating genomic organization and function in the cell nucleus. J. Cell Biochem. *47*, 109-123.
6. Berezney, R. and Coffey, D.S. (1974). Identification of a nuclear protein matrix. Biochem. Biophys. Res. Commun. *60*, 1410-1417.
7. Benham, C., Kohwi-Shigematsu, T., and Bode, J. (1997). Stress-induced duplex DNA destabilization in scaffold/matrix attachment regions. J. Mol. Biol. *274*, 181-196.
8. Berezney, R. and Wei, X. (1998). The new paradigm: integrating genomic function and nuclear architecture. J. Cell Biochem. Suppl *30-31*, 238-242.
9. Boulikas, T. (1993). Nuclear localization signals (NLS). Crit Rev. Eukaryot. Gene Expr. *3*, 193-227.
10. Boulikas, T. (1995). Chromatin domains and prediction of MAR sequences. Int. Rev. Cytol. *162A*, 279-388.
11. Cai, S. and Kohwi-Shigematsu, T. (1999). Intranuclear relocalization of matrix binding sites during T cell activation detected by amplified fluorescence *in situ* hybridization. Methods *19*, 394-402.
12. Cai, S., Han, H.J., and Kohwi-Shigematsu, T. (2003). Tissue-specific nuclear architecture and gene expression reguiated by SATB1. Nat. Genet. *34*, 42-51.
13. Cai, S., Lee, C.C., and Kohwi-Shigematsu, T. (2006). SATB1 packages densely looped, transcriptionally active chromatin for coordinated expression of cytokine genes. Nat. Genet. *38*, 1278-1288.
14. Carter, D., Chakalova, L., Osborne, C.S., Dai, Y.F., and Fraser P. (2002). Long-range chromatin regulatory interactions in vivo. Nat. Genet. *32*, 623-626.

15. Case, S.S., Huber, P., and Lloyd, J.A. (1999). The gammaPE complex contains both SATB1 and HOXB2 and has positive and negative roles in human gamma-globin gene regulation. DNA Cell Biol. *18*, 805-817.
16. Chambeyron, S. and Bickmore,W.A. (2004). Does looping and clustering in the nucleus regulate gene expression? Curr. Opin. Cell Biol. *16*, 256-262.
17. Chan, H.M., and Thangue, N.B.L. (2001). p300/CBP proteins: HATs for transcriptional bridges and scaffolds. J.Cell. Science. *114*, 2363-2373.
18. Chattopadhyay, S., Whitehurst, C.E., Schwenk, F., and Chen, J. (1998). Biochemical and functional analyses of chromatin changes at the TCR-beta gene locus during CD4-CD8- to CD4+CD8+ thymocyte differentiation. J. Immunol. *160*, 1256-1267.
19. Chimera, J.A., and Musich, P.R. (1985). The association of the interspersed repetitive KpnI sequences with the nuclear matrix. J. Biol.Chem. *260*, 9373-9379.
20. Ciejek, M.E., Tsai, M., and O'Malley, B.W. (1983). Actively transcribed genes are associated with nuclear matrix. Nature *306*, 607-609.
21. Cockerill, P.N., and Yuen, M.H., and Garrard, W.T. (1987).The enhancer of the immunoglobulin heavy chain locus is flanked by presumptive chromosomal loop anchorage elements. J. Biol. Chem. *262*, 5394-5397.
22. Cunningham, J.M., Purucker, M.E., Jane, S.M., Safer, B., Vanin, E.F., Ney, P.A., Lowrey, C.H., and Nienhuis, A.W. (1994). The regulatory element 3′ to the A gamma-globin gene binds to the nuclear matrix and interacts with special A-T-rich binding protein 1 (SATB1), an AR/MAR-associating region DNA binding protein. Blood *84*, 1298-1308.
23. Davido, T. and Getzenberg, R.H. (2000) Nuclear matrix proteins as cancer markers. J. Cell Biochem. Suppl. *35*, 136-141.
24. de Laat, W. and Grosveld, F. (2003). Spatial organization of gene expression: the active chromatin hub. Chromosome. Res. *11*, 447-459.
25. deBelle, I, Cai, S., and Kohwi-Shigematsu, T. (1998). The genomic sequences bound to special AT-rich sequence-binding protein 1 (SATB1) *in vivo* in Jurkat T cells are tightly associated with the nuclear matrix at the bases of the chromatin loops. J. Cell Biol. *141*, 335-348.
26. Dekker, J. (2003). A closer look at long-range chromosomal interactions. Trends Biochem. Sci. *28*, 277-280.
27. Dickinson, L.A. and Kohwi-Shigematsu, T. (1995). Nucleolin is a matrix attachment region DNA-binding protein that specifically recognizes a region with high base-unpairing potential. Mol. Cell Biol. *15*, 456-465.
28. Dickinson, L.A., Dickinson, C.D., and Kohwi-Shigematsu, T. (1997). An atypical homeodomain in SATB1 promotes specific recognition of the key structural element in a matrix attachment region. J. Biol. Chem. *272*, 11463-11470.
29. Dickinson, L.A., Joh, T., Kohwi, Y., and Kohwi-Shigematsu, T. (1992). A tissue-specific MAR/SAR DNA-binding protein with unusual binding site recognition. Cell *70*, 631-645.
30. Dillon, N., and Grosveld, F. (1991). Human γ-globin genes silenced independently of other genes in the β-globin locus. Nature. *350*, 252-254.
31. Dostie, J., Richmond, T.A., Arnaout, R.A., Selzer, R.R., Lee, W.L., Honan, T.A., Rubio, E.D., Krumm, A., Lamb, J., Nusbaum, C., Green, R.D., and Dekker, J. (2006). Chromosome Conformation

Capture Carbon Copy (5C): a massively parallel solution for mapping interactions between genomic elements. Genome Res. *16*, 1299-1309.

32. Fernandez, L.A., Winkler, M., and Grosschedl, R. (2001). Matrix attachment region-dependent function of the immunoglobulin mu enhancer involves histone acetylation at a distance without changes in enhancer occupancy. Mol. Cell Biol. *21*, 196-208.
33. Fey, E.G., and Penman, S. (1988). Nuclear matrix proteins reflect cell type of origin in cultured human cells. Proc. Natl. Acad. Sci. *USA. 85*, 121-125.
34. Fields, P.E., Kim, S.T., and Flavell, R.A. (2002). Cutting edge: changes in histone acetylation at the IL-4 and IFN-gamma loci accompany Th1/Th2 differentiation. J. Immunol. *169*, 647-650.
35. Forrester, W.C., Epner, E., Driscoll, M.C., Enver, T., Brice, M., Papayannopoulou, T., and Groudine, M. (1990). A deletion of the human beta-globin locus activation region causes a major alteration in chromatin structure and replication across the entire beta-globin locus. Genes Dev. *4*, 1637-1649.
36. Forrester, W.C., van Genderen, C., Jenuwein, T., and Grosschedl, R. (1994). Dependence of enhancer-mediated transcription of the immunoglobulin mu gene on nuclear matrix attachment regions. Science *265*, 1221-1225.
37. Francastel, C., Schubeler, D., Martin, D.I., and Groudine, M. (2000). Nuclear compartmentalization and gene activity. Nat. Rev. Mol. Cell Biol. *1*, 137-143.
38. Fujii, Y., Kumatori, A., and Nakamura, M. (2003). SATB1 makes a complex with p300 and represses gp91(phox) promoter activity. Microbiol. Immunol. *47*, 803-811.
39. Galande, S. (2002). Chromatin (dis)organization and cancer: BUR-binding proteins as biomarkers for cancer. Curr. Cancer Drug Targets. *2*, 157-190.
40. Galande, S. and Kohwi-Shigematsu, T. (1999). Poly(ADP-ribose) polymerase and Ku autoantigen form a complex and synergistically bind to matrix attachment sequences. J. Biol. Chem. *274*, 20521-20528.
41. Galande, S. and Kohwi-Shigematsu, T. (2000). Caught in the act: binding of Ku and PARP to MARs reveals novel aspects of their functional interaction. Crit Rev. Eukaryot. Gene Expr. *10*, 63-72.
42. Galande, S. and Kohwi-Shigematsu, T. (2000). Linking chromatin architecture to cellular phenotype: BUR-binding proteins in cancer. J. Cell Biochem. Suppl *35*, 36-45.
43. Galande, S., Dickinson, L.A., Mian, I.S., Sikorska, M., and Kohwi-Shigematsu, T. (2001). SATB1 cleavage by caspase 6 disrupts PDZ domain-mediated dimerization, causing detachment from chromatin early in T-cell apoptosis. Mol. Cell Biol. *21*, 5591-5604.
44. Galande, S., Purbey, P.K., Notani, D., and Kumar, P.P. (2007). The third dimension of gene regulation: organization of dynamic chromatin loopscape by SATB1. Curr. Opin. Genet. Dev.
45. Gasser, S.M.; Laemmli, U.K. (1987) A glimpse at chromosomal order. Trends Genet., 3, 16-22.
46. Gasser, S.M. and Laemmli, U.K. (1987). Improved methods for the isolation of individual and clustered mitotic chromosomes. Exp. Cell Res. *173*, 85-98.
47. Gheldof, N., Tabuchi, T.M., and Dekker, J. (2006). The active FMR1 promoter is associated with a large domain of altered chromatin conformation with embedded local histone modifications. Proc. Natl. Acad. Sci. U.S.A *103*, 12463-12468.
48. Goldberg, G.I., Collier, I., and Cassel, A. (1983). Specific DNA sequences associated with the nuclear matrix in synchronized mouse 3T3 cells. Proc. Natl. Acad. Sci. *USA. 80*, 6687-6691.

49. Greaves, D.R., Antoniou, M., van Assendelft, G.B., Collis, P., Dillon, N., Hanscombe, O., Hurst, J., Lindenbaum, M., Talbot, D., and Grosveld,F. (1989). The beta-globin dominant control region. Prog. Clin. Biol. Res. *316A*, 37-46.

50. Hart, C.M. and Laemmli, U.K. (1998). Facilitation of chromatin dynamics by SARs. Curr. Opin. Genet. Dev. *8*, 519-525.

51. He, L., Liu, J., Collins, I., Sanford, S., O'Connell, B., Benham, C.J., Levens, D. (2000). Loss of FBP function arrests cellular proliferation and extinguishes c-myc expression. EMBO J. *19*, 1034-1044.

52. Hentzen, P.C., Rao, J.H., and Bekhor, I. (1984).Nuclear matrix DNA from chicken erythrocytes contains beta-globin gene sequences. Proc. Natl. Acad. Sci. USA. *81*, 304-307.

53. Herrscher, R.F., Kaplan , M.H., Lelsz, D.L., Das, C. Scheuremann, R., and Tucker, P.W. (1995). The immunoglobulin heavy-chain matrix-associating regions are bound by Bright: a B cell-specific *trans*-activator that describes a new DNA-binding protein family. Genes and Dev. *9*, 3067-3082.

54. Horike, S., Cai, S., Miyano, M., Cheng, J.F., and Kohwi-Shigematsu, T. (2005). Loss of silent-chromatin looping and impaired imprinting of DLX5 in Rett syndrome. Nat. Genet. *37*, 31-40.

55. Hwang, S.E., Szabo, S.J., Schwatzberg, P.L., and Glimcher, L.H. (2005). T Helper Cell Fate Specified by Kinase-Mediated Interaction of T-bet with GATA-3. Science. *307*, 430-433.

56. Jackson, D.A., Caton, A.J., Mccready, S.J., and Cook, P.R. (1982). Influenza virus RNA is synthesized at fixed sites in the nucleus. Nature. *296*, 366-368.

57. Jenuwein, T., Forrester, W.C., Ferrnandez-Herrero, L.A., Laible, G., Dull, M., and Grosschedl, R. (1997).Extension of chromatin accessibility by nuclear matrix attachment regions. Nature. 385, 269-272.

58. Jenuwein, T. and Allis, C.D. (2001). Translating the histone code. Science *293*, 1074-1080.

59. Jost, J.P., and Seldran, M. (1984) Association of transcriptionally active vitellogenin II gene with the nuclear matrix of chicken liver. EMBO J. *3*, 2005–2008.

60. Kirillov, A., Kristler, B., Mostoslavsky, R., Cedar, H., Wirth, T., and Bergman, Y. (1996). A role for nuclear NF-κB in B-cell-specific demethylation of the Igκ locus. Nature Genet. 13, 435-441.

61. Kioussis, D., Vanin, E., deLange, T., Flavell, R.A., and Grosveld, F.G. (1983). Beta-globin gene inactivation by DNA translocation in gamma beta-thalassaemia. Nature *306*, 662-666.

62. Kohwi-Shigematsu, T., Maass, K., and Bode, J. (1997). A thymocyte factor SATB1 suppresses transcription of stably integrated matrix-attachment region-linked reporter genes. Biochemistry *36*, 12005-12010.

63. Kumar, P.P., Bischof, O., Purbey, P.K., Notani, D., Urlaub, H., Dejean, A., and Galande, S. (2007a). Functional interaction between PML and SATB1 regulates chromatin-loop architecture and transcription of the MHC class I locus. Nat. Cell Biol. *9*, 45-56.

64. Kumar, P.P., Mehta, S., Purbey, P.K., Notani, D., Jayani, R.S., Purohit, H.J., Raje, D.V., Ravi, D.S., Bhonde, R.R., Mitra, D., and Galande, S. (2007b). SATB1-binding sequences and Alu-like motifs define a unique chromatin context in the vicinity of human immunodeficiency virus type 1 integration sites. J. Virol. *81*, 5617-5627.

65. Kumar, P.P., Purbey, P.K., Ravi, D.S., Mitra, D., and Galande, S. (2005). Displacement of SATB1-bound histone deacetylase 1 corepressor by the human immunodeficiency virus type 1 transactivator induces expression of interleukin-2 and its receptor in T cells. Mol. Cell Biol. *25*, 1620-1633.

66. Lambert, P.F., Kashanchi, F., Radonovich, M.F., Shiekhattar, R., and Brady, J.N. (1998). Phosphorylation of p53 Serine 15 Increases Interaction with CBP. J. Biol.Chem. 49, 33048-33053.
67. Langmore, J.P., and Paulson, J.R. (1983). Low angle x-ray diffraction studies of chromatin structure *in vivo* and in isolated nuclei and metaphase chromosomes. J. Cell Biol. *96*, 1120-1131.
68. Lee, G.R., Fields, P.E., Griffin, T.J., and Flavell, R.A. (2003). Regulation of the Th2 cytokine locus by a locus control region. Immunity. *19*, 145-153.
69. Lee, G.R., Kim, S.T., Spilianakis, C.G., Fields, P.E., and Flavell, R.A. (2006). T helper cell differentiation: regulation by cis elements and epigenetics. Immunity. *24*, 369-379.
70. Li, Q., Peterson, K.R., Fang, X., and Stamatoyannopoulos G. (2002). Locus control regions. Blood. *100*, 3077-3086.
71. Lichtenstein, M., Keini, G., Cedar, H. and Bergman, Y.B. (1994) Cell-Specific Demethylation: a Novel Role for the Intronic k Chain Enhancer Sequence. Cell *76*, 913-923.
72. Ling, J.Q., Li, T., Hu, J.F., Vu, T.H., Chen, H.L., Qiu, X.W., Cherry, A.M., and Hoffman, A. R. (2006). CTCF mediates interchromosomal colocalization between Igf2/H19 and Wsb1/Nf1. Science *312*, 269-272.
73. Liu, J., Barnett, A., Neufeld, E.J., and Dudley, J.P. (1999). Homeoproteins CDP and SATB1 interact: potential for tissue-specific regulation. Mol. Cell Biol. *19*, 4918-4926.
74. Liu, J., Bramblett, D., Zhu, Q., Lozano, M., Kobayashi, R., Ross, S.R., and Dudley, J.P. (1997). The matrix attachment region-binding protein SATB1 participates in negative regulation of tissue-specific gene expression. Mol. Cell Biol. *17*, 5275-5287.
75. Liu, Z. and Garrard, W.T. (2005). Long-range interactions between three transcriptional enhancers, active Vkappa gene promoters, and a 3′ boundary sequence spanning 46 kilobases. Mol. Cell Biol. *25*, 3220-3231.
76. Luger, K., Mader, A.W., Richmond, R.K., Sargent, D.F., and Richmond, T.J. (1997). Crystal structure of the nucleosome core particle at 2.8 A resolution. Nature *389*, 251-260.
77. Lund, R., Ahlfors, H., Kainonen, E., Lahesmaa, A.M., Dixon, C ., and Lahesmaa, R. (2005). Identification of genes involved in the initiation of human Th1 or Th2 cell commitment. Eur. J. Immunol. *35*, 3307-3319.
78. Ma, C., Zhang, J., Durrin, L.K., Lv, J., Zhu, D., Han, X., and Sun, Y. (2007). The BCL2 major breakpoint region (mbr) regulates gene expression. Oncogene *26*, 2649-2657.
79. Messi, M., Giacchetto, I., Nagata, K., Lanzavecchia, A., Natoli, G., and Sallusto, F. (2003). Memory and flexibility of cytokine gene expression as separable properties of human T(H)1 and T(H)2 lymphocytes. Nat. Immunol. *4*, 78-86.
80. Mirkovitch, J., Mirault, M.E., and Laemmli, U.K. (1984). Organization of the higher-order chromatin loop: specific DNA attachment sites on nuclear scaffold. Cell *39*, 223-232.
81. Murrelle, A., Heeson, S., Cooper, W.N., Douglas, E., Apostilodou, S., Moore, G.E., Maher, F.R., and Reik, W. (2004). An association between variants in the *IGF2* gene and Beckwith–Wiedemann syndrome: interaction between genotype and epigenotype. Hum. Mol. Genet. *13*, 247-255.
82. Nelson, W.G., Pienta, K.J., Barrack, E.R., and Coffey, D.S.(1986). The Role of the Nuclear Matrix in the Organization and Function of DNA. Annual Review of Biophysics and Biophysical Chem. *15*, 457-475.

83. Nickerson, J.A., Krockmalnic. G., Wan, K.M., AND Penman, S. (1997). The nuclear matrix revealed by eluting chromatin from a cross-linked nucleus. Proc. Natl. Acad. Sci. *USA. 9*, 4446-4450.

84. Oei, L.S., Griesenbeck, J., Schweiger, M., and Ziegler, M. (1998).Regulation of RNA Polymerase II-dependent Transcription by Poly(ADP-ribosyl)ation of Transcription Factors.J. Biol.Chem. 48, 31644-31647.

85. Opstelten, R.J.G., Clement, J.M.E., and Wanka, F. (1989) Direct repeats at nuclear matrix-associated DNA regions and their putative control function in the replicating eukaryotic genome. Chromosoma. *98*, 422-427.

86. Palstra, R.J., Tolhuis, B., Splinter, E., Nijmeijer, R., Grosveld, F., and de Laat, W. (2003). The beta-globin nuclear compartment in development and erythroid differentiation. Nat. Genet. *35*, 190-194.

87. Patrinos, G.P., de Krom, M., de Boer, E., Langeveld, A., Imam, A.M., Strouboulis, J., de Laat, W., and Grosveld, F.G. (2004). Multiple interactions between regulatory regions are required to stabilize an active chromatin hub. Genes Dev. *18*, 1495-1509.

88. Pavan, K.P., Purbey, P.K., Sinha, C.K., Notani, D., Limaye, A., Jayani, R.S., and Galande, S. (2006). Phosphorylation of SATB1, a global gene regulator, acts as a molecular switch regulating its transcriptional activity *in vivo*. Mol. Cell *22*, 231-243.

89. Peschon, J.J., Morrissey, P.J., Grabstein, K.H., Ramsdell, F.J., Maraskovsky, E., Gliniak, B.C., Park, L.S., Ziegler, S.F., Williams, D.E., Ware, C.B., Meyer, J.D., and Davison, B.L. (1994). Early lymphocyte expansion is severely impaired in interleukin 7 receptor-deficient mice. J. Exp. Med. *180*, 1955-1960.

90. Peterson, C.L. and Tamkun, J.W. (1995). The SWI-SNF complex: a chromatin remodeling machine? Trends Biochem. Sci. *20*, 143-146.

91. Poux, A.N. and Marmorstein, R. (2003). Molecular basis for Gcn5/PCAF histone acetyltransferase selectivity for histone and nonhistone substrates. Biochemistry. *42*, 14366-14374.

92. Purbey, P.K., Singh, S., Kumar, P.P., Mehta, S., Ganesh, K.N., Mitra, D., and Galande, S. (2008). PDZ domain-mediated dimerization and homeodomain-directed specificity are required for high affinity DNA binding by SATB1. Nucleic Acids Res. *36*, 2107-2122.

93. Puri, P., Iezzi, S., Steigler, P., Chen, T., Schiltz, R., Muscat, G., Giordano, A., Kedes, L., Wang, J., and Sartorelli, V. (2001). Class I Histone Deacetylases Sequentially Interact with MyoD and pRb during Skeletal Myogenesis. Mol. Cell. *8*, 885-897.

94. Replogle-Schwab, T.S.. Getzenberg, R.H., Donat, T.L. and Pienta, K.J. (1996) Effect of organ site on nuclear matrix protein composition. J. Cell Biochem., *62*, 132-141.

95. Reyes, J.C., Muchardt, C., and Yaniv, M. (1997). Components of the human SWI/SNF complex are enriched in active chromatin and are associated with the nuclear matrix. J. Cell Biol. *137*, 263-274.

96. Robinson, S.I., Nelkin, B.D., and Vogelstein, B. (1982). The ovalbumin gene is associated with the nuclear matrix of chicken oviduct cells. Cell. *28*, 99-106.

97. Rose, S.M., and Garrard, W.T. (1984).Differentiation-dependent chromatin alterations precede and accompany transcription of immunoglobulin light chain genes. J. Biol.Chem. *259*, 8534-8544.

98. Singh, G.B., Kramer, J.A., and Krawetz, S.A. (1997). Mathematical model to predict regions of chromatin attachment to the nuclear matrix. Nucl. Acids. Res. *25*, 1419-1425.

99. Schubeler, D., Francastel, C., Cimbora, D.M., Reik, A., Martin, D.I., and Groudine, M. (2000). Nuclear localization and histone acetylation: a pathway for chromatin opening and transcriptional activation of the human beta-globin locus. Genes Dev. *14*, 940-950.

100. Seo, J., Lozano, M.M., and Dudley, J.P. (2005). Nuclear matrix binding regulates SATB1-mediated transcriptional repression. J. Biol. Chem. *280*, 24600-24609.

101. Simonis, M., Klous, P., Erik Splinter., Moshkin, E.Y., Willemsen, R., 3, de Wit, E., van Steensel, B., and de Laat, W. (2006) Nuclear organization of active and inactive chromatin domains uncovered by chromosome conformation capture–on-chip (4C). Nature Genetics. *38*, 1348-1354.

102. Spilianakis, C.G. and Flavell, R.A. (2004). Long-range intrachromosomal interactions in the T helper type 2 cytokine locus. Nat. Immunol. *5*, 1017-1027.

103. Spilianakis, C.G., Lalioti, M.D., Town, T., Lee, G.R., and Flavell, R.A. (2005). Interchromosomal associations between alternatively expressed loci. Nature *435*, 637-645.

104. Stratling , W.H., and Yu, F. (1999) Origin and roles of nuclear matrix proteins. Specific functions of the MAR-binding protein MeCP2/ARBP. Crit Rev Eukaryot Gene Expr. *9*, 311-318.

105. Strätling, W.H., Dölle, A., and Sippel, A.E. (1986). Chromatin structure of the chicken lysozyme gene domain as determined by chromatin fractionation and micrococcal nuclease digestion. Biochemistry. *25*, 495-502.

106. Strätling, W.H. (1987). Gene-specific differences in the supranucleosomal organization of rat liver chromatin. Biochemistry. 26, 7893-7899.

107. Tintignac, L.A., Sirri, V., Leibovitch, M.P., Lecluse, Y., Castedo, M., Metivier, D., Kroemer, G., and Leibovitch, S.A. (2004). Mutant MyoD lacking Cdc2 phosphorylation sites delays M-phase entry. Mol.Cell. Biol. *24*, 1809-1821.

108. Tolhuis, B., Palstra, R. J., Splinter, E., Grosveld, F., and de Laat, W. (2002). Looping and interaction between hypersensitive sites in the active beta-globin locus. Mol. Cell *10*, 1453-1465.

109. Tucker, M.L., Whitelaw, C.A., Lyssenko, N.N., and Nath, P. (2002). Functional analysis of regulatory elements in the gene promoter for an abscission-specific cellulase from bean and isolation, expression, and binding affinity of three TGA-type basic leucine zipper transcription factors. Plant Physiol. *130*, 1487-1496.

110. Vakoc, C., Mandat, S., Olenchock, B., and Blobel, G. (2005). Histone H3 Lysine 9 Methylation and HP1γ Are Associated with Transcription Elongation through Mammalian Chromatin. Mol. Cell. *19*, 381-391.

111. Von Kries, J.P., Buhrmester, H., and Stratling, W.H. (1991). A matrix/scaffold attachment region binding protein: identification, purification, and mode of binding. Cell. *64*, 123-135.

112. Wan, K.M., Nickerson, J.A., Krockmalnic, G., and Penman, S. (1999). Matrix prepared by amine modification. Proc. Natl. Acad. Sci. *USA*.. *2*, 933-938.

113. Watson J.D. and Crick F.H.C. (1953). A Structure for Deoxyribose Nucleic Acid. Nature. *171*, 737-738

114. Webb, C.F., Smith, E.A., Medina, K.L., Buchanan, K.L., Smithson, G., and Dou, S. (1998). Expression of bright at two distinct stages of B lymphocyte development. J. Immunol. *160*, 4747-4754.

115. Wen, J., Huang, S., Rogers, H., Dickinson, L.A., Kohwi-Shigematsu, T., and Noguchi, C.T. (2005). SATB1 family protein expressed during early erythroid differentiation modifies globin gene expression. Blood *105*, 3330-3339.

116. West, A.G. and Fraser, P. (2005). Remote control of gene transcription. Hum. Mol. Genet. *14 Spec No 1*, R101-R111.

117. Will, K., Warnecke, G., Wiesmuller, L., Deppert, W. Specific interaction of mutant p53 with regions of matrix attachment region DNA elements (MARs) with a high potential for base-unpairing. Proc. Natl. Acad. Sci. U S A 1998, 95, 13681-13686.

118. Willerford, D.M., Chen, J., Ferry, J.A., Davidson, L., Ma, A., and Alt, F.W. (1995). Interleukin-2 receptor alpha chain regulates the size and content of the peripheral lymphoid compartment. Immunity. *3*, 521-530.

119. Winders, B.R., Schwartz, R.H., and Bruniquel, D. (2004). A distinct region of the murine IFN-gamma promoter is hypomethylated from early T cell development through mature naive and Th1 cell differentiation, but is hypermethylated in Th2 cells. J. Immunol. *173*, 7377-7384.

120. Xu, Z., Meng, X ., Cai, Y., Koury, M.J., and Brandt, S.J. (2006). Recruitment of the SWI/SNF protein Brg1 by a multiprotein complex effects transcriptional repression in murine erythroid progenitors. Biochem. J. *399*, 297-304.

121. Yamashita, M., Ukai-Tadenuma, M., Kimura, M., Omori, M., Inami, M., Taniguchi, M., and Nakayama, T. (2002). Identification of a conserved GATA3 response element upstream proximal from the interleukin-13 gene locus. J. Biol. Chem. *277*, 42399-42408.

122. Yanagisawa, J., Ando, J., Nakayama, J., Kohwi, Y., and Kohwi-Shigematsu, T. (1996). A Matrix Attachment Region (MAR)-binding Activity Due to a p114 Kilodalton Protein Is Found Only in Human Breast Carcinomas and Not in Normal and Benign Breast Disease Tissues. Cancer Res. *56*, 457-462.

123. Yano, S., Ghosh, P., Kusaba, H., Buchholz, M., and Longo, D.L. (2003). Effect of promoter methylation on the regulation of IFN-gamma gene during *in vitro* differentiation of human peripheral blood T cells into a Th2 population. J. Immunol. *171*, 2510-2516.

124. Yasui, D., Miyano, M., Cai, S., Varga-Weisz, P., and Kohwi-Shigematsu, T. (2002). SATB1 targets chromatin remodelling to regulate genes over long distances. Nature *419*, 641-645.

125. Yeh, T.Y., Chuang, J.Z., and Sung, C.H. (2005). Dynein light chain rp3 acts as a nuclear matrix-associated transcriptional modulator in a dynein-independent pathway. J. Cell Sci. *118*, 3431-3443.

126. Zbarskii, I.B., and Debov, S. S. (1948). On the proteins of the cell nucleus. Dokl. Acad. Nauk. SSSR 63, 795-798.

127. Zhang, F., Rincon, M., Flavell, R.A., and Aune, T.M. (2000). Defective Th function induced by a dominant-negative cAMP response element binding protein mutation is reversed by Bcl-2. J. Immunol. *165*, 1762-1770.

128. Zhao, Z., Tavoosidana, G., Sjolinder, M., Gondor, A., Mariano, P., Wang, S., Kanduri, C., Lezcano, M., Sandhu, K.S., Singh, U., Pant, V., Tiwari, V., Kurukuti, S., and Ohlsson, R. (2006). Circular chromosome conformation capture (4C) uncovers extensive networks of epigenetically regulated intra- and interchromosomal interactions. Nat. Genet. *38*, 1341-1347.

129. Zhong, H., May, M., Jimi, E., and Ghosh, S. (2002).The Phosphorylation Status of Nuclear NF-κB Determines Its Association with CBP/p300 or HDAC-1. Mol. Cell. *9*, 625-636.

Azoospermia Factor (AZF) Genes Function for Human Male Fertility

6

P. H. Vogt, U. Bender, C. Enders and J. Zimmer

Unit Molecular Genetics, Department of Gynecology, Endocrinol, University Women Hospital, Heidelberg, Germany

Introduction

Human male infertility is a major health problem worldwide with a prevalence of ~20%. Anatomic defects, endocrinopathies, immunologic problems, ejaculatory failures and environmental exposures are significant causes of this disease. Frequently, however, in the clinic no clear cause for the observed infertility could be diagnosed which coined the term "idiopathic infertility". At least in part, it reflects our still poor molecular understanding of the basic mechanisms regulating the genetic networks controlling male fertility, respectively, our poor knowledge of the genetic mutation base causing human male infertility. From mouse and *Drosophila* "gene-knock-out" experiments it has been deduced that more than 3000 genes are involved in the genetic control of male (and female) fertility (Hackstein et al., 2000; Cooke and Saunders, 2002; Matzuk and Lamb, 2002). However, the notion of a rapid evolution of just reproductive proteins (Swanson and Vacquier, 2002) especially of those involved in the postmeiotic maturation process of the male germ cell (Wyckoff, 2000) indicates that animal models may not always serve for a better functional understanding of homologous human male fertility genes. Additionally, many human male fertility genes like the *DAZ* gene family in AZFc on the long arm of the human Y chromosome are absent in the mouse genome because they were first evolved during primate evolution (Shan et al., 1996; Seboun et al., 1997).

In human, molecular identification of the major male fertility genes is hampered by the fact that human is not an experimental species. An infertile man is only recognised after the couple desiring for a child is asking for a therapeutic treatment in an infertility clinic. Moreover, current genetic screening programmes in the infertility clinic are usually restricted to an analysis of the chromosome picture in the infertile man´s lymphocytes. In case of absence of germ cells in the patients ejaculate, i.e. non-obstructive or obstructive azoospermia, only a molecular analysis for AZF microdeletion in the long arm of the human Y chromosome is performed, respectively, analysis of some CFTR gene mutations (F508del; 5T polymorphism in intron 8; R117H) causing frequently a bilateral absence of the vas deferens (CBAVD) (Oates et al., 1994; Vogt, 2004). Common insights on the frequency of other molecular genetic defects causing human male infertility are therefore not available and it has become popular to bypass the patients severe infertility in the clinic by using the ICSI (Intra-Cytoplasmatic Sperm-Injection) - protocol where just a single sperm is sufficient for a successful fertilization programme. However, recent descriptions of an increased rate of chromosome abnormalities in ICSI offspring (Bonduelle et al., 2002) and a twice as high risk for major birth defects in

IVF/ISCI offspring when compared with naturally conceived infants reported in a large Australian study (Hansen et al., 2002) now put a serious question mark behind this practical bypass decision, especially in those cases where probably a genetic defect has caused the man´s infertility.

Here we focus on the male fertility function of the human Y chromosome. We present not only to a molecular description of the Y genes encoding proteins with a function in human spermatogenesis, but also point to the locus specific repetitive sequence structures being probably a major reason for the frequent occurrence of microscopically visible Y chromosome abnormalities and the genomic Y mutations (e.g. AZF micro-deletions) causing male infertility. The involvement of the dynamic Y chromatin structure in this basic process for human reproduction are discussed.

Y Chromosome Abnormalities Visible in the Microscope

First microscopically visible deletions of the human Y chromosome were observed on its long arm (Yq-). They occurred in some men with oligozoospermia and azoospermia (Kjessler, 1965; Neu et al., 1973), but since an extremely variable length of the human Y chromosome was also detected in fertile men (Unnerus et al., 1967), nobody believed in an association of these Yq- abnormalities with the men´s infertile phenotype. By fluorescent chromosome staining experiments with quinacrine it was then recognised that most of the Y variability was caused by a polymorphic length of the distal heterochromatic part of its long arm (Yq12) (Fig. 6.1). Although length variations of the non-fluorescent euchromatic Yq11 region were reported as well (Soudek et al., 1973). These chromosome variabilities were found to be stable in family pedigrees and therefore useful to assign individuals to specific populations.

Especially interest was focussed on a mosaic karyotype 45X/46XY with a normal looking Y chromosome because this was not only associated with male infertility but also with mixed gonad dysgenesis (Chang et al., 1990). Men with this karyotype can have a streak gonad on one side and a dysgenetic gonad on the other side. Outer genitalia might be normal or presented with micropenis, hypospadia and small testis volumes. The occurrence of female phenotypes with the same karyotype pointed to a delicate balance of the male and female sexual pathway in early embryogenesis.

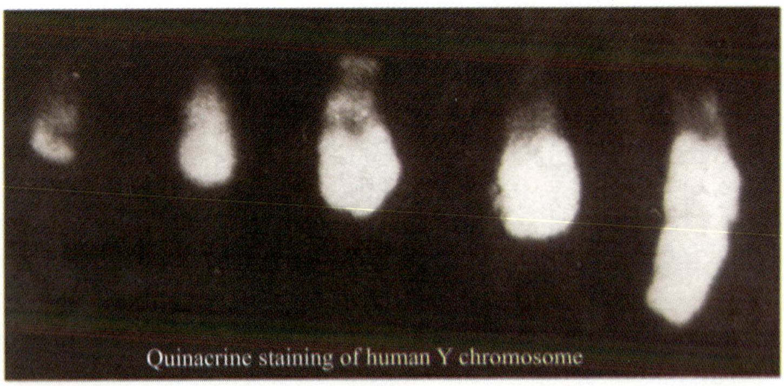

Fig. 6.1 The human Y chromosome displays a large variability of the distal long arm after staining with quinacrine. This is mainly due to variable copy number of the DYZ1 and DYZ2 repeats composing the heterochromatic Y chromosome region (Yq12).

Although the karyotype of this patient group usually displays a normal size Y chromosome, it often lacked the fluorescent staining Yq12 heterochromatin block after treatment with quinacrine. It could be shown that these "non-fluorescent" Y chromosomes (Ynf) were created after breakage in Yq11 with subsequent fusion of a second Y chromosome also broken in Yq11, both forming a dicentric iso-Yp chromosome (Daniel, 1985). In these men´s germ cells meiotic pairing of the Y to X chromosome is inhibited because of self-pairing of the iso-Yp between its doubled short arms (Chandley, 1989). Since similar pathologies were also described from patients with an AZFc microdeletion and a 45,X0 cell line (Siffroi et al., 2000; Jaruzelska et al., 2001; Papadimas et al., 2001; Patsalis et al., 2002) it is suggested that Ynf chromosomes might be one of the final rearrangements of the generally unstable Y chromosome after the occurrence of AZFc microdeletion.

In summary, seven different Yq- chromosome abnormalities could be repeatedly observed in infertile men indicating a general low stability of this chromosome region (Fig. 6.2). Today we know that major molecular reasons of this dynamic Y sequence structure are Non-Allelic Homologous

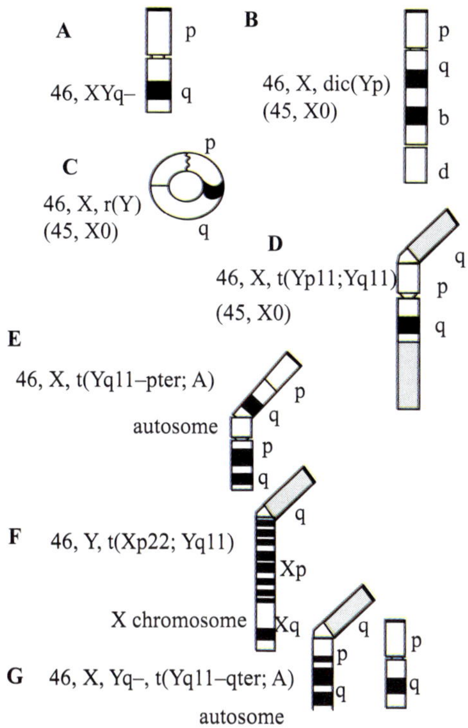

Fig. 6.2 Seven typical chromosomal rearrangements, **A-G**, in Yq11 which are associated with the occurrence of infertility, azoospermia or oligozoospermia because of disruption of the Y chromosomal Azoospermia Factor (*AZF*). Dicentric iso-Yp chromosomes (**B**) and ring-Y chromosomes (**C**) are often found together with a 45,X0 cell line in the patients lymphocyte nuclei pointing to a general instability of the Y chromosome after these breakage events. Translocations of the Y chromosome broken in Yq are possible to the X chromosome (**E**) and to autosomes (**F, G**) (*mainly the acrocentric chromosomes* 14, 15, 21).

Recombinations (NAHR) events of the locus-specific repetitive sequence blocks in distal Yp and Yq11 most of which are organised in large palindromes (Skaletsky et al., 2003).

The impact of balanced Y-A translocations on male fertility is variable because the same karyotypes were observed in fertile and sterile men sometimes even in the same pedigree (Fryns et al., 1985). In most familial cases, the distal heterochromatic part of the Y long arm (Yq12) was translocated to the short arm of an acrocentic chromosome. Since these Y-chromosomal abnormalities were all associated with male infertility a Y fertility factor probably based on a specific chromatin folding code for the generation of active X-Y pairing sites in Yq11 was suggested to be of functional importance (Vogt, 1990). Indeed, a functionally active chromatin domain in Yq11 was indicated in nuclei spreading experiments with human spermatogonia where the Y chromosome was shown to decondense in the nuclei before pairing with the X chromosome to form the condensed X-Y chromatin structure ("sex vesicle") in the spermatocyte nuclei (Speed et al., 1993) and an extension of the X-Y pairing structure beyond the Y centromere in Yq11 was shown as well (Armstrong et al., 1994). This might be functionally linked to the Azoospermia Factor (AZF) also mapped to this Y region and supposed to encode a functional protein coding gene (Tiepolo and Zuffardi, 1976).

The Azoospermia Factor (AZF) of the Human Y Chromosome in Yq11

The Azoospermia Factor (AZF) in Yq11 was first mapped in 1976 by Tiepolo and Zuffardi based on cytogenetic observations of microscopically visible deletions (i.e. macrodeletions) always affecting the distal part of the long Y arm. Since at that time nobody believed in a genetic function of the heterochromatic Yq12 chromatin region it was proposed that the Yq breakpoints should be located proximal to it in distal Yq11 (Tiepolo and Zuffardi, 1976). The six azoospermic men with a monocentric Yq11- chromosomes studied displayed a severe disruption of spermatogenesis in their testicular tissue sections. This suggested an essential function of AZF for the cellular proliferation and differentiation of the human male germ cell. Numerous similar studies then confirmed this basic assumption (Sandberg, 1985), although the individual histological evaluation of the patients testicular tubules also points to an extension of the first observed phenotypes. This promoted the basic idea that there might be a gradient of testicular pathologies observed with Yq11 macrodeletions and that the AZF locus might be composed of a number of Y genes functioning during different phases of the spermatogenic maturation process (Vogt et al., 1993).

Today, we know that there are at least fourteen protein encoding Y genes part of the AZF locus (Skaletsky et al., 2003; Vogt, 2005a) and that additionally the premeiotic pairing process of the sex chromosomes along the AZF chromatin is essential for a proper meiosis of the male germ cells (Fernandez-Capetillo et al., 2003). Moreover, with knowledge of the euchromatic Y sequence it is now also evident that two faces of this chromosome, viz, being polymorphic on one site, being functional for spermatogenesis at the other site, are structurally intermingled, i.e., functional AZF genes are structurally linked to Y-specific highly polymorphic DNA regions.

Accordingly, multiple Y chromosomes have developed during human evolution now distinguished by a rooted pedigree of 153 Y-chromosomal haplogroups (The Y chromosome Consortium [YCC], 2002; Jobling and Tyler-Smith, 2003; for more information on this subject read also the chapter of Chris Tyler-Smith of this Proceedings).

Mapping Genomic Microdeletions in Yq11 Causing Disruption of AZF

Obviously Yq11 macrodeletions were not suitable to map the molecular extension of the AZF locus in Yq11 more precisely. It was impossible to distinguish molecularly caused and chromosomally caused male infertility in this heterogeneous patient group. On the contrary because most macro-Y chromosomal rearrangements in Yq11 were expected to interfere also with the de-condensation process of the Yq chromatin in the male germ line and consequently with formation of the pairing gradient of both sex chromosomes during leptotene, nobody believed in a functional AZF gene(s) locus for a long time, and it was assumed that it would be impossible to map selectively Yq11 breakpoints which disrupt an essential AZF spermatogenesis gene.

A general breakthrough became only possible after the establishment of some detailed molecular deletion maps of the human Y chromosome in different research groups. They were first based on Southern blots and the use of Y-specific probes selected from different Y-specific clone libraries, then on the collection of a large number of Sequence Tagged Sites (STSs) selected from large clones mapped in Yq11 and application of PCR technology. With these molecular tools deletion mapping in Yq11 could be performed also with genomic DNA samples of infertile men with a normal karyotype (46,XY) and in a large screening study with 370 individuals three different microdeletions in Yq11 were detected and coined AZFa, AZFb and AZFc, respectively (Fig. 6.3).

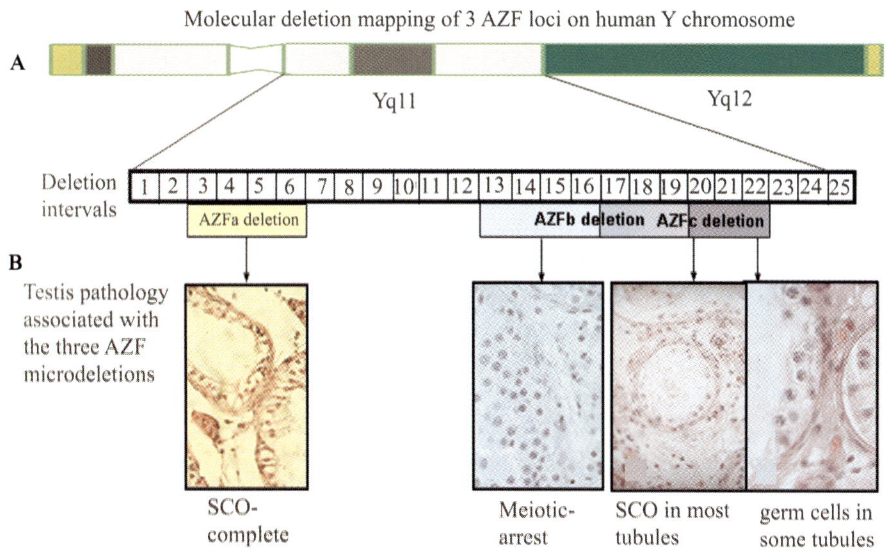

Fig. 6.3 Molecular map of the 3 AZF loci on the human Y chromosome in Yq11. (**A**): With aid of a series of Y-specific probes and STS´s the long Y arm (*Yq*) was subdivided in 25 molecular deletion intervals (D1-D25) (Vogt et al., 1996). Interval D25 correspond to the Yq12 heterochromatin. The D intervals deleted in infertile patients with complete AZFa, AZFb and AZFc deletions are drawn with a specific colour code below the interval map. (**B**): Examples of the associated testicular pathologies are displayed below the D-interval map. Patients with complete AZFa deletions have always a complete absence of germ cells. Patients with complete AZFb deletions usually show a meiotic arrest at pachytene whereas patients with complete AZFc deletions display a variable picture with a "mixed atrophy" in their testes tubules. Most tubules have a SCO, but some tubules also contain germ cells at different phases of differentiation.

Most interestingly, these AZF microdeletions interrupted spermatogenesis at three different phases: (1) a complete Sertoli-cell-only-syndrome (SCO) was observed in all patients with deletion of the complete AZFa interval, i.e. only Sertoli cells, but no germ cells were visible in the tubules of their testis tissue sections; (2) arrest at the spermatocyte stage was observed in the testis tissue of all patients with deletion of the complete AZFb interval. The populations of spermatogonia and primary spermatocytes in all tubules analysed were in the normal range; however, no postmeiotic germ cells were identified; (3) a variable testicular pathology was only found in patients with AZFc deletions in distal Yq11. In most tubules only Sertoli cells were identified but in some tubules germ cells of different developmental stages were clearly visible and the occurrence of mature sperm cells albeit in a low number was reported from different laboratories. Hypospermatogenesis seemed therefore to be the primary result of an AZFc deletion. Thus, AZF was now structurally subdivided into three distinct spermatogenesis loci: AZFa, AZFb and AZFc.

Molecular Structure of the 3 *AZF* Loci in Yq11

After the genomic human Y DNA sequence has been completed (Skaletzky et al., 2003), it could be shown that the long Y arm in Yq11 is indeed composed of numerous Y-specific repetitive sequence blocks (also called "amplicons") as suggested earlier (Kirsch et al., 1996) and especially in the AZFc deletion interval. Most interesting, amplicons between 115 kb and 678 kb length were structurally assembled in eight palindrome structures (P1-P8) with the largest palindrome (P1: 2.9 Mb) in distal AZFc (Fig. 6.4). The other palindromes, P2-P8, were mapped, from distal to proximal (towards the Y centromere) in distal AZFb (P2: 246 kb; P3: 736 kb), in proximal AZFb (P4: 419 kb; P5: 996 kb) and between the AZFa and AZFb deletion intervals (P6: 266 kb); P7: 30 kb; P8: 75 kb) respectively. These eight palindromes are highly symmetrical and comprise ~25% of the complete Y-specific sequence class, that is, 5.7 Mb of the genomic Yq11 sequence. Sequence analyses of homologous amplicons in the different palindrome arms revealed extensive homologies between 99.94–99.997% along the complete amplicon sequence (Kuroda-Kawaguchi et al., 2001).

It was found out that intrachromosomal recombination events between homologous repetitive sequence blocks have most likely caused the three genomic AZF microdeletions (Fig. 6.4): Two specific HERV15 (Human Endogenous Retroviral #15) sequence blocks in proximal Yq11 have caused the AZFa deletion (Blanco et al., 2000; Kamp et al., 2000; Sun et al., 2000). Local homologies between homologous repetitive sequence blocks in amplicons of the P1 and P5 palindromes in distal AZFc and proximal AZFb have probably led to all AZFb deletions (Repping et al., 2002), whereas all AZFc deletions were caused by homologous recombinations between the 229 kb long b2 and b4 amplicons (Kuroda-Kawaguchi et al., 2001). Similar results were also found in other research groups. A study on the sequence at the breakpoints on the Y chromosome of men with AZFb and AZFc deletions was performed in India to find clustering of L1, ERV and other retroviral repeat elements (Thangaraj et al., 2003).

However, the molecular mechanisms breaking and fusing the different recombination sites are not known. In this context it is interesting to note that short tandem repetitive simple sequence blocks like $(GATA)_{10\text{-}12}$ and $(CA)_n$ are concentrated in the AZFa and AZFb (GATA: Subramanian et al., 2003), respectively, AZFb and AZFc (CA: M. Köhler and P.H. Vogt, unpublished results).

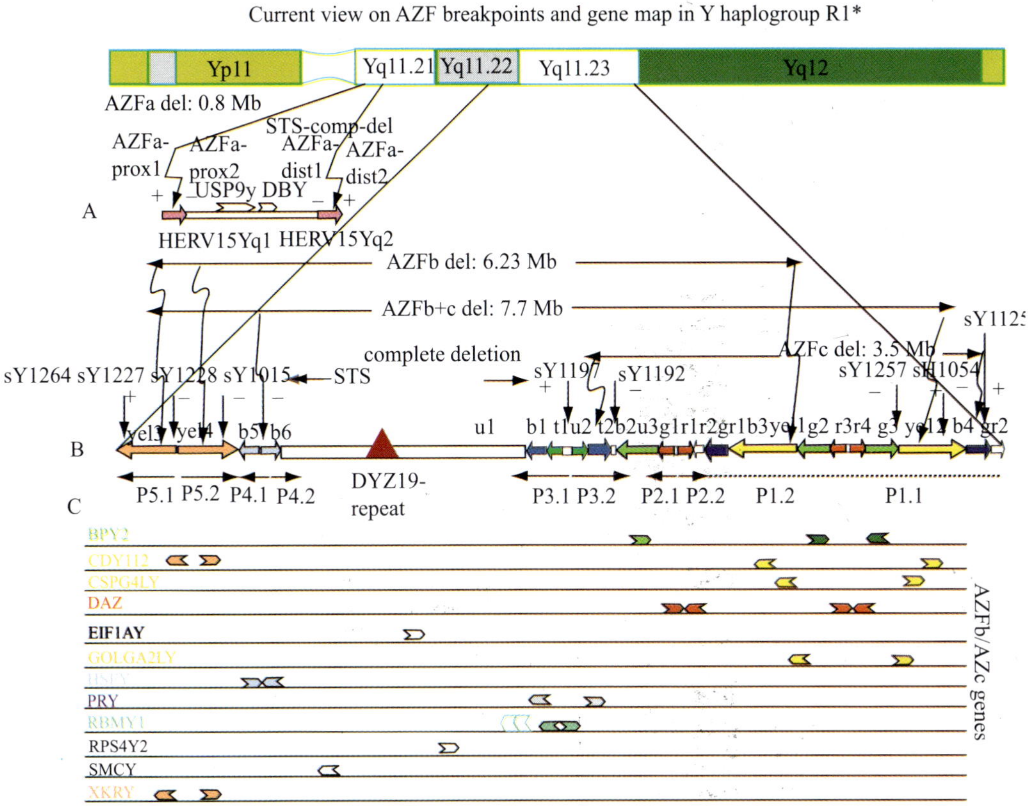

Fig. 6.4 Schematic view on AZF breakpoints and gene map in Yq11 from men with haplogroup R1*. (**A**). The AZF locus first marked by three distinct molecular deletion intervals (AZFa, AZFb, AZFc; Vogt et al., 1996) has now been sequenced. The AZFa deletion caused by recombination of two homologous HERV15 sequence blocks (HERV15Yq1/q2) is marked by a 0.8 Mb genomic DNA fragment in proximal Yq11. It contains the two AZFa genes, USP9Y and DBY. The AZFb deletion (extension 6.23 Mb) is largely overlapping with the AZFc deletion interval (*extension 3.5 Mb*) as indictated. Additionally a still larger AZFb deletion exists (coined AZFb+c del: 7.7 Mb; Repping et al., 2002), which however is not the sum of a complete AZFb + AZFc deletion interval. The indicated STS comp-del. markers can be used to diagnose complete AZFa, AZFb, AZFb+c, and AZFc deletions by a distinct deletion pattern. For example the sY1054 marker in distal AZFc is present in men with an AZFb+c deletion but absent in men with a complete AZFc deletion. (**B**). The approximate extension of the different palindromes formed by the AZFb/c amplicon structure is drawn below the amplicon blocks. Only the large P1 palindrome and the P3 palindrome include more than one amplicon type. (**C**). A schematic map of the location of all protein coding genes with a recognizable function in the AZFb and AZFc deletion interval. Their 5′-3′ polarity is indicated by direction of the arrows marking the location of the different gene copies. Pseudogene copies and the Testis-specific Transcript Y (*TTTY*) genes are listed in Table 6.1 to reduce the complexity of this map. The colour code of the genes and associated arrows indicates in which amplicon block the gene copies are located. Only the EIF1AY, RPS4Y2, and SMCY gene and two copies of the RBMY1 gene family are located outside the amplicon blocks. One copy of the duplicated CDY2 and XKRY gene is located proximal to the AZFb deletion interval (Repping et al., 2002). Please note that the ampliconic structure of the AZFb and AZFc sequence with its gene content as shown here is based on the GenBank Y chromosome sequence (http://www.ensembl.org/Homo_sapiens/ mapview?chr=Y) and that this sequence belongs to a Y chromosome of the R1* lineage. Thus it can be different in men with another Y chromosomal haplogroup (for more details see Vogt, 2005b).

Their distribution and the high matrix association potential of their flanking sequences suggests a bordering function of these simple sequence blocks for distinct decondensing chromatin domains.

Gene Content of AZFa, AZFb, AZFc Loci

The putative complete gene content of each AZF microdeletion could be established after knowledge of the euchromatic genomic Y sequence (Skaletsky et al., 2003). 31 Y genes expressed in human testis and located in one of the AZF deletion intervals were registered (Vogt, 2005a). These are 14 functional protein coding genes (Fig. 6.4) and 17 RNA genes transcribed specifically in testicular

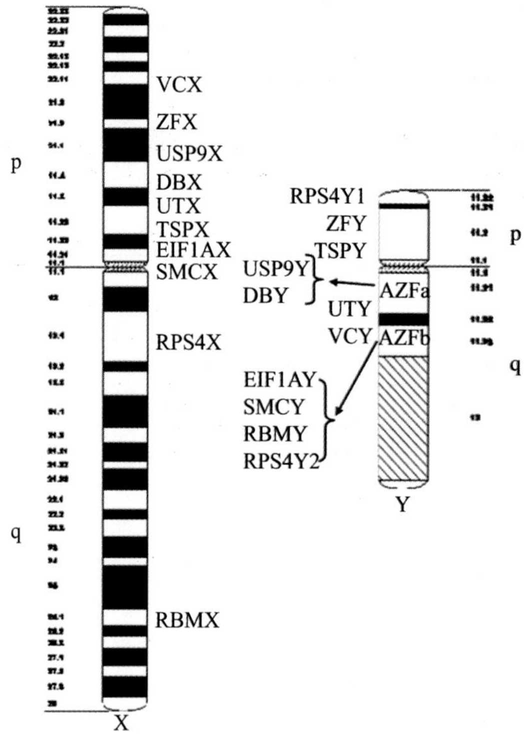

Fig. 6.5 Schematic view on location of the functional homologues of the AZFa genes, USP9Y and DBY, and AZFb genes, EIF1AY, SMCY, RBMY, RPS4Y2 on the human X chromosome. The X-Y homologous genes, RPS4Y1, TSPY/TSPX, UTY/UTX, VCY/VCX, and ZFY/ZFX are also expressed in human testicular tissue but mapped outside the AZF deletion intervals.

tissue (called *TTTY* genes) but with a non-identified function because homologous sequence domains have not yet been found in the data base (Table 6.1). Interestingly, all AZFa genes and a number of functional AZFb genes have an X chromosomal counterpart (Fig. 6.5), and some of them are conserved also on the mouse X and Y chromosome (*Dbx/Dby, Smcx/Smcy, Rbmx/Rbmy, Usp9x/Usp9y*). However, so far analysed, their function in mouse spermatogenesis looks different

Table 6.1 Human Y-genes in AZFb and AZFc with unknown Protein coding Potential[1]

Gene symbol	Gene name	Protein function homolog to	Tissue RNA expression	Copies in Yp interval	Yq11 interval	X or autosome homolog
CYorf14	Chromosome Y Openreading frame 14	unknown	multiple	no	AZFb	X; cXorf 14
CYorf15A[4]	Chromosome Y Openreading frame 15A	unknown	multiple	no	AZFb	X; cXorf 15
CYorf15B[4]	Chromosome Y Open reading frame 15B	unknown	multiple	no	AZFb	X; cXorf 15
TTTY1	Testis Transcript Y1	no-protein encod. RNA	only testis	prox. Yp11	no	no
TTTY2	Testis Transcript Y2	no-protein encod. RNA	only testis	prox. Yp11	AZFb	no
TTTY3	Testis Transcipt Y3	no-protein encod. RNA	only testis	no	AZFc	no
TTTY4	Testis Transcript Y4	no-protein encod. RNA	only testis	no	AZFc	no
TTTY5	Testis Transcipt Y5	no-protein encod. RNA	only testis	no	AZFb	no
TTTY6	Testis Transcript Y6	no-protein encod. RNA	only testis	no	AZFb	no
TTTY9	Testis Transcript Y 9	no-protein encod. RNA	only testis	no	AZFb	no
TTTY10	Testis Transcipt Y10	no-protein encod. RNA	only testis	no	AZFb	no
TTTY12	Testis Transcipt Y12	no-protein encod. RNA	only testis	no	AZFb	no
TTTY13	Testis Transcript Y13	no-protein encod. RNA	only testis	no	AZFb	no
TTTY14	Testis Transcipt Y14	no-protein encod. RNA	only testis	no	AZFb	no
TTTY15 *AZFaT1*[2] *Phex152*[3]	Testis Transcipt Y15	no-protein encod. RNA	testis, brain	no	AZFa	no
TTTY16	Testis Transcipt Y16	no-protein encod. RNA	only testis	no	AZFb	no
TTTY17	Testis Transcript Y17	no-protein encod. RNA	only testis	no	AZFc	no

[1]according to Kuroda-Kawaguchi et al., (2001); Repping et al. (2002); Skaletzky et al. (2003) Although Skaletzky et al. described 23 TTTY genes (TTTY1-TTTY23) only the 14 listed here are mapped on the Y chromosome sequence.

[2] Sargent et. al. (1999) in: J.Med. Genet. Vol. *36*: 670-677.

[3] Ditton, Hirschmann, Vogt (1999) unpublished results.

[4] degenerated by splitting and having evolved from the same ancestral CXorf15 gene.

from that in human spermatogenesis. In mouse, Usp9y is expressed in brain and testis tissue (Xu et al., 2002), whereas the human homolog *USP9Y* is expressed in all tissues analysed (Lahn and Page, 1997). In mouse, deletion of most *Rbmy* genes did not cause male infertility but only some sperms dysmorphologies (Mahadevaiah et al., 1998). In human, for patients with complete *RBMY* deletion no postmeiotic germ cells were found (Elliot et al., 1997). This suggests an early premeiotic function of the human RBMY protein in male germ cells. It can thus be assumed that the spermatogenic functions of the human AZF genes and their mouse homologues are probably divergent, similarly, as found earlier for a number of autosomal genes functioning in the genetic network controlling male fertility (Wyckoff et al., 2000). The AZFc genes can be considered evolutionarily very young spermatogenesis genes because no AZFc gene was found on the mouse Y chromosome.

The functional integrity of the AZF genes located in the genomic palindromes formed by inverted homologous amplicons is probably maintained by frequent gene conversion events between the homologous palindrome arms (Rozen et al., 2003). Assuming a mutation rate of 1.6 x 10^{-9} per nucleotide per year and 2.2 x 10^{-4} conversions per duplicate nucleotide per generation (i.e. per 20 years) it has been calculated that along the 5.4 Mb length of the eight Y palindromes (i.e. 2.7 x 10^{6} duplicated nucleotides) on average 600 duplicated nucleotides have undergone arm-to-arm gene conversion, thus probably distinguishing even the Y chromosome of father and son (Rozen et al., 2003).

Mapping AZF gene deletions in some men with idiopathic azoospermia *DBY* turned up as the most important AZFa gene (Foresta et al., 2000) and *RBMY* as the most important AZFb gene (Vogt, 2005a). In AZFc, the male fertility function of the repetitive DAZ gene family seems to be linked to their copy number with a preference for the presence of the proximal *DAZ1/DAZ2* gene doublet (Fernandes et al., 2002).

Male Fertility Function of the *DBY* AZFa Gene

The AZFa DBY gene belongs to the large family of DEAD box RNA helicase genes (Rocak and Lindner, 2004). Although all members of the DEAD box family contain eight (I-VIII) highly conserved peptide domains essential for the ATP driven RNA duplex unwinding process comparison of the peptide sequence between them and beyond these eight conserved peptide boxes revealed four human subfamilies, *DBY-*, *EIF4- P68/P72-* and *VASA*, and a strong sequence relationship between the members of each subfamily. From these sequence alignment analyses only the testis specific expressed *VASA* gene was recognised to be probably functional related to the *DBY* gene subfamily.

A systematic comparative analysis of VASA and DBX protein sequences in different metazoa including the low metazoans: sponge, hydra and planaria suggests that the ancient *vasa* gene was derived from the ancient *Dbx/PL10* gene (Mochizuki et al., 2001). This conclusion was drawn from the fact that only *PL10* gene related sequences were identified in plants and only the *PL10*-related genes have a widely tissue expression, whereas expression of the *vasa*-like genes were always restricted to the germ line from hydra to human. It was therefore concluded that VASA protein has evolved from the PL10-related protein to extend or support a specific germ line function.

The human DBY protein also expresses a germ line specific function (Ditton et al., 2004). This was surprising because transcripts of the human *DBY* gene were found in each tissue analysed and in lymphocytes. It has thus been speculated that the human Y homologue (*DBY*) of the X-chromosomal *DBX* gene has maintained functional because of an extended germ line function of the X gene in mammals. The mammalian sex chromosome structures have diverged about 100 million years ago before mammalian radiation. Usually, along such a long time interval Y gene copies have diverged to non-functional pseudo-genes (Lahn et al., 2001). However, this seems not be true for the X-Y homologous genes of the AZFa and AZFb deletion intervals (Fig. 6.5).

Indeed, the nuclear function of the DBY and DBX proteins seems to be interchangeable in the G1 phase of the cell cycle (Sekiguchi et al., 2004). Similarly, in yeast expression of *DBX* like *DBY* was able to rescue the translation initiation function of the structural homologous yeast *Ded1p* RNA helicase gene in the cytoplasm (Chuang et al., 1997). RNA helicases of the *DBX/DBY* subfamily are therefore assumed to be functional similar shuttling between both cell compartments during development.

However, this seems not to be true for the human male germ cells. With aid of two polyclonal antisera, recognising specifically only DBX protein (DBX-10 antiserum) or DBY protein (DBY-10 antiserum) on Western blots it was possible to distinguish the expression phase of DBY and DBX protein in testicular tissue sections (Ditton et al., 2004). The DBY-specific antiserum stained predominantly the cytoplasm of spermatogonia suggesting a cytoplasmic function for the DBY RNA helicase specifically in these pre-meiotic germ cells. The DBX-specific antiserum stains germ cells of a different spermatogenic phase, namely the post-meiotic spermatids (Fig. 6.6). These results explained why the lack of DBY protein in men with *DBY*- or complete AZFa- deletion cannot be rescued by expression of the DBX protein. Despite their high sequence and therefore also functional homology, the DBY and DBX RNA helicases serve their cellular function during different phases of the human male germ cell differentiation process.

DBX and DBY proteins are expressed in different male germ cells

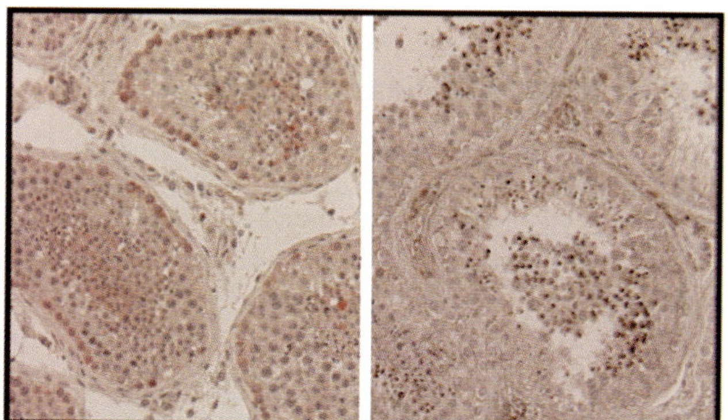

DBY is expressed in spermatogonia DBX is expressed in spermatids

Fig. 6.6 Immunohistochemical staining of the location of DBX and DBY protein in human testicular tissue sections with two polyclonal antisera specific for DBX (*DBX-10 antiserum*) and DBY (*DBY-10 antiserum*). DBY protein is predominantly expressed in spermatogonia, DBX protein in spermatids. For further description and discussion of these results see Ditton et al. (2004).

Male Fertility Function of the *RBMY* AZFb Gene

Partial AZFb deletions not including the *RBMY* genes showed variable testicular pathologies including hypospermatogenesis in four infertile men from Italy (Ferlin et al., 2003). This would suggest that the *RBMY* genes would be the major *AZFb* genes causing meiotic arrest if disrupted. The *RBMY* gene is homologous to *RBMX* on the X chromosome (Fig. 6.5) and to a functional retrogene *HNRNP G-T* on chromosome 11 (11p15) that is predominantly expressed in pachytene spermatocytes (Elliott et al., 2000). The creation of intron-less retrogenes originating from their progenitor genes by retroposition of their transcripts to an autosome seems to be a common mechanism for male germ line genes when they are located on the X chromosome and have a function at meiosis (Wang, 2004). Expression of the RBMY protein has been shown in the nuclei of both type A and B spermatogonia, spermatocytes and round spermatids. This suggests a role of this *AZFb* gene in the nuclear metabolism of newly synthesized RNAs in the germ cell at different phases (Elliot et al., 1997).

The different members of the *RBMY* gene family have obviously subdivided their germ line function to different spermatogenic phases, as was also found for the *DBY* and *DBX* genes (Ditton et al., 2004) and the *DAZ* gene family (Xu et al., 2001). This would then also explain why the lack of RBMY proteins of men with AZFb deletions cannot become compensated in spermatogenesis by the presence of RBMX and/or HNRNP G-T proteins. RBMY proteins were shown to be involved in testis specific splicing reaction of a number of still unknown testis transcripts expressed before meiosis by interaction with the splicing cofactors Sam68 and T-STAR, respectively (Elliott et al., 2000). Recent experiments identified a novel mode of interaction of the RRM domain of RBMY with its RNA target S1A which was not found with its homologous RBMX and HNRNP G-T protein (Skrisovska et al., 2007). It can therefore be concluded that a specific folding structure of the RRM domain in the RBMY protein has been selected for controlling a specific number of testis transcripts which still need to be identified.

Male Fertility Function of the *DAZ* Gene Family in AZFc

The *DAZ* gene family has been transposed from chromosome 3 (3p25; *DAZL1* locus) on to the Y chromosome after divergence of the Old world monkey line of primates, i.e., not before ~35 million years ago (Shan et al., 1996; Seboun et al., 1997). The two gene copies in proximal AZFc (*DAZ1/DAZ2*) are forming the P2 palindrome, whereas the two other gene copies (*DAZ3/DAZ4*) are located 1.47 Mb more distal in the center of the large P1 palindrome in distal AZFc (Fig. 6.4). In contrast to other AZFc gene families, *BPY2*, and *CDY*, the exon structures of the *DAZ* gene copies are rather variable especially in the number of the exon 7 variants also called *"DAZ-repeat"* (Saxena et al., 2000; Fernandes et al., 2002).

Most likely, the *DAZ* gene family has evolved and established on the Y chromosome following a duplication-diversification process similar as proposed for other duplicated genes by Hughes (1994). It would suggest that the *DAZ* genes have extended or improved the functional profile of their autosomal homologues, *DAZL1* on chromosome 3 and *BOULE* on chromosome 2. Classical theories predict that the evolutionary route towards functional divergence of two given gene copies starts first with a functional redundancy of the duplicated genes because they are still serving for identical functions. However, this state is transient because it is usually not selected for over longer evolutionary times. Functional divergence will occur by further exonic mutations then leading to a novel function of one or both duplicates (neo-functionalization) or to divison of the ancestral function among the duplicated paralogs (sub-functionalization). The last is also known as the "Duplication-Degeneration-Complementation" (DDC) hypothesis (Force et al., 1999).

We are currently trying to distinguish the spermatogenic function of the DAZ genes on the Y chromosomes from that of the autosomal DAZL1 gene by analysis of their specific promoter structures and the putative different location of DAZ and DAZL1 proteins in human male germ cells. We found DAZ2 proteins only in postmeiotic germ cells and in spermatozoa whereas DAZL1 proteins were expressed predominantly in premeiotic spermatogonia (Habermann et al., 1998; Vogt and Zimmer unpublished results). Interestingly, using a set of other polyclonal DAZ and DAZL1 antisera, Reijo and coworkers found an opposite distribution of DAZ and DAZL1 protein in human testis (Reijo et al., 2000). One can assume that some of these discordant findings can probably be explained by the use of different polyclonal DAZ-antisera. Nevertheless it is obvious, that the complete DAZ gene family is functional—essential for human spermatogenesis encoding different testis specific RNA binding proteins probably involved in the translational control of transcripts of other germ line genes.

More Y Genes Functioning for Human Male Fertility?

Prominent examples of Y genes also expressed in testicular tissue but located outside the three AZF deletion intervals are the *TSPY* genes in proximal Yp11 (Schnieders et al., 1996) and the repetitive non-coding RNA (ncRNA) gene in Yq12 (Jehan et al., 2006). This raised the question whether their function in human spermatogenesis is essential as that of at least some of the AZF genes and whether their dysfunction can be recognised by the clinical phenotype of male infertility?

The *TSPY* (Testis-Specific Protein, Y-encoded) is the product of a tandem gene cluster on human proximal Yp. Each of its 23-49 gene copies (~2.8 kb in size) is embedded in a single unit of the *DYZ5* tandem repeat array (20.4 kb sequence complexity) with a high sequence homology (>98%) within and outside the functional gene structure (Repping et al., 2006). Sequence analyses revealed that *TSPY* is a member of a protein super family including the proto-oncogene SET and the nucleosome-assembly factor NAP-1 functioning in directing spermatogonia to enter meiosis (Schnieders et al., 1996). SET/NAP proteins are generally known to be involved in cell cycle regulation, chromatin remodelling, and transcription regulation. TSPY has a structural homologue on the X chromosome (*TSPX*; Fig. 6.5) although encoding additionally an N-terminal proline-rich domain and a carboxyl bipartite acidic domain not present in TSPY proteins, and some autosomal retrogenes (*TSPY-L1,4* on chromosome 6, *TSPY-L5* on chromosome 8, *TSPY-L3* on chromosome 20, *TSPY-L6* on chromosome 2) like the RBMY gene family (Lau et al., 2007). *TSPY-L1* has been demonstrated to be responsible for the sudden infant death with a testis dysgenesis syndrome. It can therefore be assumed that the *TSPY* gene family serve for a variety of different physiological functions mediated by the conserved SET/NAP domain and gene copy specific flanking domains. In the nucleus this protein family contributes to the control of the

G2/M phase in the cell cycle. One can thus assume that *TSPY* expression in human spermatogonia might contribute to the proliferation and differentiation control of these premeiotic germ cells.

The repetitive non-coding RNA (ncRNA) gene in Yq12 was derived from the sex- and species-specific genomic clone 102d(2) mapped to the long arm of the human Y chromosome in the distal heterochromatic block (Yq12) by Fluorescence *in situ* hybridization (FISH) some years ago in the lab of Singh et al. (1994). Using it as a screening probe for homologous testis specific cDNAs the same lab was able to isolate a number of cDNA clones with a specific expression window in testis at the 36 gestation week during testicular development in the embryo (Jehan et al., 2007). Most interestingly during the same time period the heterochromatin chromosome region of the Y chromosome (Yq12) was found to be completely decondensed. Since the 102d(2) homologous cDNA clones, AK47 and AK 598346, could be also mapped to the DYZ1 repeats in Yq12 it was concluded that the tandem repetitive DYZ1 sequence structure has become transcriptionally active by de-condensation of the underlying heterochromatic chromatin structure during a distinct phase of early testicular development producing large transcripts with a heterogenous length distribution (Jehan et al., 2007).

A similar testis specific transcription profile of Y-specific satellite repeats associated with the decondensation phase of the Y chromosome also takes place in the spermatocyte nuclei of *Drosophila* (Vogt et al., 1982). Here the transcripts are extremely large accumulating large numbers of nuclear proteins and therefore becoming visible even in the light microscope. According to their visible granular structures they were also called granular Y chromosomal lampbrush loops usually found only in the oocytes of amphibians (Wu et al., 1991).

Jehan and coworkers were able to identify the DYZ1 transcripts of the human Y chromosome as testis specific functional part of the 5′ UTR sequence of the autosomal kinase gene *CDC2L2* from chromosome 1p36.3 involved in control of the cellular division cycle (Jehan et al., 2007). Obviously, these testis specific transcripts could only be generated by transplicing of the autosomal *CDC2L2* testis transcripts to the non-coding 5′ UTR transcripts encoded by the DYZ1 repeat in Yq12. Consequently, the authors suggest a translational control of these *CDC2L2* transcripts by the Y-specific 5′ UTR domain being translated only during a short period of the testicular development in the embryo (Jehan et al., 2007).

Acknowledgements

I would like to thank my friend Lalji Singh for his invitation to contribute to his 60th birthyday celebration with this lecture on human Y chromosomal male fertility function. Our long lasting friendship is based on our common interest in the fertility function of tandem repetitive locus specific sequence structures on the heteromorphic sex chromosomes of *Drosophila*, snakes, mouse and human. This research would not have been possible without the continuous support of my academic teachers, Wolfgang Hennig in Nijmegen and Friederich Vogel in Heidelberg. Thomas Strowitzki is thanked for his generous support in establishing my research projects on the Y chromosomal AZF genes in his Department of Gynecology and Reproductive Medicine in the University Hospital of Heidelberg. My studies on the AZF genes are financially supported by grants from the Deutsche Forschungsgemeinschaft (DFG: Vo403/10-2; 11-5; 14-1/2).

References

1. Armstrong, S.J., Kirkham, A.J. and Hulten, M.A. (1994). XY chromosome behaviour in the germ-line of the human male: a FISH analysis of spatial orientation, chromatin condensation and pairing. Chrom. Res. *2*, 445–452.
2. Blanco, P., Shlumukova, M., Sargent, CA., et al. (2000). Divergent outcomes of intrachromosomal recombination on the human Y chromosome: male infertility and recurrent polymorphism. Jour. Med. Genet. *37*: 752–758.
3. Bonduelle, M.; Van Assche, E.; Joris, H.; Keymolen, K.; Devroey, P.; Van Steirteghem, A. and Liebaers, I. (2002). Prenatel testing in ICSI pregnancies: incidence of chromosomal anomalies in 1586 karyotypes and relation to sperm parameters. Hum. Reprod. *17*, 2600–2614.
4. Chandley, A.C. (1989). Asymmetry in chromosome pairing: a major factor in de novo mutation and the production of genetic disease in man. J. Med. Genet., *26*, 546–552.
5. Chang H.J., Clark, R.D., Bachman ,H. (1990). The phenotype of 45X/46XY mosaicism: an analysis of 92 prenatally diagnosed cases. Amer. Jour. of Hum. Genet. *46*, 156–167.
6. Chuang, R.Y. Weaver, P.L., Liu, Z. and Chang, T.H. (1997). Requirement of the DEAD-Box protein ded1p for messenger RNA translation. Science *275*, 1468–1471.
7. Claustres, M. (2005). Molecular pathology of the CFTR locus in male infertility. Reprod. BioMed. Online, *10*, 14–41.
8. Cooke, H.J., and Saunders, P.T. (2002). Mouse models of male infertility. Nat. Rev. Genet., *3*, 790–801.
9. Daniel, A. (1985). Y isochromosomes and rings. In: Sandberg AA (eds) *The Y chromosome,* Part B: Clinical aspects of Y chromosome abnormalities. Alan R. Liss, Inc., New York, pp 105–135.
10. Ditton, H.J., Zimmer, J., Kamp, C., et al. (2004). The AZFa gene DBY (DDX3Y) is widely transcribed but the protein is limited to the male germ cells by translation control. Hum. Mol. Genet., 4, (Epub ahead of print).
11. Elliott, D.J., Millar, M.R., Oghene, K., et al. (1997). Expression of RBM in the nuclei of human germ cells is dependent on a critical region of the Y chromosome long arm. Proc. Natl. Acad. Sci. USA, *94*, 3848–3853.
12. Elliott, D.J., Venables, J.P., Newton, C.S., et al. (2000). An evolutionarily conserved germ cell-specific hnRNP is encoded by a retrotransposed gene. Hum. Mol. Genet., *9*, 2117–2124.
13. Ferlin, A., Moro, E., Rossi, A., et al. (2003). The human Y chromosome's azoospermia factor b. (AZFb) region: sequence structure and deletion analysis in infertile men. J. Med. Genet. *40*, 18–24.
14. Fernandes, S., Huellen, K., Goncalves, J., et al. (2002). High frequency of DAZ1/DAZ2 gene deletions in patients with severe oligozoospermia. Mol. Hum. Reprod., *8*, 286–298.
15. Fernandez-Capetillo, O., Mahadevaiah, K., Celeste, A., et al. (2003). H2AX is required for chromatin remodelling and inactivation of sex chromosomes in male mouse meiosis. Dev. Cell, *4*, 497–508.
16. Force, A., Lynch, M., Pickett, F.B., et al. (1999). Preservation of duplicate genes by complementary, degenerative mutations. Genetics, *151*, 1531–1445.
17. Foresta, C., Ferlin, A., Moro, E. (2000). Deletion and expression analysis of AZFa genes on the human Y chromosome revealed a major role for DBY in male infertility. Hum. Mol. Genet. *9*, 1161–1169.

18. Fryns, P., Kleczkowska, A., Van den Berghe, H. (1985). Clinical manifestation of Y/autosome translocations in man In Sandberg AA. (ed) *The Y chromosome Part B: Clinical aspects of Y chromosome abnormalities.* Alan R Liss Inc. New York, pp 213–243.
19. Habermann, B., Mi, HF., Edelmann, A., et al. 1998. DAZ. (Deleted in AZoospermia) genes encode proteins located in human late spermatids and in sperm tails. Hum. Reprod., *13*, 363–369.
20. Hackstein, J.H., Hochstenbach, R., Pearson, P.L. (2000). Towards an understanding of the genetics of human male infertility: lessons from flies. Trends Genet., *16*, 565–572.
21. Hansen, M., Kurinczuk, J.J., Bower, C., and Webb, S. (2002). The risk of major birth defects after intracytoplasmic sperm injection and *in vitro* fertilization. N. Engl. J. Med. *346*, 725–730.
22. Hughes, A.L., (1994). The evolution of functionally novel proteins after gene duplication. Proc. Royal Soc. London B Biol. Science, *256*, 119–124.
23. Jaruzelska, J., Korcz, A., Wojda, A., et al. (2001). Mosaicism for 45X cell line may accentuate the severity of spermatogenic defects in men with AZFc deletion. Journal of Medical Genetics *38*, 798–802.
24. Jehan, Z., Vallinayagam, S., Tiwari, S., et al. (2007). Novel noncoding RNA from human Y distal heterochromatic block (Yq12) generates testis specific chimeric CDC2L2. Genome Res. *17*, 433–440.
25. Jobling, M.A., Tyler-Smith, C. (2003). The human Y chromosome: an evolutionary marker comes of age. Nat. Rev. Genet., *4*, 598–612.
26. Kamp C., Hirschmann P., Voss, H., et al. (2000). Two long homologous retroviral sequence blocks in proximal Yq11 cause AZFa microdeletions as a result of intrachromosomal recombination events. Human Molecular Genetics *9*, 2563–2572.
27. Kirsch S., Keil R., Edelmann, A., et al. (1996). Molecular analysis of the genomic structure of the human Y chromosome in the euchromatic part of its long arm (Yq11). Cytogen. Cell Gene., *75*, 197–206.
28. Kjessler, B. (1965). Karyotypes of 130 childless men. Lancet, *10*, 493–494.
29. Krausz, C., Forti, G., McElreavey, K. (2003). The Y chromosome and male fertility and infertility. Int. J. Androl., *26*, 70–75.
30. Kuroda-Kawaguchi, T., Skaletzsky, H., Brown, L.G., et al. (2001). The AZFc region of the Y chromosome features massive palindromes and uniform recurrent deletions in infertile men. Nat. Genet., *29*, 279–286.
31. Lahn, B.T., Page, D.C. (1997). Functional coherence of the human Y chromosome. Science *278*, 675–679.
32. Lahn, B.T. Pearson, N.M., and Jegalian K. (2001). The human Y chromosome, in the light of evolution. Genetics *2*, 207–216.
33. Lahn, B.T., Tang, Z.L., Zhou, J., et al. (2002). Previously uncharacterized histone acetyltransferases implicated in mammalian spermatogenesis. Proc. Natl. Acad. Sci. USA, *99*, 8707–8712.
34. Lau, Y-F.C., Kido, T., Li, Y. (2007). The *TSPY* gene family. In: *The Y chromosome and Male Germ Cell Biology in Health and Diseases* (eds. Y-F. C. Lau and W-Y Chan; World Scientific Publishing Co. Pte. Ltd.), *4*, pp.73–90.
35. Mahadevaiah, S.K., Odorisio, T., Elliott, D.J. et al. (1998). Mouse homologues of the human AZF candidate gene RBM are expressed in spermatogonia and spermatids and map to a Y chromosome

deletion interval associated with a high incidence of sperm abnormalities. Hum. Mol. Genet., *7*, 715–727.

36. Matzuk, M.M., Lamb, D.J. (2002). Genetic dissection of mammalian fertility pathways. Nat.Cell. Biol., *4*, Suppl s41–49.
37. Mochizuki, K., Nishimiya-Fujisawa, C., and Fujisawa, T. (2001). Universal occurrence of the vasa-related genes among metazoans and their germline expression in Hydra. Dev. Genes. Evol. *211*, 299–308.
38. Neu, R.L., Matthew, M.S., Barlow, Jr. M.D., et al. (1973). A 46,XYq- male with aspermia. Fertil. Steril., *24*, 811–813.
39. Oates, R.D., Amos, J.A. (1994). The genetic basis of congenital bilateral absence of the vas deferens and cystic fibrosis. J. Androl., *15*, 1–8.
40. Papadimas J., Goulis DG., Giannouli, C., et al. 2001. Ambiguous genitalia 45X/46XY mosaic karyotype and Y chromosome microdeletions in a 17-year-old man. Fertility and Sterility *76*, 1261–1263.
41. Patsalis PC., Sismani C., Quintana-Mursi L et al. (2002). Effects of transmission of Y chromosome AZFc deletions. Lancet *360*, 1222–1224.
42. Quintana-Murci, L., Krausz, C., Heyer, E. et al. (2001). The relationship between Y chromosome DNA haplotypes and Y chromosome deletions leading to male infertility. Hum. Genet., *108*, 55-58.
43. Reijo, R.A., Dorfman, D.M., Slee, R. et al. (2000). DAZ family proteins exist throughout male germ cell development and transit from nucleus to cytoplasm at meiosis in humans and mice. Biol. Reprod., *63*, 1490–1496.
44. Repping, S., Skaletsky, H., Lange, J. et al. (2002). Recombination between palindromes P5 and P1 on the human Y chromosome causes massive deletions and spermatogenic failure. Am. J. Hum. Genet., *71*, 906–922.
45. Repping, S., van Daalen, S.K.M., Brown, L.G. (2006). High mutation rates have driven extensive structural polymorphism among human Y chromosomes. Nat. Genet. *38*, 463–467.
46. Rocak, S., & Linder, P. (2004). DEAD-BOX proteins: the driving forces behind RNA metabolism. Nature Reviews Cell Biology, *5*, 232–241.
47. Rozen, S., Skaletsky, H., Marszalek, J.D. et al. (2003). Abundant gene conversion between arms of palindromes in human and ape Y chromosomes. Nature, *423*, 873–876.
48. Sandberg A.A., (ed) 1985. *Clinical aspects of Y chromosome abnormalities.* Progress and topics in cytogenetics, Vol. 6, part B, Alan R. Liss, Inc, New York, USA.
49. Saxena, R., de Vries, J.W., Repping, S., et al. (2000). Four DAZ genes in two clusters found in the AZFc region of the human Y chromosome. Genomics. *67*, 256–267.
50. Schnieders, F., Dork, T., Arnemann, J., Vogel, T., Werner, M., and Schmidtke, J. (1996). Testis-specific protein, Y-encoded (TSPY) expression in testicular tissue. Hum. Mol. Genet. *5*, 1801–1807.
51. Seboun, E., Barbaux, S., Bourgeron, T., et al. (1997). Gene sequence localization and evolutionary conservation of DAZLA a candidate male sterility gene. Genomics, *41*, 227–235.
52. Sekiguchi, T., Iida, H., Fukumura, J., and Nishimoto, T. (2004). Human DDX3Y, the Y-encodedisoform of RNA helicase DDX3, rescues a hamster temperature-sensitive ET24 mutant cell line with a DDX3X mutation. Exp. Cell Res. *300*, 213–222.

53. Shan Z., Hirschmann P., Seebacher T., et al. 1996. A SPGY copy homologous to the mouse gene Dazla and the Drosophila gene boule is autosomal and expressed only in the human male gonad. Human Molecular Genetics *5*, 2005–2011.

54. Siffroi JP., Bourhis C., Krausz C., et al. 2000. Sex chromosome mosaicism in males carrying Y chromosome long arm deletions. Human Reproduction *15*, 2559–2562.

55. Singh, L., Panicker, S.G., Nagaraj, R., et al. (1994). Proc. Indian Natn. Sci. Acad. *B60*: 455–470.

56. Skaletsky, H., Kuroda-Kawaguchi, T., Minx, P.J., et al. (2003). The male-specific region of the human Y chromosome is a mosaic of discrete sequence classes. Nature, *423*, 825–837.

57. Skrisovska, L., Bourgeois, C.F., Stefl, R., et al. (2007). The testis-specific human protein recognizes RNA through a novel mode of interaction. EMBO reports *8*, 372–379.

58. Soudek, D., Langmuir, V., Stewart, D.J. (1973). Variation in the nonfluorescent segment of long Y chromosome. Humangenetik, *18*, 285–290.

59. Speed, R.M., Vogt, P., Kohler, M.R., et al. (1993). Chromatin condensation behaviour of the Y chromosome in the human testis I Evidence for decondensation of distal Yq in germ cells prior to puberty with a switch to Sertoli cells in adults. Chromosoma, *102*, 421–427.

60. Subramanian, S., Mishra, R.K., Singh, L. (2003). Genome-wide analysis of Bkm sequences (GATA repeats): predominant association with sex chromosomes and potential role in higher order chromatin organization and function. Bioinformatics *19*. 681–685.

61. Sun, C.; Skaletsky, H.; Rozen, S. et al. (2000). Deletion of azoospermia factor a (AZFa) region of human Y chromosome caused by recombination between HERV15 proviruses. Hum. Mol. Genet. *9*, 2291–2296.

62. Swanson, W.J., Vacquier, V.D. (2002). The rapid evolution of reproductive proteins. Nat. Rev. Genet. *3*, 137–144.

63. Thangaraj, K., Gupta, N.J., Pavani K., et al. (2003). Y chromosome deletions in azoospermic men in India. J. Androl. *24*, 588–597.

64. The Y Chromosome Consortium (2002). A nomenclature system for the tree of human Y- chromosomal binary haplogroups. Genome Res *12*, 339–348.

65. Tiepolo, L., Zuffardi, O. (1976). Localization of factors controlling spermatogenesis in the nonfluorescent portion of the human Y chromosome long arm. Localization of factors controlling spermatogenesis in the nonfluorescent portion of the human Y chromosome long arm. Hum. Genet. *34*, 119–124.

66. Unnérus, V., Fellmann, J., De La Chapelle, A. (1967). The length of the human Y chromosome. Cytogenet. *6*, 213–227.

67. Ursini, F., Heim, S., Kiess, M., et al. (1999). Dual function of the selenoprotein PHGPx during sperm maturation. Science, *285*, 1393–1396.

68. Vogt, P.H., Hennig, W., Siegmund, I., (1982). Identification of cloned Y chromosomal DNA sequences from a lampbrush loop of Drosophila hydei. Proc. Nat. Acad. Sci. USA. *79*, 5132–5136.

69. Vogt P.H. (1990). Potential genetic functions of tandem repeated DNA sequence blocks in the human genome are based on a higly conserved "chromatin folding code" (Review). Hum. Genet. *84*: 301–336.

70. Vogt P, Keil R, Kirsch S (1993). The AZF-function of the human Y chromosome during spermatogenesis. In: Chandley AC, Summer A (eds), *Chromosomes Today,* Vol. *11*, Capman & Hall, London, UK, pp 227–239.

71. Vogt, P.H., Edelmann, A., Hirschmann, P., et al. (1995). The azoospermia factor (AZF) of the human Y chromosome in Yq11: function and analysis in spermatogenesis. Reprod. Fertil. Dev., *7*, 685–693.

72. Vogt, P.H., Edelmann, A., Kirsch, S., et al. (1996). Human Y chromosome azoospermia factors (AZF) mapped to different subregions in Yq11. Hum. Mol. Gent. *5*, 933–943.

73. Vogt, P.H. (2004). Molecular genetic of human male infertility: from genes to new therapeutic perspectives. Current Pharmaceutical Design *10*, 471–500.

74. Vogt, P.H. (2005a). Azoospermia factor (AZF) in Yq11: towards a molecular understanding of its function for human male fertility and spermatogenesis. Reprod. BioMed. Online, *10*, 81–93.

75. Vogt, P.H. (2005b). AZF deletions in Y-chromosonal haplogroups: History and update based on sequence. Hum. Reprod. Update, *11*, 319–336.

76. Wang, P.J. (2004). X chromosomes, retrogenes and their role in male reproduction. Trends Endocrinol. Metab., *15*, 79–83.

77. Wu, Z., A., Murphy, C., Callan, H.G., Gall, J.G., (1991). Small nuclear ribonucleoproteins and heterogeneous nuclear ribonucleoproteins in the amphibian germinal vesicle: loops, spheres, and snurposomes. J., Cell Biol. *113*, 465–483.

78. Wyckoff, G.J., Wang, W., Wu, C.I. (2000). Rapid evolution of male reproduction genes in the descent of man. Nature, *403*, 304–309.

79. Xu, E.Y., Moore, F.L., Pera, R.A. (2001). A gene family required for human germ cell development evolved from an ancient meiotic gene conserved in metazoans. Proc. Natl. Acad. Sci. USA, *98*, 7414–7419.

80. Xu, J., Burgoyne, S., Arnold, A.P. (2002). Sex differences in sex chromosome gene expression in mouse brain. Hum. Mol. Genet., *11*, 1409–1419.

Hormonal Signalling via Nuclear Receptors

7

Jamshed R. Tata

National Institute for Medical Research, Mill Hill, London NW7 1AA, UK

Hormones and growth factors are primitive substances, which, during evolution, have been put to different uses. Our current concepts of hormonal signalling have arisen during the last 80 years from multiple disciplines, particularly physiology, chemistry and biochemistry (see Tata, 2005). For example, the observation by Gudernatsch in 1911 that feeding thyroid extracts to early tadpoles caused *precocious* induction of metamorphosis and was soon followed by the purification of L-thyroxine from thyroid gland, its chemical structure solved by Kendall in 1919, and confirmed by its total synthesis by Harington in 1926. Thyroxine is the precursor of the active form of the hormone, 3,3′,5-triiodothyronine (T_3) which was only identified in 1952. Thanks to the brilliant chemical studies from the laboratories of Butenandt, Reichstein, Dam and others, coupled with precise bioassays, there followed in late 1920s and 1930s the discovery and chemical identification of a number of steroid hormones, such as testosterone, oestradiol, progesterone and aldosterone. Soon, a large number of investigators were intrigued by their mechanism of action. The account below is largely restricted to steroid and thyroid hormones and nutrients, such as retinoids and fatty acids that act through nuclear receptors.

Receptors and Signalling Mechanisms

Since 1950s, the concept of a specific receptor mediating the action of a ligand has become the central dogma of every major physiological, pharmacological, endocrinological and neurobiological regulatory mechanism (Baulieu and Kelly, 1990). Three important technical developments have largely contributed to significant advances made in recent years in our current knowledge of the structure and function of receptors for extracellular signals: 1) The availability in the 1960s of radioactive ligands of sufficiently high specific activity to enable the distinction of receptor-ligand interactions of high affinity from non-specific binding to proteins and other cellular components; 2) The advent of gene cloning and sequencing, synthesis of recombinant proteins and high resolution NMR and X-ray crystallography has led to important insights into how receptor structure is linked to its function; 3) Advances in our knowledge of the role of oncogenes and transcription factors in cellular signalling mechanisms allowed us to classify different groups of receptors according to both their structure and cellular organization.

Hormone receptors have been commonly classified into two major groups: cell membrane and nuclear receptors. Receptors for protein hormones and growth factors, such as insulin, epidermal growth factor, growth hormone and prolactin, as well as many neurotransmitters, are all located in the target cell membrane; most of them are products of the oncogenes v-erbB, v-ros and v-mpl. Many membrane receptors are closely linked to adenylyl cyclase and G-proteins and through these, to cytoplasmic protein phosphokinases, which has led to a very precise defintion of extracellular signal transduction to intracellular regulatory machinery (Brivanlou and Darnell, 2002).

Hormonal Regulation of Specific Gene Expression

Early work sought to uncover a unique mode of action for all hormones and vitamins on the basis of complex physiological and biochemical changes, such as glycolysis, respiration, etc. produced by the addition of a given hormone or active principle to isolated tissues, cell homogenates and subcellular preparations (Tata, 1986). In 1956, Knox et al., demonstrated that glucocorticoids regulated hepatic enzyme tyrosine aminotransferase by selectively enhancing the synthesis of the enzyme. A few years later, the establishment of methods to study cell-free translation allowed a more precise analysis of how growth and developmental hormones, particularly steroids (oestrogen, androgen, ecdysteroids) and thyroid hormone (TH) influenced protein synthesis in their respective target cells.

A salient feature of all steroid and thyroid hormone effects on protein synthesis, when added to a cell-free amino acid incorporation system, is the failure to obtain a direct effect of the hormone on the protein synthesizing machinery *in vitro*. This single fact, and the availability of specific inhibitors of transcription (α amanitin, actinomycin D, etc.), soon shifted attention to transcriptional control by steroid and thyroid hormones.

The first indication that hormonal signals regulate transcription was however provided by gene puffing in polytenic chromosomes in larval salivary glands of insects, such as *Drosophila* and *Chironomus*, during their metamorphosis regulated by the moulting hormone ecdysone (Clever and Karlson, 1960; Ashburner et al., 1973). These puffs, which contain newly synthesised RNA from sequential activation of specific genes, can be precociously induced by incubating the target cells with ecdysteroids. Kinetics of labelling of nuclear RNA revealed that all steroid and thyroid hormones strongly influenced the formation and turnover of messenger RNA. In the mid-1960s several investigators were able to reproduce the effects of *in vivo* administration of the hormone in cell-free transcription systems using isolated nuclei and nuclear extracts from target tissues of steroids and TH (Tata, 1966).

Structure and Function of Nuclear Receptors

When Jensen and colleagues in 1961 tracked the sex hormone oestradiol-17β of high specific radioactivity to female accessory sexual tissues, such as uterus and vagina, it was found to accumulate in the nucleus as a complex with a protein which fulfilled the criteria for a receptor (Jensen, 2004). The translocation of the liganded oestradiol receptor (ER) from the cytoplasm to the nucleus is accompanied by a major shift in its size, which arises from the reversible formation of a complex with the heat-shock protein hsp90. Following the work from the laboratories of Chambon, Evans and Vennstrom in the 1980s on the cloning of ER, TR and GR, more than 35 nuclear receptors have been cloned, sequenced and obtained as pure recombinant proteins, including several termed as 'orphan' receptors, i.e. whose ligands remain unknown (see Mangelsdorf et al., 1995; Benoit et al., 2004; Chambon, 2004). All nuclear receptors, which are products of the oncogene v-*erb*A, function as ligand-activated zinc-finger transcription factors with a modular structure comprising six domains. As depicted in Fig. 7.1, the central DNA-binding domain is the most conserved, while the 3′-terminal ligand- binding domain is the most variable (not shown in Fig. 7.1). Nuclear hormone receptors can be sub-divided into two groups, according to whether they form cytoplasmic complexes with hsp90 and are active as monomer and homodimer or as heterodimers. All vertebrate steroid hormone receptors, belong to the first category, while the liganded receptors for retinoic acid, TR, vitamin D_3 (VDR) and peroxisome prolifetor (PPAR) function as heterodimers with RXR, the 9-*cis*-retinoic acid receptor (RXR does not have to be liganded to function as heterodimer). At first, only the second group of receptors were found to exist as multiple isoforms, the multiplicity

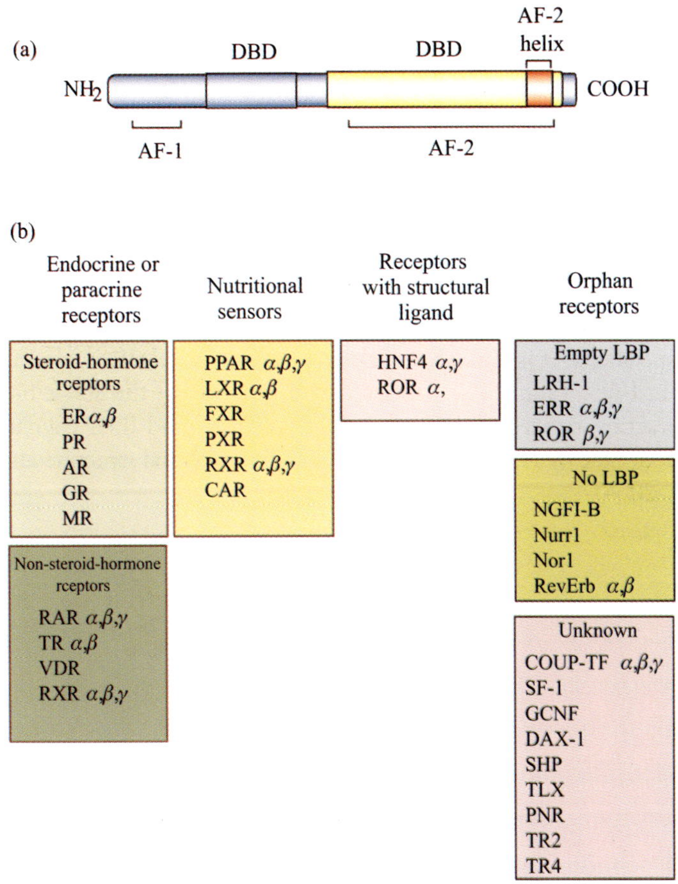

Fig. 7.1 The Nuclear Receptor Gene Superfamily. (**a**) Representation of the common structural features of all members of the supergene family as ligand-activated transcription factors. DBD, DNA-binding domain; AF, transcription activating factors. The ligand-binding domain is located at the COOH terminus.(**b**)Different categories of nuclear receptors divided by Benoit et al. (2004) into four sub-groups according to the nature of the signalling molecules. Only the two groups on the left in green boxes refer to hormonal signals and only these will be considered in this article. ER, PR, AR, GR and MR in the green box are receptors for oestrogen, progesterone, androgen, glucocoticoids and mineralocorticoids, respectively. RAR, TR, VDR and RXR in the dark green box are receptors for retinoic acid, thyroid hormone, vitamin D3 and 9-*cis* retinoic acid, respectively. For all other receptors and abbreviations refer to Benoit et al. (2004). The writer thanks these authors for use of this figure.

residing in the N-terminus of the receptor, but more recent work has shown that the same holds true for other steroid receptors.

An interesting question arises as to how the high degree of target gene specificity for a given hormone and its receptor is achieved. The answer lies in the highly precise spacing of nucleotide repeats in the hormone response element (HRE) of the promoter of the target gene and the DNA-binding domain (DBD) of the receptor which recognises it. A consensus hexanucleotide sequence, usually present as a pair, is the most common feature of all the known HREs but the sequence of each hexad, and the relative position of the two hexads, exhibit considerable variability to generate the high degree of specificity of interaction between the receptor and its target gene. The HREs recognised by the receptors of steroids (ER, PR, GR, AR, MR) in the top green box in Fig. 7.1 are characterised by a 15-nucleotide motif consisting of two hexads in a palindromic configuration separated by 3 nucleotides. The hexad sequences show a variability within this group of receptors. On the other hand, and more interestingly, the HREs of the non-steroid receptors shown in the bottom green box in Fig. 7.1 (TR, RAR, RXR, VDR, PPAR) all share the same AGGTCA hexad of ERE but are organised as direct repeats (DRs) separated by one to five nucleotides (see Fig. 7.2). This arrangement of HREs explains the fine discrimination of target genes by the heterodimers formed by each of these receptors with RXR and which confers an extraordinary hormonal specificity. It is a most remarkable example of selective transcriptional regulation (Mangelsdorf et al., 1995; Tata, 1998). Dramatic confirmation of the above biochemical findings regarding the interaction between DBD of nuclear receptors and their cognate HREs has been provided in the last decade by NMR spectroscopy and X-ray crystal structure analysis (Chakravarti et al., 1996; Evans, 2004).

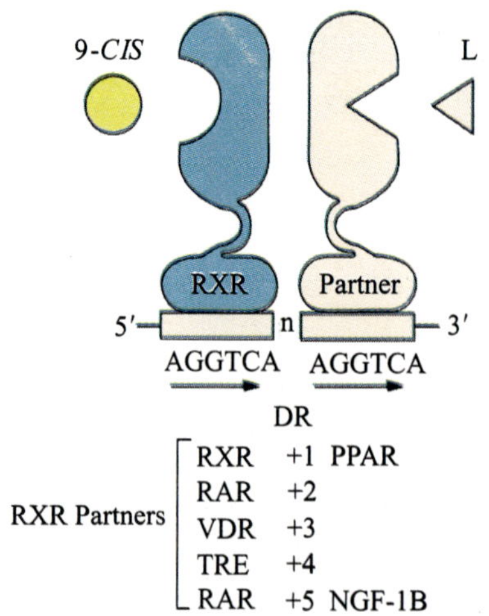

Fig. 7.2 Some examples of non-steroidal nuclear receptors that function as heterodimers with RXR to recognize the direct repeat (DR) of the hexad AGGTCA separated by 1-5 nucleotides (n) in the respective hormone response element recognized by the relevant DNA-binding domain of the receptor. RAR, retinoic acid receptor; VDR, vitamin D receptor; TR, thyroid hormone receptor. 9-*cis*, 9-*cis* retinoic acid; L, ligand; RXR, 9-*cis* retinoic acid receptor. See Mangelsdorf et al. (1995) for a detailed account of RXR heterodimers.

Nuclear Receptors, Transcriptional Control and Chromatin Structure

An ever-increasing number of non-receptor proteins are now known to interact with nuclear receptors and transcription factors, these complexes may function synergistically or in a mutually antagonistic manner. Based on such techniques as two-hybrid analysis, recent studies on transcription factors have focused on networking and integration of extracellular and intracellular signalling in order to explain how the activity of transcription complexes itself could be regulated. As shown in the over-simplified model in Fig. 7.3, large protein molecules termed CBP (CREB binding protein) and p300 are thought to form bridges between nuclear hormone receptors and other transcription factors. This conclusion was arrived at by unexpected observation that nuclear receptors inhibit the activity of the non-receptor transcription factor AP-1 by competing for limited amounts of CBP/p300 normally present in cells. Other important elements of the complex are the p160 nuclear receptor co-activator and the 270 kDa nuclear receptor co-repressor (N-CoR), which have been purified and their functions tested. Evans and collaborators (Chakravarti et al., 1996) undertook a structural and thermodynamic analysis of the interaction domains of CBP and the p160 coactivators

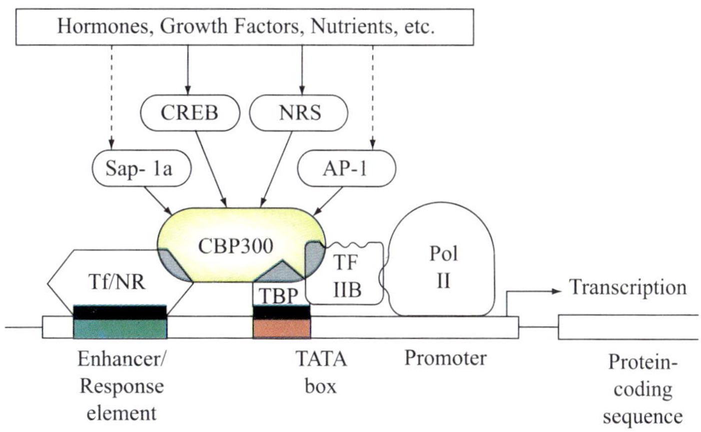

Fig. 7.3 A simplified representation of how a nuclear receptor forms complexes with elements of the transcriptional machinery. A bridging protein such as CBP/p300 is thought to be in close contact with nuclear receptors (NRS) and transcription factors (TF) that recognize specific DNA sequences, TATA box binding protein (TBP) and transcription factor IIB (TFIIB), which would form a complex with RNA polymerase II. CBP/p300 is thought to form complexes with other transcription factors without involving DNA, such as CREB, AP-1 and Sap-1a. The activities of many of these components are modified by phosphorylation (see Evans [2004] for more details).

of TR and RAR to elucidate the assembly of the multiprotein activation complex. They describe a mechanism of mutual synergistic folding whereby the co-activators recruit CBP/p300 to facilitate the transmission of the hormonal signal to the transcriptional machinery. Thus, the CBP/p300 complex serves to integrate multiple signalling pathways in the cell nucleus.

The higher order of organization of genes within the nucleus and the growing number of cross-interactions among regulatory factors have focused attention on the important role of chromatin structure in hormonal regulation of gene expression. Studies from Chambon's laboratory in the mid-1980s on the DNase I-hypersensitive regions of the chicken ovalbumin gene promoter led to the first understanding of how oestrogen reversibly modifies the chromatin structure of this target gene in a tissue-specific manner (see Chambon, 2004). Later studies provided evidence that the function of glucocorticoid receptor in regulating the target gene promoter is determined by the manner in which it itself is organized within the chromatin structure (Beato, 1996). According to a simple model, the HREs are highly organised in phased nucleosomes and the binding of the hormone to its nuclear receptor causes an alteration in the chromatin structure such that it will induce the binding of non-receptor transcription factors, such as NGF-1 and OTF-1, and thus allow the transcription of the hormone-regulated gene promoter. Similarly, as depicted in Fig. 7.4, Wolffe's group has suggested that both the silencing and activation of the Xenopus TRβ gene is determined by processes controlling nucleosome assembly (see Wolffe, 2000). The conclusions of such chromatin studies have largely been inferred indirectly from techniques such as cross-linking to determine points of contact between DNA and protein. More direct information of the spatial organization

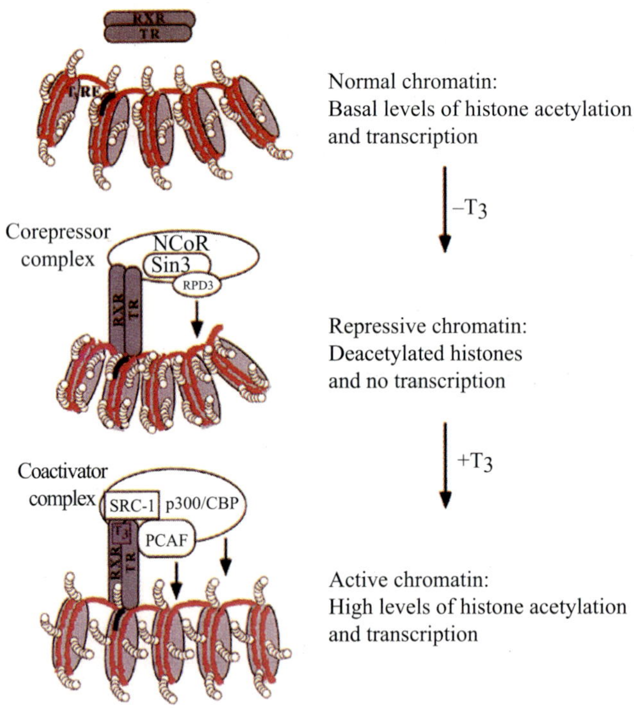

Fig. 7.4 Wolffe's view of how a ligand-activated nuclear receptor could modify the higher-order structure of chromatin. The packaging of DNA into chromatin is visualized in three transcriptionally active states: normal, repressive and active. In this example, the region of chromatin chosen contains the thyroid hormone receptor (TR)/RXR heterodimer, with or without its ligand triiodothyronine (T_3), bound to the thyroid responsive element (TRE) in the target gene. In normal chromatin, histone acetylation is at its basal level and so is the transcriptional activity. In the absence of T_3 (as during early stages of development), chromatin exists in its condensed and transcriptionally repressive form whereby the histones are in a largely deacetylated state with no transcripion of the TR's target gene. In the presence of T_3 the chromatin is now active with elevated levels of histone acetylation and transcription. The other components are proteins that form 'corepressor' and 'coactivator' complexes with complexes with the TR/RXR receptor heterodimer. For more details see Wolffe (2000).

and mobility of receptors is needed before we can draw precise conclusions as to the role played by chromatin rearrangements within the nucleus *in vivo*.

Nevertheless, it is important to emphasize that the functions of many transcription factors, nuclear receptors and chromatin proteins involved in such integration are themselves regulated by post-translational modifications, such as protein phosphorylation (Jackson, 1992) and histone acetylation (Wolffe, 2000), thus setting up a network linking membrane and nuclear receptor signal transduction pathways. In other words, the functions of many transcription factors and nuclear receptors involved in such integration are themselves regulated by protein phosphorylation and other post-translational modifications via mechanisms located in the cell membrane. As illustrated in Fig. 7.5, which is based on recent work from Darnell's laboratory (Brivanlou and Darnell, 2002),

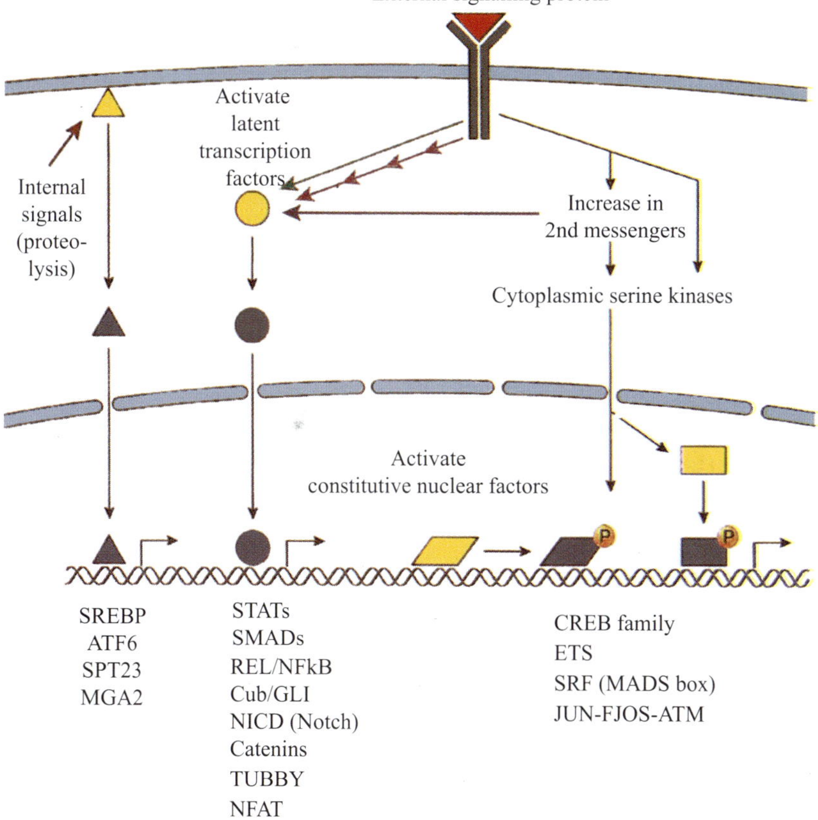

Fig. 7.5 Convergence of signals operating through mechanisms at the level of cell membrane and nucleus. According to this model, proposed by Brivanlou and Darnell (2002), latent intracellular regulatory factors, which include nuclear receptors, transcription factors and other modulators of nuclear functions, are activated or inhibited by post-translational modifications such as phosphorylation, proteolysis or through 'second messenger'-dependent functions. The latter would be regulated by signals impinging on the cell membrane. Thus, for example, the activities of nuclear proteins, receptors and transcription factors can be modified by phosphorylation by a convergence of functions operating at the levels of cell membrane and nucleus. For definitions of the abbreviations and other details see Brivanlou and Darnell (2002).

these complexes serve to bring together nuclear and extra-nuclear regulatory processes and thus allow the setting up a network linking membrane and nuclear receptor signal transduction pathways. The importance of the JAK/STAT pathway and the control of gene expression in general by transcription factors is a good example of this networking.

Expression of Nuclear Receptors

More than 30 years ago, Ashburner, while studying chromosomal puffing in *Drosophila* during ecdysone-induced metamorphosis, came to the conclusion that an early action of ecdysone is to upregulate the expression of its own receptor (EcR), to be followed by its downregulation (see Ashburner et al., 1973). What is remarkable is that the model was derived entirely from results of incubating larval salivary glands at different stages of hormone-induced puffs with inhibitors of protein and RNA synthesis, well before nuclear receptors were discovered. The validity of Ashburner model has been confirmed later by work from other laboratories (see, for example, Thummel, 1995) who were able to show quite clearly the burst of EcR expression at the onset of larval metamorphosis which anticipated that of other regulatory gene products, the early auto-induction of EcR being essential for the later expression of other hormone-induced gene products.

It is perhaps useful to consider other examples of receptor auto-induction, since these seem to be important during hormone-dependent development (Tata, 1998; see Table 7.1). Among the well-documented examples is the activation of receptor genes preceding the induction of target genes is that of ER and PR prior to the induction of egg proteins (e.g. ovalbumin and vitellogenin) upon administration of oestrogen to male and female oviparous animals. For example, the lag period between the administration of oestrogen to chickens and the detection of egg protein mRNA in immature oviducts was shown to be due to a requirement for additional ER and PR. An even more dramatic auto-induction of nuclear receptor has been recorded in virtually every tissue of the early Xenopus tadpole induced to metamorphose precociously with exogenous T_3. It is now clear that the receptor interacts with the promoter of its own gene in intact tadpole cells as well in tissue culture cells. This conclusion is validated by the findings that the promoters of both isoforms of TR comprise more than one functional DR+4-type TRE which strongly binds the TR-RXR heterodimer, and that dominant negative mutants of TRβ inhibit the auto-induction of TRβ (Ulisse et al., 1996; Tata, 2002).

Table 7.1 Examples of Auto-induction of Nuclear Receptors of Hormones During Development

Hormone and Receptor	Species and tissue	Developmental action
Thyroid hormone	All amphibian tissues Rodent astrocytes	Metamorphosis Brain development
Oestrogen	Avian and amphibian liver	Vitellogenesis
Ecdysone	All insect cells	Metamorphosis
Retinoic acid	Rodent and avian brain, eyes and skin	Development
Androgen	Mammals; male accessory sexual tissues	Sexual maturation
Vitamin D	All vertebrates	Bone and limb development

What can be the significance of auto-induction of nuclear receptors? There is now good evidence to suggest that receptor up-regulation is linked to the biological activity of the ligand, at least where it concerns developmental action. One possible explanation is that up-regulation of a given receptor is a pre-requisite for the sequential activation of its target genes that specify its biological action. It predicts a double threshold of receptor numbers, the first for the auto-induction of the receptor and the second for the activation of target genes. It also implies that the gene encoding a given receptor is constitutively expressed to produce a very low level of functional receptor in the target tissue at very early stages of development. Upon the onset of secretion of the hormonal ligand, the low concentrations of the liganded receptor would only suffice to activate its own receptor gene, but not of its target genes which would require a significantly higher concentration of the hormone and receptor. Measurement of the relative affinities of interaction between the DNA-binding domain of the receptor and the response element in the promoters of the receptor and target genes would be a first step to validating or discarding this model. An intriguing question, as yet unanswered, then arises as to what are the factors that maintain the low levels of the constitutively expressed receptor prior to the onset of the developmental process which is accompanied by the release of the hormone.

Perspectives

Receptors occupy a central position in our current thinking of signalling mechanisms, via membrane or nuclear receptors. Both the membrane and nuclear receptors are derivatives of oncogenes, which points to the special position of oncogenes in the evolution of cellular signalling processes. The applications of techniques of gene cloning, cell transfection, X-ray crystallography and NMR, analysis of DNA-protein and protein-protein interactions, all have greatly advanced our understanding of nuclear receptors as ligand-activated transcription factors. The discovery of factors such as co-activators, co-repressors and integrators have further defined how nuclear receptors regulate gene expression; it is predictable that many more such modulators will be discovered in the future and that these will be found to form even more complex structures with nuclear receptors. Although for several years many investigators had accepted the importance of considering the role of nuclear receptors in the context of their organisation into the higher order of chromatin structure, it is only in the last decade that the true significance of enzymes that add or remove acetyl and methyl groups in histones and other chromosomal proteins has been realized in the context of a network of cellular signals impinging on nuclear and extra-nuclear sites. As yet we have only looked at a fraction of chromatin-associated proteins, which means that much has yet to be learned about how hormones regulate specific gene expression. The recent findings on histone modifications emphasize the nucleosome and chromatin as dynamic entities that make up the transcriptional machinery and which, in turn, will surely lead to a major shift in our thinking on how steroid and other hormones control gene expression.

Our views on mechanisms of signalling mechanisms over the last several decades have essentially followed the progression of knowledge that is being built up about regulatory biological processes. Thus, starting with whole cellular physiological processes, such as respiration and glycolysis, these have moved on to direct interaction with enzymes and then to subcellular biochemical functions, such as oxidative phosphorylation, protein synthesis and transcription, to be followed by cell biology, structural biology and genomics at this time. This evolution in our search for the mechanisms involved is clearly a function of the availability of new technologies — from subcellular fractionation to availability of specific inhibitors of complex processes, from monoclonal antibodies to defining cell ultrastructures and from nucleic acid hybridization to three-dimensional visualization of interactions between macromolecules and the organization of genes. It is therefore not too difficult to predict that future developments in the science of signalling mechanisms will depend on the newly emerging technologies of proteomics and structural biology, while receptor pathology, RNA interference and gene therapy will expand the frontiers of clinical practice and commercial exploitation.

References

1. Ashburner, M., Chihara, C. Meltzer, P. and Richards, G. (1973). Temporal control of puffing activity in polytene chromosomes. Cold Spring Harb. Symp. Quant. Biol. *38*, 655–662.
2. Baulieu, E.-E., and Kelly, P.A. eds. (1990). Hormones. From Molecules to Disease. Hermann, Paris.
3. Beato, M. (1996). Chromatin structure and the regulation of gene expression: remodelling at the MMTV promoter. J Mol Med *74*: 711-724.
4. Benoit, G., Malewicz, M. and Perlmann, T. (2004). Digging deep into the pockets of orphan nuclear receptors: insights from structural studies. Trends Cell Biol. *14*, 369–376.
5. Brivanlou, A.H., and Darnell, J.E. Jr. (2002) Signal transduction and the control of gene expression. Science *295*, 813–818.
6. Chakravarti, D., LaMorte, V. J., Nelson, M. C., Nakajima, T., Schulman, I. G., Juguilon, H., Montminy, M. and Evans, R. M. (1996). Role of CBP/P300 in nuclear receptor signalling. Nature, *383*, 99–103.
7. Chambon, P. (2004). How I became one of the fathers of a superfamily. Nature Medicine *10*, 1027–1031.
8. Clever, U. and Karlson, P. (1960). Induktion von Puff-Veranderungen in den Speicheldrusen-chromosomen von Chironomus tentans durch Ecdyson. Exp. Cell Res. *20*, 623–626.
9. Evans, R. (2004). A transcriptional basis for physiology. Nature Medicine *10*, 1022–1026.
10. Jackson, S.P. (1992). Regulating transcription factor activity by phosphorylation. Trends Cell. Biol. *2*, 104–108.
11. Jensen, E. V. (2004). From chemical warfare to breast cancer mangement. Nature Medicine *10*, 1018–1021.
12. Knox, W.E., Auerbach, V.H., and Lin, E.C.C. (1956). Enzymatic and metabolic adaptations in animals. Physiol Rev. *36*, 164–254.
13. Mangelsdorf, D.J., Thummel, C., Beato, M., Herrlich, P., Schulze, G., Umesono, K., Blumberg, B., Kastner, P., Mark, M. and Chambon, P. (1995). The nuclear receptor superfamily: The second decade. Cell *83*, 835–839.
14. Tata, J.R. (1966). Hormones and the synthesis and utilization of ribonucleic acids. Prog. Nuc. Acid Res. Mol. Biol . *5*, 191–250.

15. Tata, J.R. (1986). The search for the mechanism of hormone action. Pers. Biol. Med. *29*, 184–204.

16. Tata, J.R. (1998). Hormonal signalling and postembryonic development. Springer, Heidelberg, Germany.

17. Tata, J.R. (2002). Signalling through nuclear receptors. Nature Rev. Mol. Cell. Biol. *3*, 702–710.

18. Tata, J.R. (2005). One hundred years of hormones. EMBO Rep. 490–496, 2005.

19. Thummel, C.S. (1995). From embryogenesis to metamorphosis: the regulation and function of Drosophila nuclear receptor superfamily members. Cell. *38*, 871–877.

20. Ulisse, S., Esslemont, G., Baker, B.S., Chatterjee, V.K.K. and Tata, J.R. (1996). Dominant-negative mutant thyroid hormone receptors prevent transcription from Xenopus thyroid hormone receptor beta gene promoter in response to thyroid hormone in Xenopus tadpoles *in vivo*. Proc. Natl .Acad. Sci. U.S.A. *93*, 1205–1209.

21. Wolffe, A. P. (2000). Chromatin: structure and function. Academic Press, San Diego, CA, U.S.A.

Genomics of Human Sarcomas

8

Subbaya Subramanian[1], Robert B. West[2] and Matt van de Rijn[2]

[1]Department of Laboratory Medicine and Pathology, University of Minnesota, Minneapolis, MN-55455, USA
[2]Department of Pathology, Stanford University, Stanford, CA-94305, USA

Introduction

Sarcomas (soft tissue sarcoma) are highly heterogeneous group of mesenchymal tumors that account for less than 1% of the overall malignant tumor types in humans. Soft tissue sarcomas are locally aggressive and invasive. They have significant risk of distant metastasis depending on the tumor type and grade. Sarcomas can occur anywhere in the body but majority of them occurs in the extremities. Many types of sarcomas are named after the type of cell, or tissue involved, as in angiosarcoma, chondrosarcoma, fibrosarcoma, liposarcoma and osteosarcoma. There are over 50 histological subtypes of sarcomas which pose a major diagnostic challenge (Weiss S, 2001). Many a times the histological features of sarcomas are not sufficient enough for the correct diagnosis of the tumor type. Incidentally, several soft tissue tumors are characterized by chromosomal translocations resulting in tumor specific fusion transcripts (reviewed in Antonescu, 2006). Cytogenetic analyses of specific chromosomal translocations are widely used as genetic markers for several sarcomas. In addition, DNA copy number changes such as gene amplification or deletion are also currently being used in the diagnosis of several sarcomas (Ferreira et al., 2007; Linn et al., 2003). A greater understanding of the biology of sarcomas is needed in order to increase the potential for identifying new therapeutic targets and treatment strategies. Integrative genomic approaches such as gene expression profiling, array comparative genomic hybridization, tissue microarrays and microRNA expression analysis are very attractive tools to study and understand the complexity of human sarcomas and cancer, in general.

In the recent years, gene signature for several of these sarcoma types have been determined- (Linn et al., 2003; Subramanian et al., 2004; Subramanian et al., 2005; van de Rijn and Fletcher, 2006). Microarray technology permits genome-wide expression analysis of thousands of genes simultaneously and has proven to be useful for further molecular characterization of tumors. These genome-wide gene expression signatures have significantly enhanced our understanding of the biology and classification of various sarcoma types. Gene array studies quickly established that tumor classification correlated well with gene expression patterns (Tschoep et al., 2007). Not surprisingly, tumor types such as gastrointestinal stromal tumor (GIST) and desmoid-type fibromatosis that cluster tightly together based on gene expression patterns also tend to be easily identified and classified by histological and/or immunohistochemical techniques. However, tumors like malignant fibrous histiocytomas have complex gene expression patterns and are difficult to classify by traditional techniques (Nielsen et al., 2002).

Sarcomas are characterized by genetic alterations such as chromosomal translocations, DNA copy number changes, mutations and aberrant gene expression profiles. Identification of mutations and translocations associated with these tumors has revealed aberrant signaling pathways that cause these diseases. Recurrent chromosomal translocations are characteristic features of several sarcoma types. In most sarcomas, translocations are balanced i.e., there are reciprocal exchange between two chromosomes with no net gain or loss of chromosomal material. The fusion gene products that are generated by these chromosomal translocations act as oncoproteins. Many of these fusion transcripts function as transcription factors that are generated by fusion of the DNA binding domain derived from one gene and transcriptional regulatory domain(s) derived from the other gene. These novel transcription factors inappropriately regulate the expression of their target genes and may interact with other transcription factors to regulate expression of novel target genes. The resulting changes in gene expression modify key signalling pathways involved in major cellular functions such as cell division and differentiation and thereby contribute to tumor initiation and/or tumor progression (Castillero-Trejo et al., 2005; Davis and Meltzer, 2007; Mercado and Barr, 2007). In this chapter we will discuss our current understanding about the genomic aspects of sarcomas which we have been studying extensively for the past several years.

Genomic Tools to Study Sarcomas

Gene Microarrays

DNA microarrays are widely used to measure gene expression across thousands of genes in parallel. Current technologies use a high density array spotter that prints the complete transcriptome of the human genome which allows for global gene expression profiling (Hoheisel, 2006). A major goal of many DNA microarray studies has been the classification of cancer. Recently, considerable efforts have been made to utilize gene array technology to improve our understanding of cancer and to identify novel diagnostic/prognostic markers and therapeutic targets (Sotiriou and Piccart, 2007). Experimental variations such as method of RNA isolation, array platform and type, etc., may influence the microarray data analysis. Gene expression profiling performed on different array platforms or laboratories were difficult to compare, but recent advances in statistical methods have demonstrated that a remarkable similarity exists between data obtained from different laboratories or array platforms (Barnes et al., 2005; Kuo et al., 2006; Yauk et al., 2004). Microarray data analysis can be divided into supervised and unsupervised methods. Supervised analyses make upfront use of sample identity, whereas unsupervised clustering is based on overall similarity in expression patterns without prior knowledge of sample identity (Tusher et al., 2001). Another major variable is the selection of number of samples/cases used for the analysis. Gene expression profiling usually looks at variation in levels of mRNA for many genes between different tumors. The conclusions are generally dependent on the type and number of sarcomas included for the analysis. The minimum information about a microarray experiment (MIAME) outlines the minimum information required to publish and public availability of the microarray data (Brazma et al., 2001).

Based on gene expression analysis a number of diagnostic markers for various sarcomas have been discovered. These include *NMB* for extraskeletal myxoid chondrosarcoma (EMC) (Subramanian et al., 2005), *DOG1* in the case of gastrointestinal stromal tumors (GISTs)(West et al., 2004) and *TLE1* in synovial sarcomas (SS)(Terry et al., 2007). Gene expression profiles can also identify potential therapeutic gene targets that are expressed particularly at high levels. These gene products can be targeted using small molecule inhibitors for treatment purpose (Khandurina and Guttman, 2002). For example, receptor tyrosine kinases such as EGFR in malignant peripheral nerve sheath tumors (MPNSTs) and synovial sarcoma (Nielsen et al., 2002); *CSF1R* in pigmented villonodular synovitis (PVNS) and tenosynovial giant cell tumor (TSGCT) (West et al., 2006) etc., can be used as novel therapeutic targets. The identification and characterization of the above mentioned diagnostic markers and therapeutic targets are detailed below in their respective sections. One of the high-throughput approaches to validate the gene expression data is tissue microarrays analysis.

Tissue Microarrays (TMAs)

Tissue microarrays are useful in evaluating protein and/or RNA expression levels in large series of tumors and in complementing gene array findings. TMA analysis generates hundreds of tumor expression data points for a single gene on a single paraffin block (Marinelli et al., 2007; West and van de Rijn, 2006). Any promising new marker that is identified through gene expression profiling can now be quickly validated across a wide variety of sarcomas using TMAs. These data are multidimensional: both relative expression levels and subcellular/tissue localization can also be determined using TMAs. A wide variety of samples can be incorporated into a TMA block, including traditional cored samples, cell pellets and embedded tissues. TMAs are widely used to assess the quantity of both RNA and protein. The RNA abundance in a tumor sample is measured by *in situ* hybridization (ISH) techniques using riboprobe generated against specific gene of interest (Subramanian et al., 2004; West et al., 2004). Likewise the abundance of the protein is measured by immunohistochemical (IHC) techniques using antibodies against the protein. TMA scoring uses the following criteria: a score of -2 is given for negative staining, defined as lesser than 5% of tumor cells showing staining at or minimally above background. A score of 1 (weak positive staining) is given for light brown staining in greater than 5% of tumor cells. A score of 2 (strong positive staining) is given for dark brown staining in greater than 50% of tumor cells and no score is given for staining in non-tumor cells and cells of unknown origin. Though tissue microarrays can be used in conventional DNA copy number identification studies, a high density tilling array comparative genomic hybridization approaches are used currently for mapping the chromosomal amplifications and deletions with precision in the cancer genome.

Array Comparative Genomics (aCGH)

In array comparative genomic hybridization (aCGH), analyses the copy number for thousands of genes is determined in a manner similar to that used to look at mRNA levels (Pollack et al., 1999). Both cDNA and oligonucleotide arrays can be used for aCGH studies (Pollack and Iyer, 2002). In array-based comparative genomic hybridization, tumor and normal genomic DNA are differentially labelled and compared by hybridization to microarrays comprising DNA probes of defined human genome map position. The resultant tumor/normal ratios can be mapped onto the genome sequence to reveal in tumor genomic DNA amplifications and deletions in tumor, which drive the altered expression of cancer genes (Pollack et al., 1999). Patterns of DNA in tumor copy number alteration, analogous to gene expression signatures, which has been proposed for cancer classification and outcome predictions.

MicroRNA Expression Signature

Recent advances in the understanding of post-transcriptional regulation of genes by microRNA have increased our understanding of the biology of several sarcomas (Subramanian et al., 2007). Gene regulation by miRNAs is an important cellular regulatory mechanism; therefore, it is plausible that abnormalities in microRNA dependent post-transcriptional regulation of genes may play important roles in the development and progression of sarcoma and cancer, in general. miRNAs are evolutionarily conserved non-coding RNAs of approximately 22 nucleotide length that can inhibit protein synthesis from a wide variety of mRNAs through either by inhibition of translation by imperfect base-pairing with their target mRNA at the 3′UTR or by inducing degradation of target mRNA through perfect complementary sequence base-paring (Bartel, 2004). In humans, about 500 miRNAs have been discovered so far, many of which are located in genomic loci that are fragile sites on chromosomes (Calin et al., 2004). DNA copy number abnormalities in these fragile sites can cause a direct impact on the miRNA expression levels but levels of miRNA can also vary in the absence of miRNA copy number changes. miRNAs can influence a wide variety of biological processes, including development, proliferation, and differentiation (Chapman and Carrington, 2007). Recent studies have shown that miRNAs can act as either tumor suppressors or oncogenes (oncomiRs) and also can reflect the developmental lineage and differentiation stage of the tumors (Calin and Croce, 2006; Esquela-Kerscher and Slack, 2006). Expression profiles of miRNAs have been studied in several carcinomas and in hematological tumors (Calin et al., 2005; Lu et al., 2005). In contrast to carcinomas, that are mixtures of epithelial cells and varying amounts of connective tissue cells, sarcomas have little supporting stroma and can be seen as relatively pure populations of transformed mesenchymal cells.

Novel Diagnostic Markers in Sarcomas

Gene microarray analysis of soft tissue tumors allows for a genome-wide search for novel diagnostic markers that classify tumors in different diagnostic groups and distinguish subsets of tumors within a particular sarcoma type.

DOG1 in GIST

Gastrointestinal stromal tumor (GIST) is the most common mesenchymal tumor arising in the gastrointestinal tract and have been proposed to arise from the interstitial cells of Cajal (Sircar et al., 1999). Most GISTs contain an activating mutation in either *KIT* or *PDGFRA* resulting in constitutive, ligand-independent activation of these receptor tyrosine kinases (RTK) (Heinrich et al., 2003b; Hirota et al., 1998; Kinoshita et al., 2003). In the vast majority of GISTs, high levels of KIT expression are accompanied by a *KIT* gene mutation in any of the exons 9, 11, 13, or 17 (Hirota et al., 1998). Imatinib mesylate (Gleevec), a small molecule inhibitor against KIT and PDGFRA, is currently an effective treatment strategy for metastatic or unresectable GIST (Judson and Demetri, 2007). In general, a diffuse or strong KIT immunoreactivity is widely used in the diagnosis of GIST, but a subset of GIST cases is immunonegative while many other types of sarcomas may also be immunoreactive for this marker.

A novel gene *TMEM16A* aka *DOG1* identified by a gene microarray analysis is highly expressed in both *KIT*- and *PDGFRA*-mutant GISTs. DOG1 is localized on chromosome 11 (11q13) and encodes for a protein of unknown function (Katoh and Katoh, 2003). It contains 26 exons and encodes for a 960 amino acid protein. The protein has 8 transmembrane domains. Monoclonal antibody raised against DOG1 yielded positive staining for over 90% of the GIST cases irrespective of their mutational status showing that DOG1 can be used as a novel diagnostic marker for GIST (West et al., 2004).

NMB in Extraskeletal Myxoid Chondrosarcoma

Extraskeletal myxoid chondrosarcoma (EMC) is a malignant soft tissue tumor that occurs primarily in the extremities. EMC is most frequently diagnosed in middle-aged and elderly men, but has also been reported in children (Weiss S, 2001). EMC is characterized by a balanced translocation. The most common breakpoint involves the *EWS* gene on chromosome 22q12 and the *TEC* gene on chromosome 9q22 (Labelle et al., 1999). The fusion protein is a potential transcriptional activator (Ohkura et al., 2002). In most cases, EMC is diagnosed based on histological examination. Gene microarray experiments identified significant number of genes to be differentially expressed in EMC (Subramanian et al., 2005). One of these is *Neuromedin B* (*NMB*), a mammalian homologue of amphibian bombesin and a secreted neuropeptide involved in stimulation of smooth muscle contraction (Minamino et al., 1983). NMB is a potent mitogen and growth factor for normal and neoplastic lung and gastrointestinal epithelial tissue. *NMB* was among the first neuropeptides to be implicated as an autocrine growth factor in lung cancer cells (Matusiak et al., 2005). *In situ* hybridization using *NMB* on TMAs validated the predominant expression of *NMB* in 87% of EMC cases (Subramanian et al., 2005). Both the gene expression data and tissue array data indicate that NMB is a useful diagnostic marker for EMC. In addition, NMB being a secreted protein, can be clinically useful. If it is found in higher levels in the serum of patients, it may serve as a useful serum marker in the diagnosis of recurrence of EMC. This could be of clinical relevance as EMC can recur, sometimes after very long periods of time.

TLE1 in Synovial Sarcoma

Synovial sarcoma is a soft tissue malignancy typically occurring in the limbs of young adults (Soule, 1986). Correct diagnosis of synovial sarcoma on the basis of histology alone can be challenging. Synovial sarcoma is characterized by a t(X;18) chromosomal translocation. This translocation most commonly fuses either the *SSX1* or *SSX2* gene on chromosome X to the *SYT* gene on chromosome 18, resulting in the production of an SYT-SSX fusion protein (Nilsson et al., 1999). Confirming the presence of t(X;18) by cytogenetics, fluorescence *in situ* hybridization (FISH) or reverse-transcriptase polymerase chain reaction is considered the gold standard for the diagnosis of synovial sarcoma (Terry et al., 2005).

The consistent identification of strong *TLE1* expression in synovial sarcoma, from several gene expression profiling studies (Allander et al., 2002) in different laboratories using different DNA microarray platforms led to investigate its value as a diagnostic immunohistochemical marker. TLE1 is one of the best discriminators for synovial sarcoma when compared with histologically and/or biologically similar sarcomas such as MPNST. Immunohistochemical detection of TLE1 is not only highly sensitive for synovial sarcoma, but is also specific in the context of other mesenchymal neoplasms (Terry et al., 2007).

Distinct Gene Expression Profiles of GIST with KIT and PDGFRA Mutations

The treatment of sarcomas will be more effective with an individualized treatment strategy for each patient. In sarcomas such as GIST, the *KIT* or *PDGFRA* gene is mutated. It becomes imperative to identify tumor subtypes and the signalling pathways that may be affected by these different gene mutations. Most GISTs (67%) are characterized by mutations in exon 11 of *KIT* gene, which causes ligand-independent activation of its tyrosine kinase function (Hirota et al., 1998). Mutations in *KIT* exon 9 are present in about 11% of GISTs (Lux et al., 2000). More rarely, mutations are found in exon 13 or 17 (Kinoshita et al., 2003) . Although most GISTs have activating *KIT* mutations, a subset is *KIT* wild type (Hirota et al., 1998). Approximately 35% of these GISTs have intragenic activating mutations in the related receptor tyrosine kinase, platelet-derived growth factor receptor alpha (*PDGFRA*) (Heinrich et al., 2003b). The biological significance of different mutations in GIST is evident from its clinical relevance. While *KIT* exon 11 mutants were highly sensitive to the drug, imatinib mesylate, GISTs with *KIT* exon 9 mutations—and those without demonstrable *KIT* and *PDGFRA* mutations – often failed to respond to this drug. *KIT* and *PDGFRA* mutations are mutually exclusive oncogenic mechanisms in GISTs, and the biological outcome—of these alternate mechanisms appear to be similar (Bauer et al., 2007; Subramanian et al., 2004).

Distinct gene expression profiles can distinguish most GISTs with mutations in different kinases and also among those with different types of mutations in the *KIT* gene (Subramanian et al., 2004). Studies have shown that both *KIT* and *PDGFRA* mutant GISTs feature functional activation of the *PI3K/AKT* and *MAPK* pathways (Porter and Vaillancourt, 1998). A considerable overlap of expressed genes with different kinase mutation types reflects the close biological relationship between GISTs, irrespective of their kinase oncogenic mechanisms.

Tumor Landscaping and Novel Drug Targets

Kinase gene family with over 500 genes is one of the large gene families in humans. Receptor tyrosine kinases (RTKs) are involved in regulating diverse cellular processes such as growth, cell cycle, cell migration, differentiation, and survival (Manning et al., 2002). RTK genes or their ligands are frequently altered by translocation or mutation in several tumors. Several of these genes are susceptible to treatment with small molecule inhibitors (Andrews et al., 2007; Heinrich et al., 2003a). Signalling pathways mediated by kinase genes are implicated in several cancer types. Clinical trials suggest that tumors harboring mutations involving these RTK genes are particularly susceptible to small-molecule inhibitors. These small molecule drugs effectively inhibit the oncogenic signalling through these kinase pathways. An example of this is the gastrointestinal stromal tumor (GIST), a tumor which has activating mutations in *KIT* or in *PDGFRA* (Heinrich et al., 2003b; Hirota et al., 1998). With the use of Gleevec, a small molecule inhibitor of these type III RTK, three therapeutic categories for GIST can be recognized: tumors that respond to Gleevec therapy, tumors that do not respond, and tumors that develop resistance (Fletcher and Rubin, 2007). Moreover, a number of studies have demonstrated clinicopathological trends associated with the type of mutation in *KIT* (Antonescu et al., 2003; Ladanyi, 2001; Miettinen and Lasota, 2005).

TMA analysis of tenosynovial giant cell tumors (TGCT), a benign tumor, demonstrated high levels of *CSF1R*. TGCT are reactive tumors rather than neoplastic and are morphologically similar to a more aggressive tumor, pigmented villonodular synovitis (PVNS). Both these tumors are composed of mononuclear and multinucleated cells. High levels of *CSF1R* are also noticed in PVNS. TGCT and PVNS are characterized by cytogenetic abnormalities, commonly involving a balanced translocation of 1p11and 2q37 chromosomal loci (Dal Cin et al., 1994; Mertens et al., 1993). The genes *CSF1* and *COL6A3* involved in these chromosomal abnormalities have been recently identified (West et al., 2006). *CSF1*, encoding the ligand of CSF1R, is translocated in majority of TGCT and PVNS, but only a minority of tumor cells (2–16%) carry the translocation and express *CSF1*(West et al., 2006). These findings imply that only a minority of cells in TGCT and PVNS are neoplastic and that the majority of cells in these tumors are non-neoplastic cells that are recruited by the local over-expression of CSF1.

The *CSF1–COL6A3* translocation in TGCT and PVNS is reminiscent of the translocation that defines dermato fibrosarcoma protuberans (DFSP). In DFSP, t(17;22) brings *PDGF-B* under the control of the strong *COL1A1* promoter. The post-translationally processed form of the fusion protein is a fully functional PDGF-B protein that may stimulate oncogenesis through its receptor, PDGFRB. The *PDGFRB* receptor is also up-regulated in DFSP, suggesting an autocrine loop. However, TGCT differs from DFSP and other soft-tissue tumors in that only a minority of cells in the tumor contains the translocation and expresses *CSF1* RNA, but most of the cells within the tumor express *CSF1R*. The cells with the *CSF1* translocation most likely recruit CSF1R-expressing macrophages and may induce the formation of multinucleated giant cells (West et al., 2006). Thus, the neoplastic cells create a tumor landscape comprised non-neoplastic cells responding to secreted CSF1. The translocation involving *COL6A3* and *CSF1* results in high levels of *CSF1* expression in TGCT and PVNS. TGCT is a small benign, localized tumor and is easily controlled by surgery. PVNS is a more aggressive tumor and causes significant morbidity. The evidence suggesting a central role for *CSF1* in the pathogenesis of these tumors suggests that they might respond to treatment with a specific inhibitor such as SU11248 (Murray et al., 2003).

MicroRNAs in Sarcomas

MicroRNA expression profiles in a series of sarcomas demonstrated that the tumor clusters and sub-grouping were based largely on tumor types as well as on lines of differentiation (Subramanian et al., 2008). The oncogenesis of various sarcomas is quite different. Synovial sarcoma (SS) is characterized by a translocation event that leads to the formation of a fusion gene *SYT-SSX* and the resulting aberrant transcriptional activity exerted by the fusion oncoprotein is believed to underlie its malignant behaviour (Ladanyi, 2001). GIST, on the other hand is driven in the majority of cases by an activating mutation in KIT or PDGFRA while no known molecular event has been uncovered for leiomyosarcoma. Phenotypically, there is evidence to indicate that GIST is derived from interstitial cells of Cajal in the myenteric plexus while LMS are thought to arise from smooth muscle cells in the uterus from extra-uterine sites. In contrast, SS belongs to a class of sarcomas in which the cell of origin is unknown and it can demonstrate a mixture of mesenchymal and epithelial histology and immunophenotypes. However, recent report has identified myogenic origin of synovial sarcomas (Haldar et al., 2007). It is therefore plausible that the differential microRNA expressions that are observed among the sarcoma types are purely reflective of the differences in cell lineage or differentiation. Some of the upregulation or downregulation of microRNA observed for SS, LMS and GIST may be related to the oncogenic processes that are occurring in each of these tumor types. Rhabdomyosarcomas such as alveolar rhabdomyosarcomas (ARMS) and pleomorphic rhabdomyosarcomas (PRMS) showing histologic, ultrastructural and/or immunophenotypic evidences of skeletal muscle differentiation possess microRNA expression signatures that are sufficiently different from each other. The oncogenic mechanisms for ARMS and PRMS again show considerable differences in that *PAX7/FKHR* or *PAX3/FKHR* translocation is critical for the development of ARMS while no recurrent molecular event has been found for PRMS (Gallego Melcon and Sanchez de Toledo Codina, 2007; Lae et al., 2007). ARMS and PRMS both show primarily downregulation of multiple microRNAs that are typically expressed at a relatively high level in skeletal muscle.

Significance of microarray analysis quantifies the relative abundance of miRNA in specific sarcoma types. In GISTs, *miR-221* and *miR-222* were expressed at lower levels. Both these miRNAs are known to target 3′ UTR of *KIT*, thereby preventing its translation (Felli et al., 2005; Poliseno et al., 2006). Given that activating mutation in *KIT* is present in ~80% of GIST, downregulation of *miR-221* and *miR-222* may allow for increased translation of *KIT* and further enhance its oncogenic influence on the cell. In contrast, overexpression of *miR-21* has been found in several tumor types including leiomyosarcomas. *miR-21* has been demonstrated to regulate cell growth both *in vitro* and *in vivo* and can function as an oncogene to promote tumorigenesis. In addition, miRNAs such as *miR-133* and *miR-1* involved in muscle differentiation shows differential expression in LMS compared to normal smooth muscle tissues (Subramanian et al., 2008).

In synovial sarcoma, *miR-143* was expressed at very low levels relative to other sarcomas, as demonstrated by microarray, cloning and Northern analyses. Similarly, *miR-143* is also downregulated in several cancer types, including colorectal neoplasia (Michael et al., 2003) and cervical cancer (Lui et al., 2007). *ERK5* (also known as *MAPK7*) is the only experimentally verified target for *miR-143*, which is known to promote cell growth and proliferation in response to tyrosine kinase signalling (Esau et al., 2004). However, its role in sarcomagenesis remains unclear. Based on previously published mRNA expression array data for sarcomas (Baird et al., 2005), *ERK5* mRNA seems to be low in SS, indicating that it is unlikely to be the target of *miR-143*. On the other hand, *SSX1*, a common 3′ fusion partner gene resulting from a t(X;18) in synovial sarcoma, is predicted to be the target for *miR-143* by miRBase (http://microrna.sanger.ac.uk/targets/v4/) and Target Scan (http://www.targetscan.org/). It may be possible that downregulation of *miR-143* in SS tumor cells enable the production of the SYT-SSX1 oncoprotein.

The differentially expressed miRNAs in rhabdomyosarcomas include *miR-1* and *miR-133* which have been implicated in the differentiation of skeletal muscle. *miR*-206 which is implicated in myogenic differentiation is also highly expressed in both ARMS and PRMS. This indicates that the alterations in microRNA expression profiles in ARMS and PRMS compared to normal skeletal muscle are likely related to the different oncogenic processes that are occurring in the respective tumor types.

In RMS, two of the muscle-specific miRNAs (*miR-1* and *miR-133*) were downregulated compared to normal skeletal muscle tissues. Several myogenic genes are regulated by these two miRNAs (Clop et al., 2006). Therefore, low expression of *miR-1* may lead to increased expression of myogenin. Although many of the significantly downregulated miRNAs are common in both ARMS and PRMS, some differences observed. *miR-335* seems to be of particular interest as it is upregulated in ARMS. Notably, *miR-335* resides in the intron 2 of *MEST* (also known as *PEG1*). *MEST* is an imprinted gene expressed only from the paternal allele during development (Yan et al., 2003). Interestingly, *MEST* has been indicated to play a role in muscle differentiation. Its expression is high in ARMS (Baer et al., 2004) and it is also one of the downstream target genes of *PAX3* (Mayanil et al., 2001). The presence of the PAX3-FKHR fusion oncoprotein in ARMS may induce the expression of *MEST*, and subsequently the production of *miR-335*. It is tempting to speculate that the target genes of *miR-335* might play important role in the pathogenesis of ARMS. Several interesting targets were predicted by Target Scan 3.1, such as *CHFR*, *HAND1* and *SP1*. The expression of CHFR (checkpoint with forkhead and ring finger domains) is lost in a wide range of tumors, furthermore, its downregulation has been shown to contribute directly to tumorigenesis in mice (Yu et al., 2005) . The *HAND1* (heart and neural crest derivatives expressed 1) gene encodes a basic helix-loop-helix (bHLH) transcription factor that has been demonstrated to play important roles in extra-embryonic mesoderm and heart development (Firulli et al., 1998) as well as to regulate cardiomyocyte proliferation and differentiation in the developing heart (Risebro et al., 2006). The transcription factor SP1 is a DNA-binding protein that interacts with a variety of gene promoters containing GC-box elements. Interestingly, SP1 has been demonstrated to play a role in muscle cell differentiation (Guo et al., 2003) and the role of SP1-mediated transcription control is important in RMS (Yu et al., 2004). Further investigations of these targets are warranted to elucidate the molecular mechanism of pathogenesis in ARMS.

Novel MicroRNAs in Sarcomas

Identification of a moderate number of novel miRNAs and other small RNAs from sarcomas using a cloning approach indicates that some of these small RNAs are unique to this tumor type or they might be expressed in cells (or at levels), which escaped detection in previous cloning experiments. Although one might expect some novel miRNAs to be expressed specifically in one tumor type, it is conceivable that others would be expressed in a range of tumor types. Small RNA cloning and sequencing in sarcomas identified fourteen novel candidate miRNAs (Subramanian et al., 2007). Interestingly, one of the novel miRNA candidate is mapped to a bidirectional transcript pair, *i.e.*, sense strand of intron 22 of coagulation factor VIII (F8; NM_000132) and antisense of coding-strand of coagulation factor VIII-associated protein (F8A1; NM_012151).

Prognostic Markers

Gene vectors from sarcoma predicts outcome in patients with breast cancer

A gene expression gene list (gene vectors) for a soft tissue tumor can be projected into other oncogenic processes, such as breast carcinoma, to understand the significance of stromal expression patterns that may be a minor component of the overall process (West et al., 2005). Measurement of the global expression pattern of a stromal cell can be difficult. However, tumors with different fibroblastic phenotypes might represent different activation states or different subtypes of normal fibroblasts or stromal cells. Gene expression patterns soft tissue tumor such as desmoid-type fibromatosis (DTF) and solitary fibrous tumor (SFT) help to define normal stromal cell counterparts.

Expression profiling studies have shown remarkable differences in expression profiles between cultured fibroblasts isolated from different sites (Chang et al., 2002; 2004). The gene expression profile of a soft tissue tumor represents primarily a single cell type because the majority of the cells in the lesion are tumor cells. The gene expression patterns of DTF and SFT are distinguished by differences in expression of a variety of functional groups of genes: DTF expresses a fibrotic response and SFT tumors express genes that are typically found in the basement membrane. The cells positive for DTF markers are found in reactive tissues such as inflammatory granulation tissue and scar tissue, whereas, cells positive for SFT markers tend to be found in normal tissue (West et al., 2005). The stromal cells surrounding breast lobules were strongly reactive for SFT markers and negative for DTF genes. Analyses of the gene expression signatures (gene vectors) of DTF and SFT in a breast carcinoma expression dataset suggest that there may be distinct molecular patterns of stromal reaction in breast cancer. These stromal reaction patterns appear to be correlated with the clinical outcome. Breast cancer patients who had predominant DTF signature showed better prognosis than those who have SFT gene signature (West et al., 2005).

Summary

Integrative approach of using gene expression profiling, DNA copy number analysis, microRNA expression analysis and tissue microarrays have greatly advanced our understanding of sarcomas. Knowledge of sarcomagenesis also helps understanding several carcinoma types. The novel drug targets identified through sarcomas studies can also be used as therapeutic targets for other epithelial and hematological tumor types. Establishing cell lines for most of the sarcoma types would greatly advance the field of sarcoma biology. In addition, studying the mRNA and miRNA association in context to gene regulation and tumor development/progression could establish novel treatment strategies.

Acknowledgements

The authors would like to thank Drs. Shirley Zhu, Inigo Espinosa, Cheng-Han Lee, Reena Kartha and Kelli Montgomery for support and helpful comments.

References

1. Allander, S.V., Illei, P. B., Chen, Y., Antonescu, C.R., Bittner, M., Ladanyi, M., and Meltzer, P. S. (2002). Expression profiling of synovial sarcoma by cDNA microarrays: association of ERBB2, IGFBP2, and ELF3 with epithelial differentiation. Am J Pathol *161*, 1587–1595.
2. Andrews, S., Stephens, L.R., and Hawkins, P.T. (2007). PI3K class IB pathway. Sci STKE *2007*, cm2.
3. Antonescu, C.R. (2006). The role of genetic testing in soft tissue sarcoma. Histopathology *48*, 13–21.
4. Antonescu, C.R., Sommer, G., Sarran, L., Tschernyavsky, S. J., Riedel, E., Woodruff, J. M., Robson, M., Maki, R., Brennan, M. F., Ladanyi, M., et al., (2003). Association of KIT exon 9 mutations with nongastric primary site and aggressive behavior: KIT mutation analysis and clinical correlates of 120 gastrointestinal stromal tumors. Clin Cancer Res *9*, 3329–3337.

5. Baer, C., Nees, M., Breit, S., Selle, B., Kulozik, A. E., Schaefer, K. L., Braun, Y., Wai, D., and Poremba, C. (2004). Profiling and functional annotation of mRNA gene expression in pediatric rhabdomyosarcoma and Ewing's sarcoma. Int J Cancer *110*, 687–694.
6. Baird, K., Davis, S., Antonescu, C.R., Harper, U.L., Walker, R. L., Chen, Y., Glatfelter, A.A., Duray, P.H., and Meltzer, P. S. (2005). Gene expression profiling of human sarcomas: insights into sarcoma biology. Cancer Res *65*, 9226–9235.
7. Barnes, M., Freudenberg, J., Thompson, S., Aronow, B., and Pavlidis, P. (2005). Experimental comparison and cross-validation of the Affymetrix and Illumina gene expression analysis platforms. Nucleic Acids Res *33*, 5914–5923.
8. Bartel, D. P. (2004). MicroRNAs: genomics, biogenesis, mechanism, and function. Cell *116*, 281–297.
9. Bauer, S., Duensing, A., Demetri, G.D., and Fletcher, J.A. (2007). KIT oncogenic signaling mechanisms in imatinib-resistant gastrointestinal stromal tumor: PI3-kinase/AKT is a crucial survival pathway. Oncogene *26*, 7560–7568.
10. Brazma, A., Hingamp, P., Quackenbush, J., Sherlock, G., Spellman, P., Stoeckert, C., Aach, J., Ansorge, W., Ball, C. A., Causton, H. C., et al. (2001). Minimum information about a microarray experiment (MIAME)-toward standards for microarray data. Nat Genet *29*, 365–371.
11. Calin, G. A., and Croce, C. M. (2006). MicroRNA signatures in human cancers. Nat Rev Cancer *6*, 857–866.
12. Calin, G.A., Ferracin, M., Cimmino, A., Di Leva, G., Shimizu, M., Wojcik, S. E., Iorio, M.V., Visone, R., Sever, N.I., Fabbri, M., et al. (2005). A MicroRNA signature associated with prognosis and progression in chronic lymphocytic leukemia. N. Engl. J. Med. *353*, 1793–1801.
13. Calin, G.A., Sevignani, C., Dumitru, C. D., Hyslop, T., Noch, E., Yendamuri, S., Shimizu, M., Rattan, S., Bullrich, F., Negrini, M., and Croce, C. M. (2004). Human microRNA genes are frequently located at fragile sites and genomic regions involved in cancers. Proc Natl Acad Sci U.S.A *101*, 2999–3004.
14. Castillero-Trejo, Y., Eliazer, S., Xiang, L., Richardson, J.A., and Ilaria, R.L., Jr. (2005). Expression of the EWS/FLI-1 oncogene in murine primary bone-derived cells Results in EWS/FLI-1-dependent, ewing sarcoma-like tumors. Cancer Res *65*, 8698–8705.
15. Chang, H. Y., Chi, J.T., Dudoit, S., Bondre, C., van de Rijn, M., Botstein, D., and Brown, P.O. (2002). Diversity, topographic differentiation, and positional memory in human fibroblasts. Proc Natl Acad Sci U S A *99*, 12877–12882.
16. Chang, H.Y., Sneddon, J.B., Alizadeh, A.A., Sood, R., West, R.B., Montgomery, K., Chi, J.T., van de Rijn, M., Botstein, D., and Brown, P.O. (2004). Gene expression signature of fibroblast serum response predicts human cancer progression: similarities between tumors and wounds. PLoS Biol *2*, E7.
17. Chapman, E.J., and Carrington, J.C. (2007). Specialization and evolution of endogenous small RNA pathways. Nat Rev Genet *8*, 884–896.
18. Clop, A., Marcq, F., Takeda, H., Pirottin, D., Tordoir, X., Bibe, B., Bouix, J., Caiment, F., Elsen, J. M., Eychenne, F., et al. (2006). A mutation creating a potential illegitimate microRNA target site in the myostatin gene affects muscularity in sheep. Nat Genet *38*, 813–818.
19. Dal Cin, P., Sciot, R., Samson, I., De Smet, L., De Wever, I., Van Damme, B., and Van den Berghe, H. (1994). Cytogenetic characterization of tenosynovial giant cell tumors (nodular tenosynovitis). Cancer Res *54*, 3986–3987.

20. Davis, S.R., and Meltzer, P.S. (2007). Modeling synovial sarcoma: timing is everything. Cancer Cell *11*, 305–307.

21. Esau, C., Kang, X., Peralta, E., Hanson, E., Marcusson, E.G., Ravichandran, L.V., Sun, Y., Koo, S., Perera, R. J., Jain, R., et al. (2004). MicroRNA-143 regulates adipocyte differentiation. J Biol Chem *279*, 52361–52365.

22. Esquela-Kerscher, A., and Slack, F.J. (2006). Oncomirs - microRNAs with a role in cancer. Nat Rev Cancer *6*, 259–269.

23. Felli, N., Fontana, L., Pelosi, E., Botta, R., Bonci, D., Facchiano, F., Liuzzi, F., Lulli, V., Morsilli, O., Santoro, S., et al. (2005). MicroRNAs 221 and 222 inhibit normal erythropoiesis and erythroleukemic cell growth via kit receptor down-modulation. Proc Natl Acad Sci U S A *102*, 18081–18086.

24. Ferreira, B.I., Alonso, J., Carrillo, J., Acquadro, F., Largo, C., Suela, J., Teixeira, M.R., Cerveira, N., Molares, A., Gomez-Lopez, G., et al. (2007). Array CGH and gene-expression profiling reveals distinct genomic instability patterns associated with DNA repair and cell-cycle checkpoint pathways in Ewing's sarcoma. Oncogene.

25. Firulli, A.B., McFadden, D. G., Lin, Q., Srivastava, D., and Olson, E. N. (1998). Heart and extra-embryonic mesodermal defects in mouse embryos lacking the bHLH transcription factor Hand1. Nat Genet *18*, 266–270.

26. Fletcher, J. A., and Rubin, B. P. (2007). KIT mutations in GIST. Curr Opin Genet Dev *17*, 3–7.

27. Gallego Melcon, S., and Sanchez de Toledo Codina, J. (2007). Molecular biology of rhabdomyosarcoma. Clin Transl Oncol *9*, 415–419.

28. Guo, C.S., Degnin, C., Fiddler, T. A., Stauffer, D., and Thayer, M.J. (2003). Regulation of MyoD activity and muscle cell differentiation by MDM2, pRb, and Sp1. J Biol Chem *278*, 22615–22622.

29. Haldar, M., Hancock, J. D., Coffin, C. M., Lessnick, S. L., and Capecchi, M. R. (2007). A conditional mouse model of synovial sarcoma: insights into a myogenic origin. Cancer Cell *11*, 375–388.

30. Heinrich, M.C., Corless, C.L., Demetri, G.D., Blanke, C. D., von Mehren, M., Joensuu, H., McGreevey, L.S., Chen, C.J., Van den Abbeele, A.D., Druker, B. J., et al. (2003a). Kinase mutations and imatinib response in patients with metastatic gastrointestinal stromal tumor. J Clin Oncol *21*, 4342–4349.

31. Heinrich, M.C., Corless, C.L., Duensing, A., McGreevey, L., Chen, C.J., Joseph, N., Singer, S., Griffith, D.J., Haley, A., Town, A., et al. (2003b). PDGFRA activating mutations in gastrointestinal stromal tumors. Science *299*, 708–710.

32. Hirota, S., Isozaki, K., Moriyama, Y., Hashimoto, K., Nishida, T., Ishiguro, S., Kawano, K., Hanada, M., Kurata, A., Takeda, M., et al. (1998). Gain-of-function mutations of c-kit in human gastrointestinal stromal tumors. Science *279*, 577–580.

33. Hoheisel, J.D. (2006). Microarray technology: beyond transcript profiling and genotype analysis. Nat Rev Genet *7*, 200–210.

34. Judson, I., and Demetri, G. (2007). Advances in the treatment of gastrointestinal stromal tumours. Ann Oncol *18 Suppl 10*, x20–24.

35. Katoh, M., and Katoh, M. (2003). FLJ10261 gene, located within the CCND1-EMS1 locus on human chromosome 11q13, encodes the eight-transmembrane protein homologous to C12orf3, C11orf25 and FLJ34272 gene products. Int J Oncol *22*, 1375–1381.

36. Khandurina, J., and Guttman, A. (2002). Microchip-based high-throughput screening analysis of combinatorial libraries. Curr Opin Chem Biol *6*, 359–366.
37. Kinoshita, K., Isozaki, K., Hirota, S., Nishida, T., Chen, H., Nakahara, M., Nagasawa, Y., Ohashi, A., Shinomura, Y., Kitamura, Y., and Matsuzawa, Y. (2003). c-kit gene mutation at exon 17 or 13 is very rare in sporadic gastrointestinal stromal tumors. J Gastroenterol Hepatol *18*, 147–151.
38. Kuo, W. P., Liu, F., Trimarchi, J., Punzo, C., Lombardi, M., Sarang, J., Whipple, M. E., Maysuria, M., Serikawa, K., Lee, S. Y., et al. (2006). A sequence-oriented comparison of gene expression measurements across different hybridization-based technologies. Nat Biotechnol *24*, 832–840.
39. Labelle, Y., Bussieres, J., Courjal, F., and Goldring, M. B. (1999). The EWS/TEC fusion protein encoded by the t(9;22) chromosomal translocation in human chondrosarcomas is a highly potent transcriptional activator. Oncogene *18*, 3303–3308.
40. Ladanyi, M. (2001). Fusions of the SYT and SSX genes in synovial sarcoma. Oncogene *20*, 5755–5762.
41. Lae, M., Ahn, E.H., Mercado, G.E., Chuai, S., Edgar, M., Pawel, B.R., Olshen, A., Barr, F.G., and Ladanyi, M. (2007). Global gene expression profiling of PAX-FKHR fusion-positive alveolar and PAX-FKHR fusion-negative embryonal rhabdomyosarcomas. J Pathol *212*, 143–151.
42. Linn, S.C., West, R.B., Pollack, J.R., Zhu, S., Hernandez-Boussard, T., Nielsen, T.O., Rubin, B.P., Patel, R., Goldblum, J. R., Siegmund, D., et al. (2003). Gene expression patterns and gene copy number changes in dermatofibrosarcoma protuberans. Am. J. Pathol. *163*, 2383–2395.
43. Lu, J., Getz, G., Miska, E.A., Alvarez-Saavedra, E., Lamb, J., Peck, D., Sweet-Cordero, A., Ebert, B.L., Mak, R.H., Ferrando, A.A., et al. (2005). MicroRNA expression profiles classify human cancers. Nature *435*, 834–838.
44. Lui, W.O., Pourmand, N., Patterson, B.K., and Fire, A. (2007). Patterns of known and novel small RNAs in human cervical cancer. Cancer Res. *67*, 6031–6043.
45. Lux, M.L., Rubin, B.P., Biase, T. L., Chen, C.J., Maclure, T., Demetri, G., Xiao, S., Singer, S., Fletcher, C.D., and Fletcher, J.A. (2000). KIT extracellular and kinase domain mutations in gastrointestinal stromal tumors. Am. J. Pathol. *156*, 791–795.
46. Manning, G., Whyte, D.B., Martinez, R., Hunter, T., and Sudarsanam, S. (2002). The protein kinase complement of the human genome. Science *298*, 1912–1934.
47. Marinelli, R.J., Montgomery, K., Liu, C.L., Shah, N.H., Prapong, W., Nitzberg, M., Zachariah, Z.K., Sherlock, G.J., Natkunam, Y., West, R.B., et al. (2007). The Stanford Tissue Microarray Database. Nucleic Acids Res.
48. Matusiak, D., Glover, S., Nathaniel, R., Matkowskyj, K., Yang, J., and Benya, R.V. (2005). Neuromedin B and its receptor are mitogens in both normal and malignant epithelial cells lining the colon. Am J Physiol Gastrointest Liver Physiol *288*, G718–728.
49. Mayanil, C.S., George, D., Freilich, L., Miljan, E.J., Mania-Farnell, B., McLone, D.G., and Bremer, E.G. (2001). Microarray analysis detects novel Pax3 downstream target genes. J Biol Chem *276*, 49299–49309.
50. Mercado, G.E., and Barr, F.G. (2007). Fusions involving PAX and FOX genes in the molecular pathogenesis of alveolar rhabdomyosarcoma: recent advances. Curr Mol Med *7*, 47–61.
51. Mertens, F., Orndal, C., Mandahl, N., Heim, S., Bauer, H.F., Rydholm, A., Tufvesson, A., Willen, H., and Mitelman, F. (1993). Chromosome aberrations in tenosynovial giant cell tumors and nontumorous synovial tissue. Genes Chromosomes Cancer *6*, 212–217.

52. Michael, M.Z., SM, O.C., van Holst Pellekaan, N.G., Young, G.P., and James, R.J. (2003). Reduced accumulation of specific microRNAs in colorectal neoplasia. Mol Cancer Res *1*, 882–891.
53. Miettinen, M., and Lasota, J. (2005). KIT (CD117): a review on expression in normal and neoplastic tissues, and mutations and their clinicopathologic correlation. Appl Immunohistochem Mol Morphol *13*, 205–220.
54. Minamino, N., Kangawa, K., and Matsuo, H. (1983). Neuromedin B: a novel bombesin-like peptide identified in porcine spinal cord. Biochem Biophys Res Commun *114*, 541–548.
55. Murray, L.J., Abrams, T.J., Long, K.R., Ngai, T. J., Olson, L.M., Hong, W., Keast, P.K., Brassard, J.A., O'Farrell, A.M., Cherrington, J.M., and Pryer, N.K. (2003). SU11248 inhibits tumor growth and CSF-1R-dependent osteolysis in an experimental breast cancer bone metastasis model. Clin Exp Metastasis *20*, 757–766.
56. Nielsen, T.O., West, R.B., Linn, S.C., Alter, O., Knowling, M.A., O'Connell, J.X., Zhu, S., Fero, M., Sherlock, G., Pollack, J. R., et al. (2002). Molecular characterisation of soft tissue tumours: a gene expression study. Lancet *359*, 1301–1307.
57. Nilsson, G., Skytting, B., Xie, Y., Brodin, B., Perfekt, R., Mandahl, N., Lundeberg, J., Uhlen, M., and Larsson, O. (1999). The SYT-SSX1 variant of synovial sarcoma is associated with a high rate of tumor cell proliferation and poor clinical outcome. Cancer Res *59*, 3180–3184.
58. Ohkura, N., Yaguchi, H., Tsukada, T., and Yamaguchi, K. (2002). The EWS/NOR1 fusion gene product gains a novel activity affecting pre-mRNA splicing. J Biol Chem *277*, 535–543.
59. Poliseno, L., Tuccoli, A., Mariani, L., Evangelista, M., Citti, L., Woods, K., Mercatanti, A., Hammond, S., and Rainaldi, G. (2006). MicroRNAs modulate the angiogenic properties of HUVECs. Blood *108*, 3068–3071.
60. Pollack, J.R., and Iyer, V.R. (2002). Characterizing the physical genome. Nat Genet *32 Suppl*, 515–521.
61. Pollack, J. R., Perou, C.M., Alizadeh, A.A., Eisen, M.B., Pergamenschikov, A., Williams, C.F., Jeffrey, S.S., Botstein, D., and Brown, P.O. (1999). Genome-wide analysis of DNA copy-number changes using cDNA microarrays. Nat Genet *23*, 41–46.
62. Porter, A.C., and Vaillancourt, R.R. (1998). Tyrosine kinase receptor-activated signal transduction pathways which lead to oncogenesis. Oncogene *17*, 1343–1352.
63. Risebro, C.A., Smart, N., Dupays, L., Breckenridge, R., Mohun, T.J., and Riley, P.R. (2006). Hand1 regulates cardiomyocyte proliferation versus differentiation in the developing heart. Development *133*, 4595–4606.
64. Sircar, K., Hewlett, B.R., Huizinga, J.D., Chorneyko, K., Berezin, I., and Riddell, R.H. (1999). Interstitial cells of Cajal as precursors of gastrointestinal stromal tumors. Am J Surg Pathol *23*, 377–389.
65. Sotiriou, C., and Piccart, M.J. (2007). Taking gene-expression profiling to the clinic: when will molecular signatures become relevant to patient care? Nat Rev Cancer 7, 545–553.
66. Soule, E.H. (1986). Synovial sarcoma. Am J Surg Pathol *10 Suppl 1*, 78–82.
67. Subramanian, S., Lui, W. O., Lee, C.H., Espinosa, I., Nielsen, T.O., Heinrich, M.C., Corless, C. L., Fire, A. Z., and van de Rijn, M. (2007). MicroRNA expression signature of human sarcomas. Oncogene 27, 2015–2026.
68. Subramanian, S., West, R.B., Corless, C.L., Ou, W., Rubin, B.P., Chu, K.M., Leung, S.Y., Yuen, S. T., Zhu, S., Hernandez-Boussard, T., et al. (2004). Gastrointestinal stromal tumors (GISTs) with KIT and PDGFRA mutations have distinct gene expression profiles. Oncogene *23*, 7780–7790.

69. Subramanian, S., West, R.B., Marinelli, R.J., Nielsen, T.O., Rubin, B.P., Goldblum, J.R., Patel, R.M., Zhu, S., Montgomery, K., Ng, T.L., et al. (2005). The gene expression profile of extraskeletal myxoid chondrosarcoma. J Pathol *206*, 433–444.

70. Terry, J., Barry, T.S., Horsman, D.E., Hsu, F. D., Gown, A.M., Huntsman, D.G., and Nielsen, T.O. (2005). Fluorescence *in situ* hybridization for the detection of t(X;18)(p11.2;q11.2) in a synovial sarcoma tissue microarray using a breakapart-style probe. Diagn Mol Pathol *14*, 77–82.

71. Terry, J., Saito, T., Subramanian, S., Ruttan, C., Antonescu, C.R., Goldblum, J. R., Downs-Kelly, E., Corless, C. L., Rubin, B.P., van de Rijn, M., et al. (2007). TLE1 as a diagnostic immunohistochemical marker for synovial sarcoma emerging from gene expression profiling studies. Am J Surg Pathol *31*, 240–246.

72. Tschoep, K., Kohlmann, A., Schlemmer, M., Haferlach, T., and Issels, R.D. (2007). Gene expression profiling in sarcomas. Crit Rev Oncol Hematol *63*, 111–124.

73. Tusher, V.G., Tibshirani, R., and Chu, G. (2001). Significance analysis of microarrays applied to the ionizing radiation response. Proc Natl Acad Sci U S A *98*, 5116–5121.

74. Van de Rijn, M., and Fletcher, J.A. (2006). Genetics of soft tissue tumors. Annu Rev Pathol *1*, 435–466.

75. Weiss S, Goldblum. J. (2001). Soft tissue tumors, Elsevier, Mosby, St Louis. West, R.B., Corless, C.L., Chen, X., Rubin, B.P., Subramanian, S., Montgomery, K., Zhu, S., Ball, C.A., Nielsen, T.O., Patel, R., et al. (2004). The novel marker, DOG1, is expressed ubiquitously in gastrointestinal stromal tumors irrespective of KIT or PDGFRA mutation status. Am J Pathol *165*, 107–113.

76. West, R.B., Nuyten, D. S., Subramanian, S., Nielsen, T.O., Corless, C.L., Rubin, B. P., Montgomery, K., Zhu, S., Patel, R., Hernandez-Boussard, T., et al. (2005). Determination of stromal signatures in breast carcinoma. PLoS Biol *3*, e187.

77. West, R.B., Rubin, B.P., Miller, M.A., Subramanian, S., Kaygusuz, G., Montgomery, K., Zhu, S., Marinelli, R.J., De Luca, A., Downs-Kelly, E., et al. (2006). A landscape effect in tenosynovial giant-cell tumor from activation of CSF1 expression by a translocation in a minority of tumor cells. Proc Natl Acad Sci U.S.A. *103*, 690–695.

78. West, R.B., and van de Rijn, M. (2006). The role of microarray technologies in the study of soft tissue tumours. Histopathology *48*, 22–31.

79. Yan, Z., Choi, S., Liu, X., Zhang, M., Schageman, J.J., Lee, S.Y., Hart, R., Lin, L., Thurmond, F.A., and Williams, R.S. (2003). Highly coordinated gene regulation in mouse skeletal muscle regeneration. J Biol Chem *278*, 8826–8836.

80. Yauk, C.L., Berndt, M.L., Williams, A., and Douglas, G.R. (2004). Comprehensive comparison of six microarray technologies. Nucleic Acids Res *32*, e124.

81. Yu, S.J., Zheng, L., Ladanyi, M., Asa, S. L., and Ezzat, S. (2004). Sp1-mediated transcriptional control of fibroblast growth factor receptor 4 in sarcomas of skeletal muscle lineage. Clin. Cancer Res. *10*, 6750–6758.

82. Yu, X., Minter-Dykhouse, K., Malureanu, L., Zhao, W.M., Zhang, D., Merkle, C.J., Ward, I.M., Saya, H., Fang, G., van Deursen, J., and Chen, J. (2005). Chfr is required for tumor suppression and Aurora A regulation. Nat Genet *37*, 401–406.

Telomere Dynamics in Aging and Cancer

9

Sen Pathak and Asha S. Multani

Department of Genetics, The University of Texas, M. D. Anderson Cancer Center, Houston, Texas 77070, USA

Introduction

"Chromosomes never tell a lie" Sen Pathak, 1970. Birth and death, the two extremities of life, are bridged by the process of aging. Aging and cancer development, which occur in living organisms, seem to be related. Multiple molecular genetic and physiological modifications occur at a cellular and tissue level as we grow old (Kanungo, 1980). Neoplastic transformation is caused by multiple gene mutations and complex biochemical alterations. The processes of aging and carcinogenesis has many cellular alterations in common (Pathak, 2001; Pathak et al., 2002). Human life expectancy is determined by a number of genetic factors that frequently interact with the environment. It is said that the human genome is a loaded gun and environment pulls the trigger. There is plenty of evidence to support the concept that cancer is the result of an interaction between the environment and the genetic make-up of a person (Pathak and Dhaliwal, 1991; Ames and Gold, 1991). Mitotic catastrophe in general and telomere dynamics in particular are among the most common genetic determinants of aging and the development of cancer, whether cellular or organismal (Shay et al., 1993; Bank and Fossel, 1997; Multani et al., 2002).

Telomeres are specialized DNA-protein structures present at the termini of every eukaryotic chromosome and are required to stabilize individual chromosomes. Telomerase, first identified in the ciliate, *Tetrahymena* (Greider and Blackburn, 1985), is a ribonucleoprotein which adds on nucleotides at the ends of chromosomes to keep telomeres from being dysfunctional. The telomeric DNA sequence, (TTAGGG)n, is conserved in all vertebrate species, from fish to human, and can be identified in chromosomes and interphase cytological preparations by a technique called *fluorescence in situ hybridization* (FISH). Cytogenetic studies have shown that true terminal deletions including the telomere do not exist in altered chromosomes capable of surviving. Telomere repeat sequences are characteristic for each organism and, for that matter, the number of repeats vary from organ to organ within the same organism. In addition, telomeres undergo attrition with each cell division because of the end replication problem in somatic cells (Watson, 1972; Olovnikov, 1973). Attrition of telomeres resulting from somatic cell divisions ultimately leads to dysfunctional telomeres and the cessation of normal cell division—characteristic of aging and, in some cases, the induction of genetic instability. Telomere attrition might be more relevant for the aging process in humans and other long-lived organisms (Blasco et al., 1997). Since telomere length is highly conserved, heritable and genetically determined, its role in human aging is of special importance (Shay et al., 1993; Slagboom et al., 1994; Blasco, 2003). Lansdorp and associates have shown that the 46 human chromosomes within a cell do not have similar telomere repeat numbers; in some cases the short arm of human chromosome 17 has the least number of repeats (Lansdorp et al., 1996). The biological characteristics of cancer cells are largely influenced by telomeric dynamics. Multani and associates, using a quantitative FISH (Q-FISH) technique, have shown amplification of telomeric DNA in human and mouse invasive and metastatic cancers of different histopathologic origin (Multani et al., 1999). A similar report was published on human prostate cancer cells where cancer cells showed more telomeric DNA compared to the surrounding normal prostate cells (Ozen et al., 1998). Similarly, in the somatic cells of aged humans, the amount of telomeric DNA is reduced as compared to neonates (Harley et al., 1990; Vaziri et al., 1994; Bank and Fossel, 1997; Ly et al., 2000). These observations provide strong support for the hypothesis that cancer is a group of complex genetic diseases of old age, not the age of the person (phenotypic age), but of the somatic cells of the body (genotypic age) (Pathak et al., 2002).

The purpose of this review is to discuss the relationship between telomere dynamics, cellular aging and the development and metastasis of human cancer. A brief mention will also be made of the future directions of cancer treatment, research and prevention and the role of stem cells in cancer development.

Telomere Concept

In the eukaryotic interphase nucleus, each chromosome has its own domain, which is determined by the telomere. The word telomere originates from the Greek words, *telo,* meaning "ends", and *meros,* meaning "part." It was coined by Hermann J. Muller (1938), who believed that telocentric chromosomes could not exist in nature. According to Muller's hypothesis, a chromosome must be guarded by a telomere at both ends to maintain its integrity during the subsequent cell divisions. This concept was passed on to many early geneticists and cytogeneticists, which included my teacher, the late Professor S. P. Ray-Chaudhuri, a former student of Muller. It supplanted belief in the existence of a "true" telocentric chromosome, which has a centromere at one end and a telomere at the other. Telomeres present on both ends of a chromosome act as "body guards" for the genome. They protect chromosomes from fragmentation, determine domains of individual chromosomes within the interphase nucleus, and help in the pairing of homologous chromosomes during meiosis (Pathak et al., 1996). Barbara McClintock (1941a and b), while working on maize chromosomes, provided experimental evidence to show that without telomeres, chromosomes translocate with one another or with themselves, forming dicentric and ring configurations. These follow breakage-fusion-bridge-cycles during subsequent cell divisions and finally, resulting in either cell death or abnormal genotype (Pathak et al., 1994a and b). Fig. 9.1 shows the areas of intense research where telomere dynamics is heavily involved.

Telomere and Telomerase

Telomeres are nucleoprotein structures consisting of repetitive TTAGGG sequences that terminate in a 3′ single-stranded G-rich overhang in all vertebrates. Telomeres are maintained by the enzyme telomerase, a specialized ribonucleoprotein complex that includes an RNA template (*Terc)* and a reverse transcriptase catalytic subunit (*Tert)* (Smogorzewska and deLange, 2004). In humans, RNA template (hTERC) is located on chromosome 3q26 and the catalytic subunit (hTERT) on chromosome 5p15 (see review, Pathak and Multani, 2006). Telomeres in human

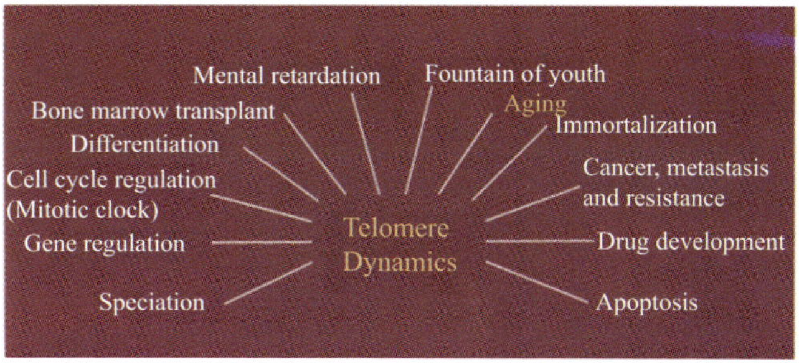

Fig. 9.1 Some areas of active research where telomere dynamics is implicated.

somatic cells comprises the TTAGGG sequences that are repeated 1,000 to 2,000 times (Moyzis et al., 1988). At the time of birth, in humans, telomeres consist of about 15,000 base pairs (bp) of TTAGGG repeat sequences. Each time a human cell divides, it loses 25-200 DNA base pairs from the telomeric ends. When this pruning takes place about 100 times, a cell ages and stops dividing. Telomere attrition is directly implicated in cellular and individual aging processes. This will be discussed in detail later in this review. Telomeres are present in two conformations: a "closed" one, in which the 3′ single-stranded overhang is folded forming a structure called the T-loop (Griffith et al., 1999, Stansel et al., 2001), and another "open" telomeric conformation which helps in replication during late S phase. Telomerase is generally not active in most human somatic cells; progressive telomere attrition takes place after each cell division since DNA polymerases cannot fully replicate the chromosomal termini. As a result of telomere shortening, dysfunctional telomeres are not able to protect chromosomal ends, which results in ring, dicentric and multicentric. During ensuing cell divisions, such complex chromosomes will undergo breakage-fusion-bridge-cycles, ultimately giving rise to genomic instability. This may lead to loss of heterozygosity (LOH) and gene amplification, the hallmarks of cancer cells (Lengauer, et al., 1998). A number of proteins that are directly or indirectly associated with telomeres are now recognized and listed in a recent book chapter (Pathak and Multani, 2006). The term "Telosome" or "Shelterin" has been used for the telomere loop and the associated six telomere-specific proteins, TRF1, TRF2, TIN2, RAP1, TPP1 and POT1 (Hockemeyer et al., 2005). Three of these six proteins, TRF1,TRF2 and POT1 directly bind the TTAGGG repeats and are connected by the remaining proteins to form a complex that protects the telomeres. TRF2 and POT1 prevent end-to-end fusions of chromosomes by interacting with DNA-damage-signaling proteins and DNA repair machinery. This entire complex is called *Shelterin* or *Telosome*; a detailed description of which along with the protein complex that shapes and safeguards human telomeres was recently reviewed (de Lange, 2005).

Telomere Attrition Causes Genomic Instability

Chromosome end fusions are the hallmarks of telomere dysfunction (Pathak et al., 1994a and b). Barbara McClintock's work on irradiated maize chromosomes showing breakage-fusion-bridge-cycles provided the first evidence that proper maintenance of telomeres is essential for chromosomal integrity and stability (McClintock, 1941a and b). Attrition of telomeres, which is highly recombinogenic, may lead to chromosome fusions in the form of dicentrics, new markers and ring conformations (Fig. 9.2). During cell divisions, a dicentric or a ring chromosome may form a bridge at anaphase, depending on the assembly of microtubules on the daughter kinetochores. Subsequent chromosome breaks will generate broken chromosome ends in the daughter cells, resulting in the breakage-fusion-bridge-cycles, as mentioned above. This type of genomic instability, which arises either due to telomere shortening from the end of a chromosome or from breaks in the chromosomal DNA, may lead to loss of heterozygosity (LOH) of tumor suppressor gene(s) in one daughter cell and amplification of a chromosomal segment in another. Most human epithelial malignancies are characterized by LOH and gene amplification (Tlsty et al., 1989). Genetic instability, the hallmark of most cancers, could be due to variation in the telomere repeats among different chromosomes (Lansdorp et al., 1996; Charames and Bapat, 2003; Bisoffi et al., 2006). The differential loss of telomere repeats can easily explain why only specific chromosomes often participate in translocations or other structural alterations in a given species or cancer type (Pathak et al., 1989). For example, in the human karyotype, D/D, G/G, or D/G translocations are very common (Hurley and Pathak, 1977; Pathak, 1986). These Robertsonian translocations are mainly due to the fusion of short arms of D and G group chromosomes. Similar translocations involving short arms of mouse chromosomes have been reported in late generation mTERC-/- animals (Blasco et al., 1997; Artandi et al., 2000).

Broken telomeric DNA, as in the case of broken chromosomes, may be repaired by nucleotide excision repair, base excision repair, and mismatch repair cell machineries. Some of these repair results may have detrimental effects on cell survival. Both homologous recombination (HR) and non-homologous end-joining (NHEJ) of two telomeres may result in abnormal telomere length due to recombination with intercalary telomeric DNA and the formation of dicentric chromosomes, respectively. Three types of HR are identified with known detrimental effects at chromosomal termini: (1) Telomere-Sister Chromatid Exchange (T-SCE), which is caused by homologous recombination between sister telomeres. The unequal exchange between sister chromatids can be detected by the chromosome orientation-fluorescent *in situ* hybridization (CO-FISH) technique,

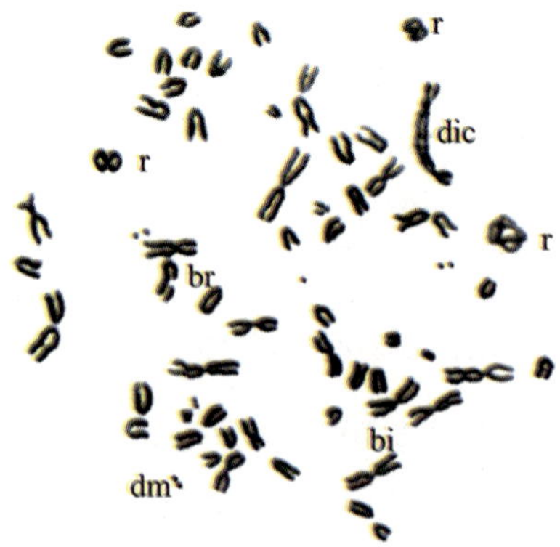

Fig. 9.2 A Giemsa-stained metaphase spread from a tumor formed by injecting late generation (*G5*) immortalized WRN -/-, mTERC -/- mouse embryonic fibroblast cells in nude mouse showing genetic instability which includes ring (*r*), dicentric (*dic*), chromatid break (*br*), double minute (*dm*), and biarmed chromosomes (*bi*).

first described by Susan Bailey (Bailey et al., 2004). A typical example of CO-FISH preparation from a late generation (G5) mTERC-/-, WRN-/- mouse cell is shown in Fig. 9.3. In normal human and murine cells T-SCE is rare, but in telomerase-independent or ALT (alternative lengthening of telomere) cells, T-SCE frequency is high (Bechter et al., 2004; Wu et al., 2006). (2) T-loop HR, discovered through a separation of a functional mutant of TRF2, protects chromosomes from NHEJ. The control of T-loop HR is further relaxed in ALT cells, which contain many telomeric circles (Wang et al., 2004). (3) The third type of HR with aberrant outcomes is the recombination between a terminal telomere and intercalary telomeric DNA. In human chromosomes, intercalary telomeric DNA is rare, except on chromosome 2. However, many vertebrate species have been described with interstitial telomeric DNA in their chromosomes (Meyne et al., 1990; Pathak et al., 1998a; Multani et al., 2001). Recombination between terminal and interstitial telomeres may generate extrachromosomal fragments that contain a stretch of telomeric DNA in the form of Telomeric DNA-containing Double Minutes (TDMs). The presence of TDMs is characteristic of mouse ALT cells (Laud, et al., 2005). These observations highlight the crucial role played by dysfunctional telomeres in the production of genetic instability.

Telomere Erosion, Cellular and Organismal Aging

In most human normal somatic cells, telomere length provides a "molecular clock" that regulates the lifespan of cells. Leonard Hayflick and Paul Moorehead (1961) of the Wistar Institute in Philadelphia were the first to report that human fibroblast cells in culture undergo a finite number of doublings, approximately 50, before they enter a nondividing stage of life, termed senescence (Hayflick, 1965). The senescent cells alter their personality and look very different from the normal

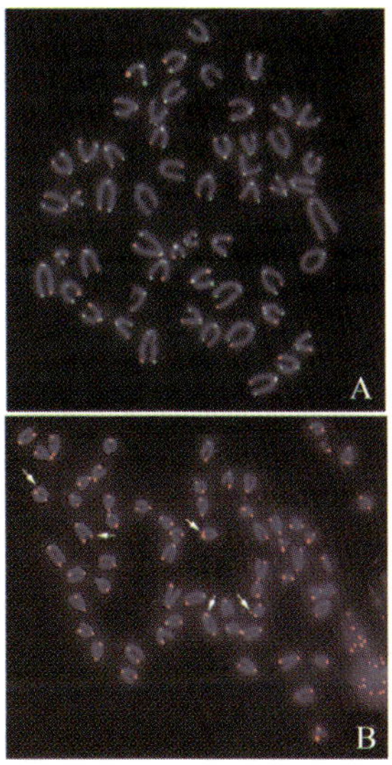

Fig. 9.3 CO-FISH preparation of mouse metaphase spreads showing: (**A**) telomeres labelled with Tam-OO-(*CCCTAA*)4 PNA probe (*red*) to detect lagging strand and FITC-OO-(*TTAGGG*)4 probe (*green*) to detect leading strand telomeres for T-SCE analysis; and (**B**) showing elevated T-SCE (*arrows*) in mouse ALT tumor cell.

cells. Before senescent cells stop dividing, their metaphases show increased frequency of dicentric configurations. In 1970, a Russian scientist, Alexy M. Olovnikov, heard a talk by Hayflick and later speculated that the gradual pruning of chromosomal ends (telomeres) might be the regulator for the number of cell divisions (Olovnikov, 1973). It was eventually discovered that telomere shortening triggers cell senescence both *in vitro* as well as *in vivo*. A similar phenomenon was reported by us in fibroblast cultures from normal individuals and patients with Li-Fraumeni syndrome (Bischoff et al., 1990). Later on, it was shown that telomeric association of chromosomes could serve as a biomarker for telomere attrition and programmed cell death (Pathak et al., 1994a and b).

Most human syndromes having characteristics of premature aging, such as Bloom syndrome (BS), Down syndrome (DS), Fanconi's anemia (FA), xeroderma pigmentosum (XP), ataxia telangiectasia (AT), and Werner syndrome (WS), are known to develop various types of malignancies (German, 1983; Salk et al., 1985). In a subsequent report, Multani and associates demonstrated that lymphocytes of individuals with such syndromes show spontaneous chromosome aberrations caused by telomere attrition (Multani et al., 2002). It was further reported by the same group that in human populations, there are individuals who are born with reduced telomeric DNA in one or more of their chromosomes (Pathak et al., 2004). Such asymptomatic and so-called normal individuals, who can be identified by studying their telomeric DNA by Q-FISH technique in their peripheral blood lymphocytes, may develop cancer and in some cases, early cancer diagnosis can be done. In many cases, such individuals also show salient characteristics of aging. One such study of a Caucasian family involving three generations with birth defects, sensitivity to mutagens and having cancer was recently published from the state of Texas (Pathak et al., 2004). Such studies should be performed on different ethnic populations for the early identification of family members who could be predisposed to certain types of cancer so that early primary and secondary interventions can be done in a timely manner.

Werner Syndrome (WS) is an autosomal recessive disorder showing premature aging and early onset of cancer (sarcomas). WS patients initially present normal phenotypes but in later life develop alopecia, osteoporosis, bilateral cataracts, type II diabetes, hypogonadism and ischemic heart disease (Thannhauser, 1945). The fibroblast cells of WS patients senesce prematurely in culture and show increased mitotic catastrophe (Faragher et al., 1993; Salk et al., 1981; Melcher et al., 2000). WS is caused by mutation of a single gene, *WRN*, which encodes a protein containing a highly conserved 3′ to 5′ DNA helicase of the RecQ family (Martin and Oshima, 2000). This gene is localized on human chromosome 8 (Goto et al., 1992). In recent publications Sandy Chang's group, using double knockout TERC -/- and WRN -/- mice have shown that late generation animals with dysfunctional telomeres exhibit classical Werner -like premature aging, graying of hairs, osteoporosis, alopecia, type II diabetes, cataracts and development of cancer. Even the fibroblast cultures of such animals displayed accelerated replicative senescence and increased rate of mitotic catastrophe (Chang et al., 2004; Multani and Chang, 2007). As a result of these observations, we conclude that telomere dysfunction is required for the manifestation of age-related pathologies in WS patients. Evidence presented here further enforces the evidence that telomere attrition plays an important role not only in cellular senescence but also in organismal aging. A more detailed account of aging and cancer in telomerase knockout mice can be found in a recent review by Sandy Chang (2005).

Telomere Dynamics in Cancer

Cancer is a genetic disease which is caused by genetic instability. It is genetic not in the sense of an inheritable disease but, in a broad fashion, a genetic disease that develops from alterations in an individual's genetic material, the DNA. Telomeres are dysfunctional in human cancer cells. Telomerase, which is inactive in most human somatic cells, may be activated in cancer cells through a gene controlling the DNA-damage response machinery. In the absence of this enzyme, human chromosomes with shortened telomeres undergo ring formation and chromosome fusions forming dicentrics, breakage-fusion-bridge cycles and profound genetic instability (also known as 'crisis'). Such a huge degree of genomic instability, through as yet unknown molecular mechanisms, is either able to activate/up-regulate the telomerase-dependent pathway in most cases or activate the ALT pathway in rare cases. Either mechanism transforms the cells into immortal entities, the precursor phenotype of all cancers (Hahn et al., 1999; Hanahan and Weinberg, 2000). Despite reactivation/up-regulation of telomerase, telomeres in primary human cancer cells are often considerably shorter than their matched adjacent normal tissue cells (Shay et al., 1993; deLange, 1995; Rhyu, 1995). However, in highly metastatic and invasive tumor cells of human and murine origin, amplification of telomeric DNA has been reported (Multani et al., 1999). This results in aneuploidy, translocation, LOH, gene amplification and, consequently, phenotypic and tumorigenic variability of different tumor clones (Nowell, 1976; Pathak, 2001, 2003 and 2004; Pathak and Multani, 2006).

Organ/Tissue-Specific Stem Cells are the Precursors of Malignant Transformation

There are basically four types of stem cells: (1) embryonic stem (ES) cells, (2) bone marrow-derived mesenchymal stem (MS) cells, (3) adult tissue-specific stem cell (ATS) and (4) cancer stem (CS) cells. These stem cells have three salient characteristics in common: (1) self renewal, (2) migration *in vivo* and *in vitro*, (3) differentiation into other cell types. Since more than 95% of human cancers caused by our "life styles" are the result of an interaction between one's genetic make-up and the exposure to one's environment, it is not unreasonable to hypothesize that all human cancers originate in the ATS cells (Pathak, 2002). This hypothesis is based on four compelling factors: (1) It is now well established that all adult organs/tissues in our body have their own stem cells which are needed for repair of damaged and aging cells; (2) only cycling (dividing) cells are susceptible to environmental insults and hence, accumulation of more mutations than quiescent (non-dividing) somatic cells; (3) only stem cells have the capability of undergoing cell divisions, whereas differentiated (mature) somatic cells do not and (4) long-term established human and murine cancer cell lines have their own cancer stem cell population, known as side population (SP) cells. Such SP cells have been isolated from long-term cancer cell lines including HeLa, MCF-7 and other human breast cancer, human brain tumor, C6 glioma and B104 neuroblastoma (Reya et al., 2001; Pardal et al., 2003; Dick, 2003; Singh et al., 2003 and 2004; Kondo et al., 2004). These CS cells isolated from the parental tumor initiate tumors upon transplantation into a secondary host (Al-Hajj et al., 2003; Singh et al., 2004). The stem cell transformation can be the underlying cause of human cancers that has been reported in leukemia, myeloma, lung, mammary gland, liver, pancreas, kidney, retina, brain, skin, intestinal polyposis and ovary (Bapat, et al., 2005; He et al., 2007; Chang, 2006; Chang et al., 2001; Singh et al., 2003; Kao et al., 1995). It is still an open question whether cancer stem cells are derived from the ATS cells, MS cells or dedifferentiated somatic cells. Evidence presented thus indicates that cancer stem cells are derived from ATS cells because the markers for CS cells are also markers for normal stem cells (Chang, 2006). More research is urgently needed to solve this burning question of the relationship between ATS and CS cells (see, book chapter by Pathak and Multani, 2006). Presently, these results strongly support the hypothesis that cancer is a disease of aging organ-specific stem cells. The American Association for Cancer Research (AACR) recently convened a conference of experts to discuss this emerging field of stem cell model for tumor development and treatment (Clarke et al., 2006).

Are Circulating Cells with Altered Chromosome Morphology Derived from Organ/ Tissue-specific Stem Cells of Cancer Patients and some of their Asymptomatic Relatives?

Cancer-initiating specific genetic (chromosomal) abnormalities may not be present in every somatic cell (Pathak et al., 1989). We have earlier proposed that predisposed individuals and cancer patients inherit susceptibility traits that make their specific chromosomes prone to breaking at a particular locus (Pathak, 1990a). The chromosomes of such individuals may undergo specific changes at a relatively low frequency in all tissues, including peripheral blood lymphocytes (PBL). Since telomeres are a vital survival factor for the cells and also serve as guardians of the individual chromosomes, it is highly likely that only the chromosomes that undergo alterations are the ones that have reduced telomeric DNA or are not protected (Pathak, 2001; our unpublished data).

Characteristic cytogenetic abnormalities of most human tumors were first identified in PBL. For example, anomaly of chromosome 13 in retinoblastoma, of chromosome 11 in Wilms' tumor, of 3 in renal cell carcinoma, of 2,5 and 12 in colorectal cancer, of 1,6 and 9 in melanoma, of 1 in endometrial and breast cancer, and of 5,7,8,10 and 16 in prostate cancer were first identified in a small number (1-3%) of phytohemagglutinin-stimulated lymphocyte metaphase spreads (see reviews, Pathak, 1990a and b, 2001 and 2003). These circulating cells with aberrant metaphase are not cancer cells *per se*, because: (a) they show mainly chromatid breaks, simple translocations and deletions, (b) such aberrations are present in the PBL of only some asymptomatic family members, and (c) when tumor cells and PBL both were analyzed from the same patient, simple chromatid or chromosome translocations were present in the PBL but the same chromosomes were involved in stable marker formation and were present in most tumor metaphases (Ghayee et al., 1997). Is it possible that these rare PBL metaphases are coming from tissue-specific stem cells? More research is needed to substantiate this hypothesis, although our observations on animal models showed very similar results. Cytogenetic analyses of late generation (G4 and G5) telomerase knockout mice have shown a very similar pattern. Our unpublished data show that when bone marrow, spleen and spontaneously-developed tumors of the same telomerase knockout G4 and G5 mice are studied cytogenetically, bone marrow spreads show 2-3 % of metaphases with one or rarely two metacentrics or a long marker chromosome, whereas all metaphases in tumors show many such abnormalities, presenting suitable support for the above hypothesis. Hagmar and associates in a population based study have also shown that chromosomal abnormalities in human PBLs predict cancer development (Hagmer et al., 1998).

Telomere Attrition and Programmed Cell Death

Telomeric DNA which is attached to the inner wall of the cell nucleus is the determinant of domains for individual chromosomes within the nucleus. When a clastogen, whether chemical, physical or biological agent that breaks the chromosome, enters the cell nucleus, it first contacts the telomere, the "body guard" of the genome (Pathak, 2003). Because of this location and contact, telomeres are the first to be hit by a clastogen entering the nucleus. It is, therefore, logical to think that telomeres must be playing a role in spontaneous or induced cell death (Pathak et al., 1994a and b; 1998b and 2000; Mukhopadhyay et al., 1998; Ishibashi and Lippard, 1998; Multani et al., 2000; Sohn et al., 2002). In a series of reports mentioned above, we and others have shown that telomeric DNA is the primary target when cancer cells are treated with chemotherapeutic drugs and radiation therapy. It is high time to target the telomeric DNA of cancer cells for the discovery of new generation anticancer drugs. Izbicka and colleagues (1999) have rightly said, "The fact that DNA (telomeric), not telomerase protein or RNA, can be a target for rationale drug design has important implications for understanding the telomerase mechanism of action." It is, therefore, necessary to focus on and discover telomere-interactive agents that can preferentially target the cancer cells and do not inflict adverse effect on normal cells (Pathak et al., 2003).

Cancer Treatment and Future Directions

Cancer patients often die due to the metastasis and invasion of cancer cells to other organs (Fidler, 1990). The prognosis and successful treatment of human cancers are dependent on early and correct diagnosis, disease stage and its metastatic potential. Examination of xenografts of human tumors into nude mice have shown that injected human tumor cells may transform the host cells, resulting in a mixture of human and murine tumor cells or the pure host tumor cells (Pathak et al., 1997). This observation may have huge clinical implications in cancer patients. It is possible that cancer cells capable of metastasizing to distant organs of a patient may also transform the local normal stromal cells into malignant cells. It is, therefore, important for physicians to consider not only the treatment of the original tumor but also the new tumor which might be formed due to transformation in the distant organ. Cancer patients rarely die due to the primary neoplasm. Only metastatic cancer cells are capable of amplifying their telomeric DNA, forming new blood vessels (angiogenesis), developing resistance to treatment, and bringing death to the hosts due to their selfish drive to survive. Cancer treatment modalities including surgery, chemotherapy, radiotherapy, immunotherapy and cell therapy should be directed more towards the cancer stem cells as compared to the differentiated cancer cells. Cancer stem cells divide more slowly than other cells in a tumor mass—though still faster than normal stem cells. Since CS cells divide slowly, conventional cancer therapies that affect the fast dividing tumor cells may miss these stowaways. Then, when the tumor is eradicated and the patient is considered cancer free, the CS cells remain alive. As long as the CS cells persist and are hidden due to the formation of complex neo-vasculature in a given organ, cancer may return. Therefore, eradication of CS cells is absolutely necessary in "**Making Cancer History**" which is the mission of the University of Texas M. D. Anderson Cancer Center in Houston, Texas. Since chemotherapy has severe side effects, it is worth considering the use of alternative therapies such as, ayurvedic and homeopathic medicines, acupuncture, yoga, hydrotherapy, massage, pranayam, kapalbhati, and other forms of ancient treatment modalities in combination with current cancer treatments. Targeting telomeres with G-quadruplex interactive drugs would be beneficial because the highest gene transcriptions, recombinations and active gene concentration in the human genome are found in the telomeric regions of chromosomes. It is absolutely essential to combine the best therapies from the West and the East to tackle these dreadful diseases called cancers. In the University of Texas M. D. Anderson Cancer Center, where one of us (S.P.) had spent more than 37 years researching, had taken a lead in this direction by establishing Global Oncology Program with other countries of the world, including India.

Conclusion

Cancers are genetic diseases of organ/tissue-specific defective (aged) stem cells. Aging and cancer initiation are interrelated in that both are caused by telomere attrition. Although invasive and metastatic cancer cells show amplification of telomeric DNA, primary tumor cells have much fewer telomere repeats as compared to normal cells of the same organ. It is our belief that successful treatments of cancer will depend on discovering new chemotherapeutic agents that are targeted to cancer stem cells and that have minimal or no side effects on normal cells. Quantification of telomeric DNA in PBL can be used to identify cancer-predisposed individuals in human populations and cancer patients whose cancer cells have potential for invasion and metastasis. Such identifications will be of crucial importance in deciding primary and secondary interventions and in early diagnosis and prevention of cancer, because **"Cancer prevention is still better than a cure".**

Cancer research is the key for the successful treatment of cancer patients and eradication of the disease. It is, therefore, extremely important to bring talented clinical researchers and basic scientists interested in cancer research to work together, a model that the University of Texas M. D. Anderson Cancer Center in Houston, Texas, has established. At this second to none in the world Cancer Center, clinicians and basic researchers work hand-in-hand. Because of this active collaboration and cooperation of clinicians and basic scientists, molecular genetic pathways of some human cancers have been dissected to gain in-depth knowledge for their successful treatment. Isolated clinical and molecular studies of cancer conducted without sufficient input and data from each other most often fail to produce the knowledge needed to have an impact on the lives of cancer patients in countries like India. Even today, there appears to be a "deep valley" between researchers studying basic cancer cell biology and the clinicians taking care of cancer patients. The time has come to study cancer by a multi-disciplinary approach in every country of the world, and India should not be left behind. Scientists with very diverse interests in genomics, proteomics, microRNA work, bio-informatics, and many other fields should work together and institutions like – Centre for Cellular and Molecular Biology (CCMB), Hyderabad, and Tata Memorial Centre (TMC) and Advanced Centre for Treatment, Research & Education in Cancer (ACTREC), Mumbai, in India can play a leading role in this direction. *This is the era of translational research* which will herald the day when clinicians will be able to prescribe treatments based on the genetic and molecular profile of a patient's cancer. This is unlikely to be achieved in a country like India where clinicians and basic scientists are "playing their own music" separately.

Acknowledgements

We would like to thank Radha Dixit, Cynthia and Burr Furlong for critical reading of the manuscript and Sandy Chang for facilities and his encouragement. This work was supported in part by the institutional T.C. Hsu Molecular Cytogenetic Core under grant NCI CA0161 established in the Department of Genetics. Our apology to all those scientists whose published work could not be cited here due to space limitations.

References

1. Al-Hajj, M., Wicha M.S., Benito-Hernandez, A., Morrison, S.J., and Clarke, M.F. (2003). Prospective identification of tumorigenic breast cancer cells. Proc. Natl. Acad. Sci. USA *100,* 3983–3988.
2. Ames, B.N., and Gold, L.S. (1991). Endogenous mutagens and the causes of aging and cancer. Mutat. Res. *250,* 3–16.
3. Artandi, S., Chang, S., Lee, S., Alson, S., Gottlieb, G., Chin, L., and DePinho, R.A. (2000). Telomere dysfunction promotes non-reciprocal translocations and epithelial cancers in mice. Nature *406,* 641–645.
4. Bailey, S.M., Brenneman, M.A., and Goodwin, E.H. (2004). Frequent recombination in telomeric DNA may extend the proliferative life of telomerase-negative cells. Nucleic Acids Res. *32,* 3743–3751.
5. Bank, D.A., and Fossel, M. (1997). Telomeres, cancer, and aging. J. Natl. Cancer Inst. *278,* 1345–1348.
6. Bapat, S.A., Mali, A.M., Koppikar, C.B., and Kurrey, N.K. (2005). Stem and progenitor-like cells contribute to the aggressive behavior of human epithelial ovarian cancer. Cancer Res. *65,* 3025–3029.
7. Bechter, O.E., Zou, Y., Walker, W., Wright, W.E., and Shay, J.W. (2004). Telomeric recombination in mismatch repair deficient human colon cancer cells after telomerase inhibition. Cancer Res. *64,* 3444–3451.
8. Bischoff, F.Z., Sun, O.Y., Pathak, S., Grant, G., Siciliano, M.J., Giovanella, B.C., Strong, L.C., and Tainsky, M.A. (1990). Spontaneous abnormalities in normal fibroblasts from patients with Li-Fraumeni cancer syndrome: Aneuploidy and immortalization. Cancer Res. *50,* 7979–7984.
9. Bisoffi, M., Heaphy, C.M., and Griffith, J.K. (2006). Telomeres: Prognostic markers for solid tumors. Int. J. Cancer *119,* 2255–2260.
10. Blasco, M.A. (2003). Telomere in cancer and aging: lessons from the mouse. Cancer Letters *194,* 183–188.
11. Blasco, M.A., Lee, H.W., Hande, M.P., Samper, E., Lansdorp, P.M., DePinho, R.A., and Greider, C.W. (1997). Telomere shortening and tumor formation by mouse cells lacking telomerase RNA. Cell *91,* 25–34.
12. Chang, C.C. (2006). Recent translational research: stem cells as the roots of breast cancer. Breast Cancer Res. *8,* 1–3.
13. Chang, C.C., Sun, W., Cruz, A., Saitoh, M., Tai, M.H., and Trosko, J.E. (2001). A human breast epithelial cell type with stem cell characteristics as target cells for carcinogenesis. Radiat. Res. *155,* 201–207.

14. Chang, S. (2005). Modeling aging and cancer in the telomerase knockout mouse. Mut. Res. *576,* 39–53.

15. Chang, S., Multani, A.S., Cabrera, N.G., Naylor, M.L., Laud, P., Lombard, D., Pathak, S., Guarente, L., and DePinho, R.A. (2004). Essential role of limiting telomeres in the pathogenesis of Werner syndrome. Nature Genet. *36,* 877–882.

16. Charames, G.S., and Bapat, B. (2003). Genomic instability and cancer. Curr. Mol. Med. *3,* 589–596.

17. Clarke, M.F., Dick, J.E., Dirks, P.B., Eaves, C.J., Jamieson, C.H.M., Jones, D.L., Visvader, J., Weissman, I.L., and Wahl, G.M. (2006). Cancer stem cells – Perspectives on current status and future directions: AACR workshop on cancer stem cells. Cancer Res. *66,* 9339–9344.

18. de Lange, T. (1995). Telomere dynamics and genome instability in human cancer. In: Telomere, Blackburn, E.H., and Greider, C.W., eds., Cold Spring Harbor Lab., New York, pp. 265–293.

19. de Lange, T. (2005). Shelterin: the protein complex that shapes and safeguards human telomeres. Genes & Development *19,* 2100–2110.

20. Dick, J.E. (2003). Breast cancer stem cells revealed. Proc. Natl. Acad. Sci. USA *100,* 3547–3549.

21. Faragher, R.G., Kill, I.R., Hunter, J.A., Pope, F.M., Tannock, C., and Shall, S. (1993). The gene responsible for Werner syndrome may be a cell division "counting" gene. Proc. Natl. Acad. Sci. USA *90,* 12030–12034.

22. Fidler, I.J. (1990). Critical factors in the biology of human cancer metastasis: Twenty-eight GHA Clowes Memorial Award Lecture. Cancer Res. *50,* 6130–6138.

23. German, J. (1983). Chromosomal Mutations and Neoplasia. New York; Wiley-Liss.

24. Ghayee, H.K., Dinney, C.P., and Pathak, S. (1997). Do lymphocytes contain chromosomal lesions that are also stable markers in cancer cells? Lymphocytes and tumor cell karyotyping in a melanoma patient. Int. J. Oncol. *11,* 681–684.

25. Goto, M., Rubenstein, M., Weber, J., Woods, K., and Drayna, D. (1992). Genetic linkage of Werner's syndrome to five markers on chromosome 8. Nature *355,* 735–738.

26. Greider, C.W., and Blackburn, E.H. (1985). Identification of a specific telomere terminal transferase activity in *Tetrahymena* extracts. Cell *43,* 405–413.

27. Griffith, J.D., Comeau, L., Rosenfield, S., Stansel, R.M., Bianchi, A., Moss, H., and de Lange, T. (1999). Mammalian telomeres end in a large duplex loop. Cell *97,* 503–514.

28. Hagmar, L., Bonassi, S., Stromberg, U., Brogger, A., Knudsen, L.E., Norppa, H., Reuterwall, C. (1998). Chromosomal aberrations in lymphocytes predict human cancer: a report from the European Study Group on Cytogenetic Biomarkers and Health (ESCH). Cancer Res. *58,* 4117–4121.

29. Hahn, W.C., Counter, C.M., Lundberg, A.S., Beijersbergen, R.L., Brooks, M.W., and Weinberg, R.A. (1999). Creation of human tumor cells with defined genetic elements. Nature *400,* 464–468.

30. Hanahan, D., and Weinberg, R.A. (2000). The hallmarks of cancer. Cell *100,* 57–70.

31. Harley, C.B., Futcher, C.A., and Greider, C.W. (1990). Telomeres shorten during ageing of human fibroblasts. Nature *345,* 458–460.

32. Hayflick, L. (1965). The limited *in vitro* lifetime of human diploid cell strains. Exp. Cell Res. *37,* 614–636.

33. Hayflick, L., and Moorhead, P.S. (1961). The serial cultivation of human diploid cell strains. Exptl. Cell Res. *25,* 585–621.

34. He, X.C., Yin, T., Grindley, J.C., Tian, Q., Sato, T., Tao, W.A., Dirisina, R., Porter-Westpfahl, K.S., Hembree, M., Johnson, T., Wiedemann, L.M., Barrett, T.A., Hood, L., Wu, H., and Li, L. (2007). PTEN-deficient intestinal stem cells initiate intestinal polyposis. Nature Genetics *39,* 189–198.

35. Hockemeyer, D., Sfeir, A.J., Shay, J.W., Wright, W.E., and de Lange, T. (2005). POT1 protects telomeres from a transient DNA damage response and determines how human chromosomes end. EMBO J. *20,* 2667–2678.

36. Hurley, J.E., and Pathak, S. (1977). Elimination of nucleolus organizers in a case of 13/14 Robertsonian translocation. Hum. Genet. *35,* 169–173.

37. Izbicka, E., Wheelhouse, R.T., Raymond, E., Davidson, K.K., Lawrence, R.A., Sun, D., Windle, B.E., Hurley, L.H., and von Hoff, D.D. (1999). Effects of cationic porphyrins as G-Quardruplex interactive agents in human tumor cells. Cancer Res. *59,* 639–644.

38. Ishibashi, T., and Lippard, S.J. (1998). Telomere loss in cells treated with cisplatin. Proc. Natl. Acad. Sci. USA *95,* 4219–4223.

39. Kanungo, M.S. (1980). Biochemistry of aging. Academic Press, New York.

40. Kao, C.Y., Nomata, K., Oakley, C.S., Welsch, C.W., and Chang, C.C. (1995). Two types of normal human breast epithelial cells derived from reduction mammoplasty: phenotypic characterization and response to SV 40 transfection. Carcinogenesis *16,* 531–538.

41. Kondo, T., Setoguchi, T., and Taga, T. (2004). Presistence of a small subpopulation of cancer stem-like cells in the C6 glioma cell line. Proc. Natl. Acad. Sci. USA *101,* 781–786.

42. Lansdorp, P.M., Verwoerd, N.P., van de Rijke, F.M., Dragowska, V., Little, M.T., Dirks, R.W., Raap, A.K., and Tanke, H.J. (1996). Heterogeneity in telomere length of human chromosomes. Hum. Mol. Genet. *5,* 685–691.

43. Laud, P.R., Multani, A.S., Bailey, S.M., Wu, L., Ma, J., Kingsley, C., Lebel, M., Pathak, S., DePinho, R.A., and Chang, S. (2005). Elevated telomere-telomere recombination correlates with increased cellular immortalization and transformation in G5 mTerc-/- Wrn-/- mouse cells. Genes & Development *19,* 2560–2570.

44. Lengauer, C., Kinzler, K.W., and Vogelstein, B. (1998). Genetic instabilities in human cancers. Nature *396,* 643–649.

45. Ly, D.H., Lockhart, D.J., Lerner, R.A., and Schultz, P.G. (2000). Mitotic misregulation and human aging. Science *287,* 2486–2492.

46. Martin, G.M., and Oshima, J. (2000). Lessons from human progeroid syndromes. Nature *408,* 263–266.

47. McClintock, B. (1941a). Spontaneous alterations in chromosome size and form in *Zea mays.* Cold Spring Harbor Symp. Quant. Biol. *9,* 72–81.

48. McClintock, B. (1941b). The stability of broken ends of chromosomes in *Zea mays.* Genetics *26,* 234–282.

49. Melcher, R., von Golitschek, R., Steinlein, C., Schindler, D., Neitzel, H., Kainer, K., Schmid, M., and Hoehn, H. (2000). Spectral karyotyping of Werner syndrome fibroblast cultures. Cytogenet. Cell Genet. *91,* 180–185.

50. Meyne, J., Baker, R.J., Hobart, H.H., Hsu, T.C., Ryder, O.A., Ward, O.G., Wiley, J.E., Wurster-Hill, D.H., Yates, T.L., and Moyzis, R.K. (1990). Distribution of non-telomeric sites of the (TTAGGG)n telomeric sequence in vertebrate chromosomes. Chromosoma *99,* 3–10.

51. Moyzis, R.K., Buckingham, J.M., Cram, L.S., Dani, M., Deaven, L.L., Jones, M.D., Meyne, J., Ratliff, R.L., and Wu, J.R. (1988). A highly conserved repetitive DNA sequence, (TTAGGG)n, present at the telomeres of human chromosomes. Proc. Natl. Acad. Sci. USA *85,* 6622–6626.

52. Mukhopadhyay, T., Multani, A.S., Roth, J.A., and Pathak, S. (1998). Reduced telomeric signals and increased telomeric associations in human lung cancer cell lines undergoing p53-mediated apoptosis. Oncogene *17,* 901–906.

53. Muller, H.J. (1938). The remaking of chromosomes. Collecting Net. *13,* 183–195.

54. Multani, A.S. and Chang, S. (2007). WRN at telomeres: implications for aging and cancer. J. Cell Sci. *120,* 713–721.

55. Multani, A.S., Narayan, S., Jaiswal, A.S., Zhao, Y.-J., Barkley, R.A., Furlong, C.L., and Pathak, S. (2002). Telomere dynamics, aging, and cancer: study of human syndromes characteristic of premature aging. J. Anti-Aging Med. *5,* 271–281.

56. Multani, A.S., Ozen, M., Furlong, C.L., Zhao, Y.-J., Hsu, T.C., and Pathak, S. (2001). Hetrochromatin and interstitial telomeric DNA homology. Chromosoma *110,* 214–220.

57. Multani, A.S., Ozen, M., Sen, S., Mandal, A.K., Price, J.E., Fan, D., Radinsky, R., Ali-Osman, F., von Eschenbach, A.C., Fidler, I.J., and Pathak, S. (1999). Amplification of telomeric DNA directly correlates with metastatic potential of human and murine cancers of various histological origin. Int. J. Oncol. *15,* 423–429.

58. Multani, A.S., Ozen, M., Narayan, S., Kumar, V., Chandra, J., McConkey, D.J., Newman, R.A., and Pathak, S. (2000). Caspase-dependent apoptosis induced by telomere cleavage and TRF2 loss. Neoplasia *2,* 339–345.

59. Nowell, P.C. (1976). The clonal evolution of tumor cell population: Acquired genetic lability permits stepwise selection of variant sublines and underlies tumor progression. Science *194,* 23–28.

60. Olovnikov, A. (1973). A theory of marginotomy: the incomplete copying of template margin in enzyme synthesis of polynucleotides and biological significance of the phenomenon. J. Theor. Biol. *41,* 181–190.

61. Ozen, M., Imam, S.A., Datar, R.H., Multani, A.S., Narayanan, R., Chung, L.W.K., von Eschenbach, A.C., and Pathak, S. (1998). Telomeric DNA: Marker for human prostate cancer development. The Prostate *36,* 264–271.

62. Pardal, R., Clarke, M., and Morrison, S. (2003). Applying the principles of stem-cell biology to cancer. Nat. Rev. Cancer *3,* 895–902.

63. Pathak, S. (1986). Constitutional translocations and predisposition to T- and B- cell malignancies. Cytogenet. Cell Genet. *43,* 223–225.

64. Pathak, S. (1990a). Chromosome alterations in speciation and neoplastic transformation: a parallelism. In: Sharma, T. (ed.), Trends in Chromosome Research. Springer-Verlag, Narosa Publishing House, New Delhi, pp.204–220.

65. Pathak, S. (1990b). Cytogenetic abnormalities in cancer: with special emphasis on tumor heterogeneity. Cancer Metast. Rev. *8,* 299–318.

66. Pathak, S. (2001). Telomeres in human cancer research. 10th All India Congress of Cytology and Genetics Award Lecture. Perspect. Cytol. Genet. *10,* 13–22.

67. Pathak, S. (2002). Organ- and tissue-specific stem cells and carcinogenesis. Anticancer Res. *22,* 1353–1356.

68. Pathak, S. (2003). Cancer Biology and genetic heterogeneity. Proc. Indian Natl. Sci. Acad. *B69,* 11–22.

69. Pathak, S. (2004). Stem cells, aging, telomere dynamics, and cancer. Perspect. Cytol. Genet. *11,* 1–10.

70. Pathak, S., and Dhaliwal, M. (1991). Understanding cancer: Myths and Facts. J. Scientific Res. *41,* 121-136.

71. Pathak, S., Dhaliwal, M.K., and Hopwood, V.L. (1989). Genetic susceptibility, somatic mosaicisms and predisposition to human cancer. Anticancer Res. *9,* 17–20.

72. Pathak, S., Multani, A.S., Furlong, C.L., and Sohn, S.H. (2002). Telomere dynamics, aneuploidy, stem cells, and cancer (Review). Int. J. Oncol. *20,* 637–641.

73. Pathak, S., Multani, A.S., and Amoss Jr., M.S. (1996). Telomere, telomerase and malignant melanomas in human and domestic mammals. Arch. Zootec. *45,* 141–149.

74. Pathak, S., and Multani, A.S. (2006). Aneuploidy, stem cells and cancer. In: Cancer: Cell, Cell Structures, Carcinogens and Genomic Instability. Bignold, L.P., eds., Birkhauser Verlag/ Switzerland, pp. 49–64.

75. Pathak, S., Multani, A.S., McConkey, D.J., Imam, A.S., and Amoss Jr, M.S. (2000). Spontaneous regression of cutaneous melanoma in Sinclair swine is associated with defective telomerase activity and extensive telomere erosion. Int J. Oncol. *17,* 1219–1224.

76. Pathak, S., Dolhonde, J.A., and Multani, A.S. (1998a). Amplification of telomeric DNA and the extent of karyotypic evolution. Cytobois *93,* 141–146.

77. Pathak, S., Multani, A.S., Ozen, M., Richardson, M.A., and Newman, R.A. (1998b). Dolastatin-10 induces polyploidy, telomeric associations and apoptosis in a murine melanoma cell line. Oncol. Report *5,* 373-376.

78. Pathak, S., Dave, B.J., and Gagos, S.H. (1994a). Chromosome alterations in cancer development and apoptosis. *In vivo 8,* 843–850.

79. Pathak, S., Risin, S., Brown, N.M., and Berry, K. (1994b). Telomere association of chromosomes is an early manifestation of programmed cell death. Int. J. Oncol. *4,* 323–328.

80. Pathak, S., Multani, A.S., Banerji, P., and Banerji, P. (2003). Ruta 6 selectively induces cell death in brain cancer cells but proliferation in normal peripheral blood lymphocytes: A novel treatment for human brain cancer. Int. J. Oncol. *23,* 975–982.

81. Pathak, S., Multani, A.S., Thalmann, G.N., von Eschenbach, A.C., and Chung, L.W.K. (1997). Can cancer cells transform normal host cells into malignant cells? Brit. J. Cancer *76,* 1134–1138.

82. Pathak, S., Multani, A.S., Narayan, S., Furlong, C.L., and Hsu, T.C. (2004). Germline telomere length dynamics and mutagen sensitivity studies in a family with acute reactions to Sun exposure: Involvement of three generations. Cancer Genomics & Proteomics *1,* 199–208.

83. Rhyu, M.S. (1995). Telomeres, telomerase, and immortality. J. Natl. Cancer Inst. *87,* 884–894.

84. Reya, T., Morrison, S.J., Clarke, M.F., and Weissman, I.L. (2001). Stem cells, cancer, and cancer stem cells. Nature *414,* 105–111.

85. Shay, J.W., Wright, W. E., and Werbin, H. (1993). Loss of telomeric DNA during aging may predispose cells to cancer (Review). Int. J. Oncol. *3,* 559–563.

86. Salk, D., Au, K., Hoehn, H., and Martin, G.M. (1981). Cytogenetics of Werner's syndrome cultured skin fibroblasts: variegated translocation mosaicism. Cytogenet. Cell Genet. *30,* 92–107.

87. Salk, D., Fujiwari, Y., and Martin, G. (1985). Werner's Syndrome and Human Aging. Plenum Press, New York.
88. Singh, S.K., Clarke, I.D., Terasaki, M., Bonn, V.E., Hawkins, C., Squire, J., and Dirks, P.B. (2003). Identification of a cancer stem cell in human brain tumors. Cancer Res. *63,* 5821–5828.
89. Singh, S.K., Hawkins, C., Clarke, I.D., Squire, J.A., Bayani, J., Hide, T., Henkelman, R.M., Cusimano, M.D., and Dirks, P.B. (2004). Identification of human brain tumour initiating cells. Nature *432,* 396–401.
90. Slagboom, P.E., Droog, S., and Boomsma, D.I. (1994). Genetic determination of telomere size in humans: a twin study of three age groups. Am. J. Hum. Genet. *55,* 876–882.
91. Smogorzewska, A., and de Lange, T. (2004). Regulation of telomerase by telomeric proteins. Ann. Rev. Biochem. *73,* 177–208.
92. Sohn, S.H., Multani, A.S., Gugnani, P.K., and Pathak, S. (2002). Telomere erosion-induced mitotic catastrophe in continuously grown Chinese hamster Don cells. Exptl. Cell Res. *279,* 271–276.
93. Stansel, R.M., de Lange, T., and Griffith, J.D. (2001). T-loop assembly *in vitro* involves binding of TRF2 near the 3′ telomeric overhang. EMBO J. *20,* E5532–E5540.
94. Thannhauser, S.J. (1945). Werner's syndrome (progeria of the adult) AND Rothmund's syndrome: two types of closely related heredfamilial atrophic dermatosis with juvenile cataracts and endocrine features, A critical study with five new cases. Ann. Intern. Med. *23,* 559–625.
95. Tlsty, T.D., Margolin, B.H., and Lum, K. (1989). Differences in the rates of gene amplification in nontumorigenic and tumorigenic cell lines as measured by Luria-Delbruck fluctuation analysis. Proc. Natl. Acad. Sci. USA *86,* 9441–9445.
96. Vaziri, H., Dragowska, W., Allsopp, R.C., Thomas, T.E., Harley, C.B., and Lansdorp, P.M. (1994). Evidence for a mitotic clock in human hematopoietic stem cells: loss of telomeric DNA with age. Proc. Natl. Acad. Sci. USA *91,* 9857–9860.
97. Wang, R.C., Smogorzewska, A., and de Lange, T. (2004). Homologous recombination generates T-loop-sized deletions at human telomeres. Cell *119,* 355–368.
98. Watson, J.D. (1972). Origin of concatemeric T7 DNA. Nature New Biol. *239,* 197–201.
99. Wu, L., Multani, A.S., He, H., Cosme-Blanco, W., Deng, Y., Deng, J.M., Bachilo, O., Pathak, S., Tahara, H., Bailey, S.M., Deng, Y., Behringer, R.R., and Chang, S. (2006). Pot1 deficiency initiates DNA damage checkpoint activation and aberrant homologous recombination at telomeres. Cell *126,* 49–62.

Genetic Basis of Tropical Calcific Pancreatitis: Similar Finds, Different Minds

10

Giriraj Ratan Chandak

Centre for Cellular and Molecular Biology, Hyderabad-500 007, India

Tropical Calcific Pancreatitis: An Etiological Enigma

Chronic pancreatitis (CP, OMIM 167800) is an inflammatory disease of pancreas characterized by irreversible destruction and fibrosis of the exocrine parenchyma, causing pain, exocrine pancreatic insufficiency and progressive endocrine failure leading to diabetes (Kloppel and Maillet, 1991). The disease is known to be highly heterogeneous and occurs due to a number of etiological factors. Alcoholic chronic pancreatitis (ALP) is the commonest type of CP that is observed in the western world but in about a third of CP patients, the etiology is not clear and they are classified under idiopathic chronic pancreatitis (ICP) (Steer and Waxman, 1995). Tropical calcific pancreatitis (TCP) is a juvenile non-alcoholic form of CP, observed almost exclusively in developing countries located in the tropical world (Geevarghese et al., 1962; Viswanathan, 1980; Abu-Bakare et al., 1986; Thomas et al., 1990; Rao, 1993; Mittal et al., 2002; Barman et al., 2003). TCP has been reported across the Indian subcontinent (Yajnik et al., 1993), but it is most prevalent in Kerala, South India (Geevarghese et al., 1962; Viswanathan, 1980; Thomas et al., 1990). The reported prevalence of TCP is 0.012% (126/10,000) in general population (Balaji et al., 1994), which is significantly higher than has been reported for CP in the West (10–15/100,000) (Steer and Waxman, 1995). The cardinal manifestations of TCP are recurrent abdominal pain in childhood, followed a few years later by the onset of insulin-requiring diabetes mellitus, also known as fibrocalculous pancreatic diabetes (FCPD) (Mittal et al., 2002). In contrast to early-onset ICP seen in the West (Layer et al., 1994), TCP is a more rapidly progressive disorder and has been described as a disease with "pain in childhood, diabetes in puberty and death at the prime of life" (Geevarghese, 1986).

The etiopathogenic mechanism of TCP is not well understood (Abu-Bakare et al., 1986; Barman et al., 2003). As is true for other complex diseases, both environmental and genetic factors are known to play important role in its causation. Evidence that environmental factors play a role in the etiology of TCP includes the fact that the disease is reported from many different third world tropical regions, where genetic backgrounds are different, but similar environmental factors (such as poor nutrition, infections) are prevalent (Abu-Bakare et al., 1986; Rao, 1993). Commonly known predisposing factors such as alcohol intake, biliary tract stones or metabolic diseases are absent in these subjects. Based on the observation that TCP almost exclusively affects the poor population of developing nations, malnutrition was strongly suspected to be a major etiologic factor (Rao, 1984, Pitchumoni, 1985; Rao and Yajnik, 1996; Mohan and Pitchumoni, 1998). However, TCP is equally prevalent in individuals with higher economic background and recent studies suggest that malnutrition could be the effect rather than cause of the disease (Barman et al., 2003). The geographical distribution of TCP coincides with the areas of consumption of cassava, which is the staple diet of poor people in Kerala, India. Toxicity due to cyanogenic alkaloids in cassava roots (Technical report series, 2004) in presence of malnutrition and antioxidant deficiency has been proposed as an ideal setting for free radical injury (McMillan et al., 1980). However, TCP is prevalent in many parts of India and Africa where cassava is not consumed and is not seen in West African population consuming a high cassava diet (Abu-Bakare et al., 1986; Tuescher et al., 1987; Yajnik et al., 1993; Barman et al., 2003; Mittal et al., 2003). A study on rats fed with cassava diet for one year did not produce either pancreatitis or diabetes (Mathangi et al., 2000). Thus, it is unlikely that cassava ingestion can explain majority of cases of TCP seen worldwide and the current opinion is that cyanogen toxicity is not relevant in its etiopathogenesis. Identification of familial aggregation of TCP suggests a genetic etiology but there is no definite mode of inheritance (Mohan et al., 1989). This observation indicates that environmental factors may be operative in the background of a genetic predisposition to pancreatitis.

In the next few pages, I will briefly summarize details of the comprehensive approach that we have undertaken to investigate the genetic basis of TCP using it as a model to understand the exocrine and endocrine dysfunction of pancreas. We have used a combination of candidate gene based study as well as functional genomic approaches including microarray analysis of the pancreatic tissue from patients and normal subjects. While candidate gene approach permits to test various hypotheses generated from the understanding of clinical picture of the disease, microarray analysis for global expression aids in generating hypotheses that can subsequently be tested in the patient cohorts to evaluate their potential applicability in predicting the clinical course of a disease.

Pancreas and its Physiology

More than a century ago in 1896, it was hypothesized by Hans Chiari that inappropriate intrapancreatic activation of pancreatic zymogens is the initiating event in the development of pancreatitis (Chiari, 1896). Three isoforms of trypsinogens have been described in human pancreatic juice and are known as cationic trypsinogen (encoded by *PRSS1*; OMIM 276000), anionic trypsinogen (encoded by *PRSS2*; OMIM 601564) and mesotrypsinogen (encoded by *PRSS3*) according to their electrophoretic mobility (Rinderknecht et al., 1978). Normally, trypsinogen becomes active by the action of intestinal enterokinase, only after it is secreted into the duodenum. It is thought that generally, about 5-10% of trypsinogens get activated within the normal pancreas. Trypsin plays a central role in pancreatic exocrine physiology by acting as a trigger enzyme leading to activation of all other pancreatic zymogens. Pancreas has several safety mechanisms to cope with the premature activation of these enzymes, which would otherwise lead to indiscriminate proteolysis, also known as autodigestion (Gaboriaud et al., 1996). Any derangement of any of the protective mechanisms may lead to CP. Trypsin is known to lose its activity spontaneously by autolysis at the initial hydrolytic point of trypsin at Arg122-Lys123, which renders it more susceptible to further degradation (Gaboriaud et al., 1996). The serine protease inhibitor, Kazal type 1 (*SPINK1*) (OMIM 167790), also known as pancreatic secretory trypsin inhibitor (PSTI), is a potent anti-protease that is thought to be a major protective mechanism of intrapancreatic trypsin activity (Laskowski and Fu, 1953). SPINK1 is present in the secretory granules of acinar cells and causes temporary inhibition of tryptic activities by binding to its active site in a 1:1 ratio (Laskowski and Fu, 1953). Other protective mechanisms include the presence of trypsin inhibitors in plasma including $alpha_1$ anti-trypsin and $beta_2$ microglobulin, which inhibit the trypsin that leaks into the interstitial space around the pancreas (Gaboriaud et al., 1996). In addition, intra-acinar calcium levels are usually kept low since trypsinogen activation and trypsin survival are known to be regulated by calcium (Whitcomb, 2004).

Mutations in Trypsinogen Gene Family are not Responsible for TCP

Nearly a century after Chiari's proposition, several groups by linkage analysis, located a gene for hereditary pancreatitis on the long arm of chromosome 7 (7q35) (LeBodic et al., 1996; Whitcomb et al., 1996a). Subsequently, a mutation in the cationic trypsinogen gene, *PRSS1*, was identified as the underlying defect. The mutation, 365G>A transition led to a substitution of arginine by histidine at residue 122 (R122H) and co-segregated with the disease (Whitcomb et al., 1996b). Subsequently, several other *PRSS1* alterations have been reported in families with suspected hereditary pancreatitis or in patients without a family history. While, R122H is the most common *PRSS1* mutation observed worldwide, mutations such as A16V, N29I, N29T, R116C, and R122C have also been identified in number of patients (Teich et al., 2006). Site-directed mutagenesis of recombinant human cationic trypsinogen revealed that while most mutations studied (e.g. N29I and R122H) significantly enhance autoactivation *in vitro*, some mutations such as R122H additionally inhibit autolysis of the active enzyme (Sahin-Toth and Toth, 2000). Although the precise disease mechanisms for many of the *PRSS1* variants have not been unraveled, it is assumed that they will fit with the gain of function nature of mutations and explain the generally accepted model that enhanced intrapancreatic trypsinogen activation may be the common initiating step of pancreatitis.

Since the clinical picture of TCP and HP resembles closely, it was assumed that similar genetic alterations might explain the genetic basis of TCP. On screening the *PRSS1* gene for commonly reported variants in 68 TCP patients and comparison with 100 healthy volunteers, we did not detect any of the reported common mutations such as R122H, N29I or A16V etc. Further, sequencing of the complete gene also failed to identify any novel variant in *PRSS1* gene in these patients (Chandak et al., 2002). Subsequently, we extended the analysis to a larger number of patients from different states including Kerala (n=88), Tamilnadu (n=91), Orissa (n=70), Maharashtra (n=113), West Bengal (n=52), Jammu & Kashmir (n=35) but did not identify any *PRSS1* variant in them. Our observations have subsequently been replicated by two other groups from North and South India [Bhatia et al., 2002; Hassan et al., 2002]. We also observed complete absence of *PRSS1* mutations in patients with other types of chronic pancreatitis including HP, ICP and ALP in Indian population (Chandak et al., 2004).

Recently, triplication of a ~605 kb segment containing *PRSS1* and *PRSS2* was identified in HP families (Le Marechal et al., 2006) and provided further evidence that gain of trypsin function through a gene dosage effect may also play a pathological role in disease pathogenesis. Since, we did not find any *PRSS1* mutations in Indian TCP patients, trypsinogen copy number variations (CNVs) could explain the role of cationic trypsinogen in them. However, analysis of 268 Dravidian patients failed to identify the causal CNV in them (Masson et al., 2008). Further, analysis of other CNVs reported in the literature showed complete absence of the three copy number polymorphisms that included a deletion of the *PRSS1* or/and *PRSS2* loci (Database of Genomic Variants) in Indian population. This not only provides support to the recent genome-wide observation that marked variation in copy number exists among populations but also suggests the presence of ethnic group-specific pancreatitis–causing or –protecting genomic factors (Masson et al., 2008). The current model of *PRSS1* mutations suggests that the identified mutations cause enhanced autoactivation of trypsinogen to trypsin or prevent prematurely activated trypsin from being inactivated by autolysis. Hence, it was surprising that no mutation could be detected in the *PRSS1* gene in these patients. Absence of any point mutations as well as the trypsinogen triplication/duplication mutations in the Indian TCP patients stresses on the lack of pathological role for cationic trypsinogen gene mutations in TCP in the Indian population.

Since increased proteolytic activity due to mutated *PRSS1* strongly predicts the risk for CP, mutations in anionic trypsinogen, which accounts for about 1/3rd of the total trypsins in the pancreatic juice, may also lead to the disease. We sequenced the complete *PRSS2* gene in 68 TCP patients but failed to detect any mutation in them and thus concluded that mutations in the trypsinogens including anionic trypsinogen are not important in the pathogenesis of TCP (Idris et al., 2005). A recent study identified a 571G>A transition leading to G191R variant in *PRSS2* and reported strong association with CP (OR=0.37, $P=1.1x10^{-8}$) in European patients (Witt et al., 2006). The risk allele was over represented in the control subjects (3.4%; 220/6459) in contrast to only 1.3% of patients (32/2466) (Witt et al., 2006). They also showed by *in vitro* analysis of the recombinant G191R protein that trypsin activity is lost due to the introduction of a novel tryptic cleavage site that renders the enzyme hypersensitive to autocatalytic proteolysis (Witt et al., 2006). Taken together, the G191R *PRSS2* variant mitigates intrapancreatic trypsin activity and thereby plays a protective role against chronic pancreatitis. Since the risk allele frequency for the G191R variant is very low, hence, we investigated whether G191R *PRSS2* variant also confers protection against TCP in a large case-control cohort comprising 700 CP patients and 1272 normal individuals. In contrast to the Europeans (Witt et al., 2006), we observed a higher prevalence of the protective allele in patients [minor allele frequency=0.7% (5/700)] vs. 0.3% in control subjects (4/1272) and no significant association with CP (OR=2.27; 95% CI 0.53-10.05; P=0.21) (Gut, in press). Further, genotyping of five SNPs around *PRSS2* revealed that the R191 allele has originated on the same haplotype background as reported in the Europeans, while the wild allele had multiple origins. A very low prevalence of G191R variant in Indians and its association with one haplotype background suggests a recent origin for the mutation. Interestingly, the frequency of 191R allele in our study is comparable to that observed in Africans (Witt et al., 2006). It is believed that migrations out of Africa, after passing through a population bottleneck has led to African-Eurasian divergence around 100,000 years ago, followed by an European Asian divergence about 40,000 years ago (http://www.actionbioscience.org/evolution/johanson.html). It has also been postulated that modern humans did not leave Africa and successfully colonize the rest of the world until as recently as 50,000 years BP, pushing back the dates for subsequent population splits (Wade, 2006). Thus, it may be surmised that the protective variant in *PRSS2* has undergone selection in the Europeans but not in the Africans and the Indians since it does not play an important role in protecting them against CP. There are 2 possible explanations; firstly, current evidence suggests that CP in Asian Indians is caused through a different mechanism (Chandak et al., 2002; Bhatia et al., 2002; Chandak et al., 2004; Idris et al., 2005; Mahurkar et al., 2006). Secondly, the protein content and the nature of amino acids in a meal contribute markedly to the levels of digestive pancreatic enzymes (Ekelund and Johnson, 1980). The total protein content of net available daily food supplies per caput (38-54g) as well as the available animal proteins (4-8g) in developing countries like India are significantly lower than those in the developed countries (82-120g and 30-75g respectively) (FAO, 1977). A recent large-scale survey has confirmed the existence of persistent protein deficiency in India (Jha, 2003). The levels of pancreatic enzymes are expected to be lower in the pancreatic juice of Indians than that of Europeans, since compared to animal foods, plant foods have lower content of essential amino acids, especially the delimiting amino acids such as methionine, lysine and tryptophan that determine the stimulatory potency of dietary proteins (Go et al., 1970). Indeed, an earlier study demonstrated a 2-4 times lower content of pancreatic enzymes like phospholipase, trypsin and chymotrypsin in the pancreatic juice of Indians compared to that in French subjects reflecting the consequences of lower protein and fat content in Indian diets (Balakrishnan et al., 1988). Hence, as a combination of the above, the overall amount of trypsinogens in the pancreatic juice of Indians is less than that in Europeans and the anionic isoform is even less since it represents only 1/3rd of the total trypsinogens. Thus, its role in pancreatic protein digestion may be miniscule and hence the G191R variant, which makes less functional anionic trypsinogen, may not make any further impact in Indians.

It is generally believed that inappropriate intrapancreatic activation of trypsinogens is a precondition for acinar cell injury and pancreatitis. Of the three isoforms of trypsinogens secreted by the human pancreas, the cationic isoform constitutes about ~70-75% and anionic trypsinogen makes up only ~25-30% (Rinderknecht et al., 1978). Hence, the larger pool of cationic trypsinogen would remain available for autoactivation, even after the autocatalytic degradation of G191R anionic trypsinogen. Intracellular calcium levels and the pH are known to influence the activation and stability of trypsinogens 2003 (Kukor et al). It has been have shown that in contrast to the characteristic stimulation of the cationic proenzyme, acidic pH (~5.0) inhibits autoactivation of anionic trypsinogen and anionic trypsinogen undergoes rapid autocatalytic degradation at pH 8.0, especially at low Ca^{2+} concentrations (Kukor et al., 2003). Both the conditions represent the state of intracellular vesicular compartments and suggest that cationic trypsinogen remains the major isoform that determines the rate of activation and the final levels of trypsin generated (Kukor et al., 2003). Taken together, anionic trypsinogen may be of little or no consequence in the initiation of CP through the premature activation of trypsinogens.

Thus, comprehensive analysis of the trypsinogen gene family suggests that trypsinogen variants do not play a vital role in TCP or other types of CP in India and alternative mechanisms may be more crucial in its pathogenesis in Indians.

A Founder Mutation in *SPINK1* Gene Predicts Susceptibility to TCP

Serine Protease inhibitor, Kazal type 1 is a potent protease inhibitor and thought to be a major protective mechanism preventing inappropriate activation of pancreatic digestive enzyme cascade by inhibiting up to 20% of potential trypsin activity (Laskowski and Fu, 1953). In view of the lack of any mutations in the *PRSS1* gene, we hypothesized that loss of function variants in pancreatic trypsin inhibitor may also cause pancreatitis. We screened the *SPINK1* gene by sequencing the coding exonic and flanking intronic sequences including the promoter region and identified a mutated *SPINK1* in 32 out of 68 patients (46%) analysed, with a founder mutation N34S in the majority, whereas the mutant allele frequency in controls was 1.5% (3/200) (Chandak et al., 2002). *SPINK1* mutations were detected in both FCPD and in TCP patients without diabetes mellitus. A novel G to T transversion at 215 bp upstream to the ATG initiation codon in the *SPINK1* promoter region (-215G>T) was also identified in 3 patients but not in the controls (Table 10.1) (Chandak et al., 2002). We identified an equally high prevalence of the risk allele in other types of chronic pancreatitis suggesting that *SPINK1* mutations rather than mutations in trypsinogen genes are associated with chronic pancreatitis in Indian population. The association between N34S *SPINK1* and CP has been confirmed by numerous others (for review see: Witt et al., 2007) and pooled analysis of 8 large studies in Europe and US showed that 12.6% of CP patients were heterozygous for N34S, and 3.6% were homozygous (Witt et al., 2007), whereas only heterozygous individuals (1.9%) were detected in the controls.

In our study cohort, 24 were FCPD patients while remaining 44 were without diabetes at the time of presentation. We observed that FCPD patients had a late age of onset (35 yrs vs. 21 yrs) and

Table 10.1 Distribution of Mutations and Polymorphisms in the SPINK1 Gene among Tropical Calcific Pancreatitis Patients and Controls

Location	Polymorphism	Patients (n = 68)	Controls (n = 100)
5′ UTR	–253T>C	6 (6 het) (0.04)	43 (42 het) (0.22)
	–215G>A	0	0
	–215G>T[a]	3 (3 het) (0.02)	0
Intron 1	IVS1–37T>C[b]	31 (8 hom) (0.29)	3 (3 het) (0.015)
Intron 2	IVS2+286A>G[b]	31 (8 hom) (0.29)	3 (3 het) (0.015)
Exon 3	101A>G (N34S)	31 (8 hom) (0.29)	3 (3 het) (0.015)
	163C>T (P55S)	02 (2 het) (0.015)	1 (1 het) (0.005)
Intron 4	IVS3-604G>A[b]	31 (8 hom) (0.29)	3 (3 het) (0.015)
	IVS3-66insTTTT[b]	31 (8 hom) (0.29)	3 (3 het) (0.015)

n, number; hom, homozygous; het, heterozygous. Allelic status and frequencies are given in the parentheses.
a, novel mutation, never reported earlier; b, detected on all alleles with N34S

majority of TCP patients had an abnormal glucose tolerance test, suggesting that FCPD may be gradually evolving diabetes in the background of TCP (Table 10.2) (Chandak et al., 2002). *SPINK1* mutation frequency in FCPD patients and TCP patients without diabetes showed similar trends suggesting that neither these mutations nor their status is directly related to the development and presentation of diabetes mellitus. There was no significant difference in parameters like the age of onset, presence or absence of diabetes mellitus and age of onset of diabetes between *SPINK1* N34S heterozygous, homozygous or compound heterozygous patients suggesting lack of genotype-phenotype correlation, the hallmark of any complex disease. Similar over-representation of the mutant allele at N34S *SPINK1* was observed in TCP patients compared to controls from different parts

Table 10.2 Clinical Data and Status of *SPINK1* Mutations in TCP Patients

	TCP with Diabetes (FCPD)	TCP without Diabetes	Total
n	24 (17M/7F)	44 (29M/15F)	68 (46M/22F)
Age at onset (in yrs) (Median, range 95% CI)	35.0 (27.6–40.2)	21.0 (19.9–27.1)	24.0 (23.8–30.5)
Age at presentation (in yrs) (Median, range 95% CI)	44.0 (39.3–49.3)	26.0 (25.7–33.0)	33.5 (31.3–38.0)
SPINK1 mutation	12	20	32
Homozygous	4	4	8
Heterozygous	8	16	24
N34S/WT	6	16	22
N34S/P55S	1	-	1
P55S/WT	1	-	1

n, number; WT, wild type; TCP; tropical calcific pancreatitis; FCPD, fibro-calculous pancreatic diabetes

of India, although the background allele frequency of the risk allele varied from 1.2% to 2.2% in different states in India. This confirms that N34S *SPINK1* acts as a susceptibility allele for TCP irrespective of the ethnic origin (unpublished data). This mutation also predicted the risk of pancreatitis in other types of chronic pancreatitis, although strength of association varied for each type (Chandak et al., 2004). Thus, irrespective of the etiology, *SPINK1* mutations predict susceptibility to CP in the Indian population.

Although, N34S *SPINK1* mutation is the strongest genetic risk factor in TCP and ICP, the pathogenic mode of action of N34S remains elusive. No alteration in trypsin inhibitor capacity or secretion has been noted on analysis of recombinant N34S mutated human SPINK1 (Kuwata et al., 2002; Kiraly et al., 2007; Bouling et al., 2007). However, targeted disruption of *SPINK3*, the mouse homologue showed acinar cell degeneration, impaired regeneration and death within 2 weeks but no evidence of enhanced acinar trypsin activity or inflammation was observed (Ohmuraya et al., 2005). Incidentally, both in Indians and Europeans, N34S is known to be in complete linkage disequilibrium (LD) with four other intronic sequence variants: –37T>C, +268A>G, –604G>A, and -65insTTTT (Witt et al., 2000), suggesting that one of these variants, and not N34S itself, may be the pathogenic relevant mutation. In addition to N34S mutation in *SPINK1* gene, another variant –215G>A identified in FCPD patients may lie in an unknown *cis*-acting element and thereby affect transcription of mutated *SPINK1* with an already compromised functional capacity. Further stress on its pathological role comes from the fact that the same nucleotide a G>A transition (–215G>A) has been described in several patients from Japan and Western countries, and is unambiguously associated with chronic pancreatitis in non-Indian populations (Kaneko et al., 2001). However, the –215G>A alteration is, in contrast to –215G>T, in complete linkage disequilibrium with another *SPINK1* variant, +2T>C, which affects the splice donor site leading to truncated SPINK1 due to skipping of exon 3 (Kaneko et al., 2001; Kume et al., 2005; Kume et al., 2006). Thus, the pathogenic significance of the –215 variants remains uncertain.

Tropical Calcific Pancreatitis has a Genetic basis, which is Different than ICP in the West

Such high frequency of *SPINK1* gene mutations in the background of complete absence of mutated *PRSS1* and *PRSS2* genes is interesting. This pancreatic protease-protease inhibitor system is very important, considering the physiological interaction between them in combating the prematurely activated trypsinogen inside the pancreas. Intrapancreatic levels of trypsin are expected to be elevated if mutations in the inhibitor molecule lead to loss of its inhibitory capacity. However, it is difficult to explain the phenotype of pancreatitis in the presence of a normal trypsin when the intact R122 autolysis "self-destruct" mechanism can take care of the prematurely activated trypsin molecule. This suggests impairment of other protective mechanisms involved in combating the prematurely activated trypsin molecule inside the pancreas. Our results suggest a common genetic basis for tropical calcific pancreatitis and there could be additional genetic/environmental factors responsible for the variability of phenotype in TCP and FCPD. Based on the the knowledge that about 50% of TCP patients have N34S mutation in *SPINK1*, patients and their relatives may be screened and counseled about the risk of developing the disease. An appropriate prevention strategy could also be suggested to help the carriers of the mutated allele in modifying the course of pancreatitis.

Thus, we showed for the first time that TCP has a genetic basis and it is different from what is reported for the western world. Based on our study, tropical calcific pancreatitis was categorized as a genetic disease and assigned a separate entry and number in the Online Mendelian Inheritance in Man (OMIM #608189).

Cathepsin B: The Link between Premature Intrapancreatic Activation of Cationic Trypsinogen and Tropical Calcific Pancreatitis

It is already established that several cationic trypsinogen mutations identified in CP patients from the western world lead to significant change in its activation (Sahin-Toth and Toth, 2000). Since no *PRSS1* mutations have been identified in TCP patients, a pathogenic role for autoactivation by trypsin is ruled out in them. Additionally, *SPINK1* mutations are prevalent in only ~50% of TCP patients and mutated *SPINK1* is hypothesized to act by reducing the capacity to block prematurely activated trypsinogen (Chandak et al., 2002). Intra-pancreatic activation of cationic trypsinogen requires a trigger (genetic or environmental) that would cause premature activation of trypsinogen leading to injury of acinar cells and recurrent attacks of pancreatitis. Intra-pancreatic activation of cationic trypsinogen and presumably other zymogens occurs within acinar cells and is known to be mediated by lysosomal hydrolase, cathepsin B (*CTSB*, OMIM 116810) (Szilagyi et al., 2001). Further, evidence comes from the fact that after induction of experimental cerulein-induced pancreatitis, Ctsb knock-out mice displayed a lowered trypsin activity within the gland and a diminished acinar necrosis compared to wild-type animals (Halangk et al., 2000). Hence, we hypothesized that CTSB-induced trypsinogen activation may play an important role in the pathogenesis of TCP and could be caused by mutations in the *CTSB* gene. Since *SPINK1* is proposed to play a modifier role, patients with mutated *SPINK1* may still need a trigger for cationic trypsinogen activation and hence *CTSB* mutations may explain the disease irrespective of the presence or absence of *SPINK1* mutations.

We initially sequenced the coding region, flanking intron-exon boundary and 5′- and 3′-UTR regions of *CTSB* gene in 51 patients and 25 healthy controls and identified 23 variants spread across the gene. Based on the initial results, we screened additional 89 patients and 130 controls for four polymorphisms, Leu26Val, C595T, T663C and Ser53Gly in the *CTSB* gene. We found a statistically significant association of Val26 allele at Leu26Val polymorphism with an OR of 2.15 (95%CI=1.60-2.90, P=0.009), which was confirmed in an independent Dravidian case-control cohort (Mahurkar et al., 2006). Overall, the Leu26Val variant was about twice as common among the TCP patients than in the controls (OR=2.10, 95%CI-1.56-2.84, P=0.013) (Table 10.3). The Val26 allele frequency was similar in N34S *SPINK1* positive and N34S *SPINK1*negative patients but higher than in controls (P=0.019 & 0.013 respectively) (Table 10.4). We did not observe any difference in the phenotype with regard to the presence or absence of N34S *SPINK1* mutation in individuals carrying the Leu26Val mutation. Thus, the proposed effects of Leu26Val mutation are unrelated to changes in the *SPINK1* gene (Mahurkar et al., 2006). Since Leu26Val affects the propeptide region of cathepsin B, crucial for its transport, it may affect CTSB trafficking as the mannose-6-phosphate dependent sorting of CTSB may mislocalize it to the zymogen granules, causing premature activation of trypsinogens. A similar mechanism may be operative at Ser53Gly polymorphism, since the hydrophobic amino acid glycine replaces the polar amino acid serine in the same region of CTSB (Mahurkar et al., 2006). Another interesting explanation regarding the role of these mutations could emerge from the fact that cathepsin B is also known to preferentially activate mesotrypsinogen, which in turn is known to degrade SPINK1 and thus lower the physiological levels of the inhibitor (Szmola et al., 2003). Hence, it may be speculated that these *CTSB* variants induce premature activation of trypsinogens especially cationic trypsinogen, which in the presence of inhibitor with reduced capacity either due to N34S *SPINK1* mutation or mesotrypsin induced inhibitor degradation leads to recurrent attacks of pancreatitis. Screening of the TCP patients for *PRSS3* gene (coding for mesotrypsinogen) may identify mutations, which may stabilize mesotrypsin or loss of function mutations that protect against pancreatitis.

Thus, we have provided the first human evidence to the 30-year-old cathepsin B hypothesis of pancreatitis (Lerch and Halangk, 2006) and propose that *CTSB* gene is the second candidate gene that can play a role in pathogenesis of TCP and predict susceptibility to the disease, independent of *SPINK1* mutations.

Current Model of TCP (CP) in the Indian Population

Based on our case-control studies on tropical calcific pancreatitis and on chronic pancreatitis of different etiology, we hypothesize a model of chronic pancreatitis in Indian population (Fig. 10.1).

Role of other Candidate Genes in explaining various Clinical Features of TCP

The clinical picture of TCP is quite different than the commonly reported ICP, such as extensive fibrosis, invariable development of diabetes, presence of large stones etc. Besides genetic alterations

Table 10.3 Association of *CTSB* Gene Polymorphisms in Tropical Calcific Pancreatitis patients and Controls

SNP[#]	**Cohort I**				**Cohort II**				**Total**			
	Allele Frequency[$]		OR (95% CI)	P value	Allele Frequency[$]		OR (95% CI)	P value	Allele Frequency[$]		OR (95% CI)	P value
	Patients (n=140)	Controls (n=155)			Patients (n=166)	Controls (n=175)			Patients (n=306)	Controls (n=330)		
Leu26Val*	0.48	0.30	2.15 (1.60–2.90)	**0.014**	0.45	0.28	2.10 (1.56–2.84)	**0.019**	0.46	0.29	2.09 (1.55–2.81)	**0.019**
C595T	0.12	0.20	0.55 (0.37–0.81)	0.177	0.10	0.21	0.41 (0.28–0.63)	0.05	0.11	0.21	0.46 (0.31–0.70)	0.083
T663C	0.52	0.58	0.78 (0.59–1.04)	0.477	0.53	0.60	0.75 (0.56–1.00)	0.392	0.53	0.59	0.78 (0.59–1.04)	0.476
Ser53Gly	0.10	0.04	2.67 (1.45–4.91)	0.166	0.09	0.04	2.37 (1.28–4.40)	0.251	0.09	0.04	2.37 (1.28–4.40)	0.251

[#], Nomenclature is represented as wild status-position-mutant status; [$], represents mutant allele frequency; n, number of individuals, OR, odds ratio; 95% CI, 95% confidence interval. The frequency of mutant allele for each SNP was compared between the patients and controls and P values were calculated along with the OR and 95% CI; *, statistically significant after allowing for multiple post hoc testing (P_{corr}=0.025, with Bonferroni correction)

Table 10.4 Distribution of *CTSB* Gene Polymorphisms in Tropical Calcific Pancreatitis Patients Based on N34S *SPINK1* Status

SNP#	Mutant allele frequency			N34S Pos vs. N34S Neg			N34S Pos vs. Controls			N34S Neg vs. Controls		
	N34S Pos (n=134)	N34S Neg (n=172)	Controls (n=330)	X^2	OR (95% CI)	P value	X^2	OR (95% CI)	P value	X^2	OR (95% CI)	P Value
Leu26Val	0.45	0.46	0.29	0.02	0.96 (0.72–0.28)	0.887	4.83	2.00 (1.49–2.70)	0.025*	5.46	2.09 (1.55–2.81)	0.019*
C595T	0.05	0.16	0.21	5.32	0.28 (0.16–0.47)	0.021*	9.95	0.20 (0.12–0.33)	0.002*	0.53	0.72 (0.50–1.03)	0.467
T663C	0.51	0.55	0.59	0.18	0.85 (0.64–0.13)	0.671	0.99	0.72 (0.54–0.96)	0.320	0.18	0.85 (0.64–1.13)	0.668
Ser53Gly	0.15	0.05	0.04	4.50	3.35 (1.96–0.74)	0.034	5.82	4.23 (2.37–7.58)	0.016*	0.12	1.26 (0.64–2.51)	0.733

N34S Pos, N34S *SPINK1* positive; N34S Neg, N34S *SPINK1* negative; n, number of individuals; X^2, chi-square value; OR, odds ratio; 95% CI, 95% confidence interval; #, Nomenclature is represented as wild status-position-mutant status, The frequency of mutant allele for each SNP was compared between three groups and Yates X^2 and P values were calculated along with the OR and 95% CI,*, statistically significant after allowing for multiple post hoc testing (P_{corr} =0.025, with Bonferroni correction)

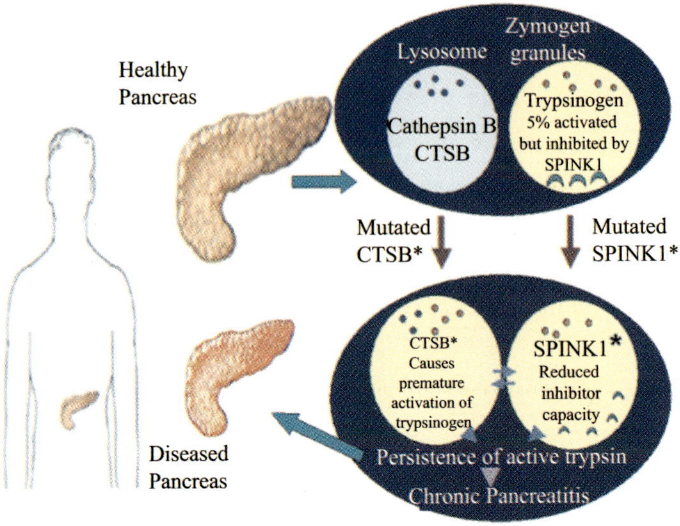

Fig. 10.1 Proposed model for tropical calcific pancreatitis in Indian population. In the healthy pancreas, cathepsin B and trypsinogens are localized in lysosomes and zymogen granules respectively and trypsin inhibitor (SPINK1) blocks the prematurely activated trypsinogen. However, mutated SPINK1 may have reduced inhibitory capacity and thus prematurely activated trypsinogen may persist. On the other hand, mutant cathepsin B may get mislocalized to the zymogen granules along with trypsinogen and act as trigger for premature intrapancreatic activation of trypsinogens. Both these conditions may co-exist but are independently capable of causing recurrent attacks of pancreatitis.

in genes encoding enzymes or enzyme inhibitors of the trypsinogen pathway, we investigated several other candidate genes, which were selected based on major clinical features in TCP.

Angiotensin Converting Enzyme (ACE) Gene I/D Polymorphism does not Predict Extensive Fibrosis in Tropical Calcific Pancreatitis

The systemic hormonal renin-angiotensin system (RAS) has important physiological functions in the regulation of blood pressure, fluid and electrolyte balance. They are mainly executed by angiotensin II, the product of hydrolysis of angiotensin I by angiotensin-converting enzyme (ACE). Apart from circulating RAS, many tissues and organs exhibit their own RAS products and activities. In the pancreas, major components of such a local RAS have been implicated in the regulation of acinar digestive enzyme secretion, islet hormonal secretion, and ductal anion secretion (Leung and Chappell, 2003). Recent literature has suggested a crucial role for pancreatic renin-angiotensin system during chronic hypoxia in acute pancreatitis and for *ACE* inhibitors in reducing pancreatic fibrosis in experimental models (Ip et al., 2003; Kuno et al., 2003). A frequent *ACE* insertion/deletion (I/D) alteration in intron 16 (g.11417_11704del287) exists that accounts for nearly half of the variation of serum ACE levels (Rigat et al., 1990). On analysis of *ACE* gene I/D polymorphism in 171 TCP patients and 99 ethnically matched healthy subjects, we did not observe any significant difference either in the 'D' allele frequency or in the genotype distribution between patients and controls (0.46 and 0.47 respectively) and between TCP and FCPD patients (0.44 and 0.48 respectively) (Bhaskar et al., 2006). No evidence of any interaction between the I/D variant in the *ACE* gene and the N34S mutation in the *SPINK1* gene in either TCP or FCPD patients was noted (Bhaskar et al., 2006). Thus, the *ACE* gene polymorphism does not show any significant association with the pathogenesis, fibrosis and progression of TCP and FCPD and hence ACE inhibitors may not play a significant role in modulating the extent of fibrosis in these patients.

Atypical Variants in *CFTR* Gene Predict Susceptibility to TCP

The clinical presentation as a result of close to a thousand mutations in cystic fibrosis transmembrane regulator (*CFTR*; OMIM 602421) gene is known to have a variable spectrum that ranges from classical cystic fibrosis (CF) to mild phenotypes which includes chronic pancreatitis (Cohn et al., 1998; Noone et al., 2001). The only published study till date has analysed 18 Indian TCP patients and identified 2 subjects with a *CFTR* variant but their pathogenic nature was doubtful (Bhatia et al., 2000). We screened by DHPLC, the complete *CFTR* gene in 35 unrelated TCP patients and nearly one-fourth of patients (9/35, 26%) showed non-synonymous SNPs including 2 novel mutations (L183R and Q1352H), whereas none were identified in any of the 100 controls (unpublished data). Other six mutations (delta F508, I807M, V920M, E410X, R75X and 5T) have been reported to be associated with CF as per the *CFTR* mutations database. Interestingly, only one patient was heterozygous for delta F508 mutation, which is the major mutation responsible for CF. Interestingly, two patients with 5T allele (commonly associated with congenital bilateral absence of vas deferens), also carried $(TG)_{12}$ in intron 9, a combination that is associated with increased *CFTR* transcript without exon 9. Out of 9 patients with *CFTR* mutations, 7 (~80%) were also heterozygous for N34S mutation in the *SPINK1* gene and the remaining two patients carrying the nonsense mutations, R75X and E410X showed wild status for *SPINK1* gene. This suggests that severe mutation in CF gene in heterozygous state or a transheterozygosity of mild *CFTR* mutations with N34S *SPINK1* mutation may lead to alternative phenotype of TCP. Thus, we have conclusively established that *CFTR* gene mutations of different nature are prevalent in TCP patients and interact with N34S *SPINK1* mutation in predicting susceptibility to the disease.

Gene Variants Predicting Endocrine and Exocrine Manifestations in TCP and FCPD Interact with Each Other

It is believed by most workers in the field that development of diabetes or FCPD is the logical end point of TCP and enough evidence exists to suggest that TCP is the pre-diabetic stage of FCPD (Mohan et al., 1983). The diabetes in TCP i.e., FCPD is thought to be type 2 diabetes (T2D) based on observations on traits including insulin resistance, although it has earlier been proposed to be diabetes secondary to destruction of pancreatic cells which is suggestive of type 1 diabetes (T1D) (Mohan et al., 1988; Mehrotra et al., 1997). Most of the patients present with pain abdomen with evidence of pancreatitis and subsequently develop diabetes at a later stage but a small proportion present with diabetes and are detected to have pancreatic stones and calcification on subsequent investigations (Reddy, 1998; Barman et al., 2003). Thus, additional diabetes susceptibility gene may account for the earlier phenotype of diabetes. We and others have established the strong association of *TCF7L2* variants with T2D in Indian population and provided evidence of its likely role in the pathogenesis of T2D by influencing both insulin secretion and insulin resistance (Chandak et al.,

2007). We analyzed variants in the *TCF7L2* gene in a cohort of TCP and FCPD patients and controls belonging to Dravidian and Indo-European ethnicities but did not observe any significant differences in allelic or genotypic relative risk in Dravidian patients vs. Dravidian controls and Indo-European patients vs. Indo-European controls (Mahurkar et al., 2008). Even, meta-analysis of all cases and control subjects from both ethnicities, showed similar results for *TCF7L2* variants. Similarly, allelic and genotypic frequencies did not differ significantly, between TCP patients and controls, FCPD patients and controls and between TCP and FCPD patients, suggesting lack of statistically significant association of *TCF7L2* polymorphisms with FCPD. Association analysis of the total cohort after dichotomization based on N34S *SPINK1* and L26V *CTSB* mutation status showed interesting results. Co-inheritance of *TCF7L2* variants with N34S *SPINK1* and L26V *CTSB* variants showed a statistically significant association between TCP and FCPD patients carrying the mutant allele compared to those without the variant (OR=1.69, 95% CI=1.11-2.56, P=0.013) (Mahurkar et al., 2008). This suggests that co-existence of *TCF7L2* variants and the variants predicting susceptibility to exocrine damage may interact to determine the onset of diabetes in TCP patients.

Our observations have important implications for diabetes in TCP since overlapping features of T1D and T2D are observed in FCPD. As *TCF7L2* is a major susceptibility gene for T2D, a lack of association of *TCF7L2* variants with TCP or FCPD observed in our study, suggests that T2D associated *TCF7L2* variants are not associated with diabetes in TCP or the diabetes in TCP patients may not be similar to T2D. Thus, although the variants in *TCF7L2* increase the risk for T2D and may affect insulin secretion, they do not alter susceptibility to FCPD, the diabetes in TCP patients.

Microarray Analysis of Tropical Calcific Pancreatitis as Model for Chronic Pancreatitis in Tropics

As illustrated above, we analysed several candidate genes and established the genetic basis for TCP, which is different from what is known for the Europeans and Americans. However, candidate gene analysis suffers from the limitation that it gives a myopic view of the pathogenesis of any disease. Hence, we also attempted to identify the differentially regulated genes in TCP and understand their expression at mRNA level by microarray approach. We have also undertaken a systems-biology approach using the differentially regulated genes and explore the role of various pathways in the pathogenesis of TCP.

We conducted the microarray experiments on pancreatic tissue from 8 patients and 5 normal individuals (analyzed in triplicates). After data normalization, genes showing 1.5-fold change and above were considered for pathway and functional categorization analysis based on the Gene Ontology description. Pathway analysis revealed the involvement of several pathways such as signaling, apoptosis, calcium binding etc. One of the pathways reported to be involved in pancreatitis and pancreatic cancer is MAP Kinase signaling pathway (Fig. 10.2). We observed down-regulation of several genes and up-regulation of few that are known to be involved in this signaling pathway. We also found the down stream effectors of calcium signaling pathway to be down regulated. Genes involved in both apoptotic pathway and necrosis were also found to be differentially regulated. This observation is in agreement with the earlier reports that pancreatic acinar cell death is a balance between apoptosis and necrosis.

Maintenance of structural and functional integrity of the intestinal epithelial cells requires highly dynamic cell-cell and cell-matrix interactions involving different types of surface receptors. Among

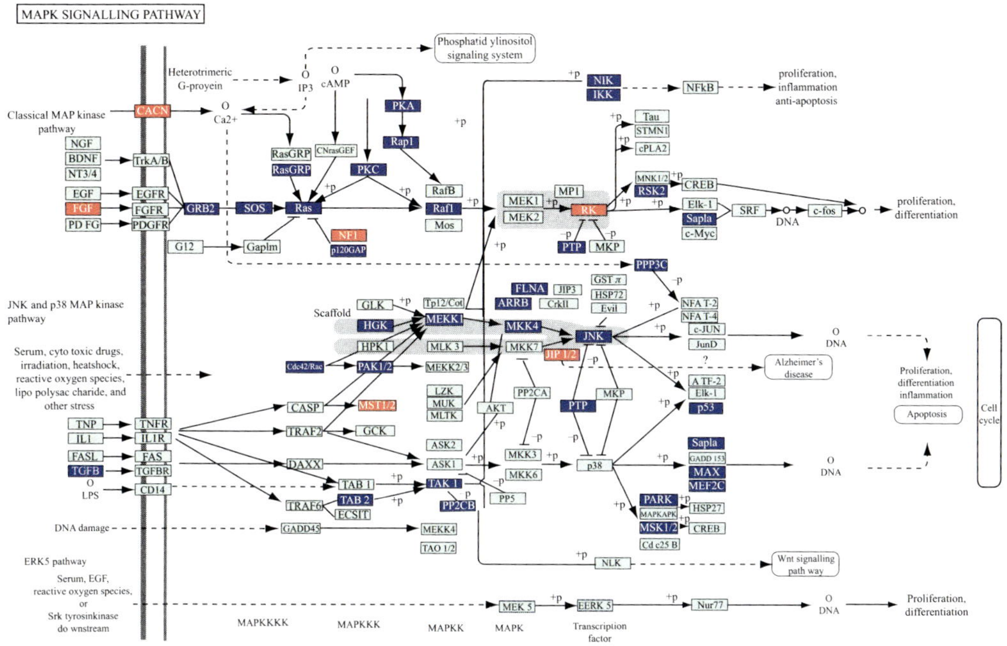

Fig.10.2 Mitogen activated protein kinase (MAPK) signaling pathway in tropical calcific pancreatitis patients. Pathway constructed from differentially expressed genes analyzed from cDNA microarray; Blue fill, down-regulation; red fill, up-regulation; green fill, no change; up-regulation of calcium channel (Calcineurin, CACN) and down-regulation of p53 are few of the interesting observations from this pathway.

these receptors, adhesion molecules, cadherins and integrins play a major role by recognizing and interacting with other cell adhesion receptors on neighboring cells and by binding components of the extra-cellular matrix (ECM). Besides providing mechanical advantage to the cell, these structures also are of functional importance; they transduce signals from the ECM and neighboring cells that are critical for survival and proliferation (Kang, et al., 2007). Loss of these signals frequently initiates apoptosis. Up-regulation of epithelial cell adhesion molecules, claudin and actin has been observed in the pancreas of TCP patients (Hofmann, et al., 2007). Our observations corroborate with the recent studies, which showed that chronic ethanol administration to rats significantly decreased pancreatic perfusion and was associated with an increase in leukocyte adhesion and increased expression of adhesion molecules and cytokines in pancreas (Werner, 2005a and b). Similarly, Ca^{2+} is a universal signal transduction element in cells modulating cell growth and differentiation. The level of free intracellular calcium ($[Ca^{2+}]_i$) is regulated and maintained as low as ~100 nM through the action of a number of binding proteins and ion exchange mechanisms (Rasmussen, et al., 1988). The calcium channels like GPCR are found to be up-regulated and the IP3R responsible for mobilization of stored calcium are down-regulated in our experiments (Clapham, 1995). These observations indicate that excessive calcium is stored in the cells, which leads to aberrant signaling and calcium related toxicity to the cells.

Overall, from a pathological point of view, CP is characterized by irregular sclerosis with destruction and loss of exocrine parenchyma, complete replacement of acinar, ductal and endocrine tissue by fibrotic tissue. Several genes involved in MAP Kinase signaling, PI3 pathway, are seen to be up- and down-regulated in our study. Pancreatic stellate cells (PSCs) are pancreatic stem cells, which have migratory and proliferative properties and are involved in fibrogenesis. Activated PSCs are characterized by the disappearance of fat globules and expression of alpha smooth muscle actin. PSCs synthesize and degrade extra cellular matrix proteins and are known to shift towards fibrogenesis if the balance is altered. Ethanol, acetaldehyde and oxidant stress activate PSCs via three MAP Kinase pathways namely, extra cellular signal kinase, p38 kinase and c-jun amino terminal kinase. p53, the prime tumor suppressor is down-regulated in TCP patients (Whitcomb, 2004; Shi, et al., 2007; Yoko et al., 2007). This could be an important determinant of progression of TCP to pancreatic cancer.

By using system biology, we are currently trying to understand various pathways projected to be involved in the pathogenesis of TCP, some details of which have been projected above while the laboratory work is focusing on validation of the differentially regulated genes inferred from microarray data by real time PCR.

Conclusions

Genetics studies conducted till date indicate that TCP and FCPD are types of chronic pancreatitis, which do not share similar genetic susceptibility loci with ICP described in the Western world. There is an attempt to categorize TCP as a type of ICP but the results till date suggest the contrary and it would be more appropriate to view TCP and ICP as being distinct entities. The atypical clinical picture of TCP, higher prevalence in the Indian subcontinent, in comparison to ICP in the West, as well as areas of endemicity, is still far from clear. Whether this is due to differences in genetic predisposition, specific environmental factors, or, as is most likely, a combination of both, still needs to be ascertained. Genetic alterations in at least two genes encoding the serine protease inhibitor, Kazal type 1, *SPINK1* and cathepsin B, *CTSB* clearly predict susceptibility to TCP. In contrast to Western countries, alterations in the cationic trypsinogen, *PRSS1* and anionic trypsinogen, *PRSS2* are of no importance in pathogenesis of TCP. Variants in the genes coding for the cystic fibrosis transmembrane conductance regulator, *CFTR* and the transcription factor-7-like 2, *TCF7L2* might play a role in explaining various manifestations of TCP. The majority of patients suffering from TCP do not harbor a mutation in one of the above-mentioned genes suggesting that further genes will be involved in their pathogenesis. Results obtained from microarray analysis on global expression of genes may help understand these etiopathogenetic pathways that may explain the peculiarities of tropical calcific pancreatitis.

References

1. Abu-Bakare, A., Taylor, R., Gill, G.V., Alberti, K.G.M. M. (1986). Tropical or malnutrition-related diabetes: a real syndrome? Lancet. **i**: 1135–1138.
2. Balaji, L.N., Tandon, B.N., Tandon, R.K., Banks, P.A. (1994). Prevalence and clinical features of chronic pancreatitis in Southern India. Int. J. Pancreatol. *15*: 29–34.

3. Balakrishnan, V., Sauniere, J.F., Hariharan, M., Sarles, H. (1988). Pancreas *3*: 30–35.
4. Barman, K.K., Premalatha, G., Mohan, V. (2003). Tropical chronic pancreatitis. Postgrad. Med. J. *79*: 606–615.
5. Bhaskar, S., Reddy, D.N., Mahurkar, S., Rao, G.V., Singh, L., Chandak, G.R. (2006). Lack of significant association of an insertion/deletion polymorphism in the angiotensin converting enzyme (ACE) gene with tropical calcific pancreatitis. BMC. Gastroenterol. *6*: 42–48.
6. Bhatia, E., Durie, P., Zielenski, J. et al. (2000). Mutations in the cystic fibrosis transmembrane regulator gene in patients with tropical calcific pancreatitis. Am. J. Gastroenterol. *95*: 3658–3659.
7. Bhatia, E., Chowdhuri, G., Sikora, S.S. et al. (2002). Tropical calcific pancreatitis: strong association with *SPINK1* trypsin inhibitor mutations. Gastroenterology *123*: 1020–1025.
8. Boulling, A., Le Maréchal, C., Trouvé, P., Raguénès, O., Chen, J. M., Férec, C. (2007). Functional analysis of pancreatitis-associated missense mutations in the pancreatic secretory trypsin inhibitor (*SPINK1*) gene. Eur. J. Hum. Genet. *15*: 936–942.
9. Chandak, G.R., Idris, M.M., Reddy, D.N. et al. (2002). Mutations in the pancreatic secretory trypsin inhibitor gene (PSTI/ *SPINK1*) rather than the cationic trypsinogen gene (*PRSS1*) are significantly associated with tropical calcific pancreatitis. J. Med. Genet. *39*: 347–351.
10. Chandak, G.R., Idris, M.M., Reddy, D.N., Mani, K.R., Bhaskar, S., Rao G.V., Singh, L. (2004). Absence of *PRSS1* Mutations and Association of *SPINK1* Trypsin Inhibitor Mutations in Hereditary and Non-Hereditary Chronic Pancreatitis. Gut *53*: 722–727.
11. Chandak, G.R., Janipalli, C.S., Bhaskar, S., Kulkarni, S.R., Mohankrishna, P. et al. (2007). Common variants in the *TCF7L2* gene are strongly associated with type 2 diabetes mellitus in the Indian population. Diabetologia *50*: 63–67.
12. Chiari, H. (1896). Ueber selbstverdauung des menschlichen pankreas. Zeitschrift fur heilkunde, *17*: 69–96.
13. Clapham, D.E. (1995). Intracellular calcium. Replenishing the stores. Nature *375*: 634–635.
14. Cohn, J.A., Friedman, K.J., Noone, P.G., Knowles, M.R., Silverman, L.M., Jowell, P.S. (1998). Relation between mutations of the cystic fibrosis gene and idiopathic pancreatitis. N. Engl. J. Med. *339*: 653–658.
15. Ekelund, K. & Johansson, C. (1980). Acta. Physiol. Scand. *110*: 161–165.
16. FAO (1977). FAO production year book, 1976. Vol. **30**. Pub. Food and Agriculture Organization of the United Nations, Rome.
17. Gaboriaud, C., Rossi, V., Bally, I. et al. (1996). Crystal structure of human trypsin 1: unexpected phosphorylation of Tyr151. J. Mol. Biol. *259*: 995–1010.
18. Geevarghese, P. J., Kumara Pillai, V., Joseph, M.P., Pitchumoni, C. S. (1962). The diagnosis of pancreatogenous diabetes mellitus. J. Assoc. Phy. India. *10*: 173–180.
19. Geeverghese, P.J. (1986). Calcific Pancreatitis. Bombay, India: Vargheese Publishing House. Rao, R.H. (1984). The role of undernutrition in the pathogenesis of diabetes mellitus. Diabetes Care. *7*: 595–600.
20. Go, V. L., Hofmann, A.F., Summerskill, W. H. (1970). J. Clin. Invest. *49*: 1558–1564.
21. Halangk, W., Lerch, M. M., Brandt-Nedelev, B., Roth, W., Ruthenbuerger, M., Reinheckel, T. et al. (2000) Role of cathepsin B in intracellular trypsinogen activation and the onset of acute pancreatitis. J. Clin. Invest. *106*: 773–781.

22. Hassan, Z., Mohan, V., Ali, L. et al. (2002). *SPINK1* is a susceptibility gene for fibrocalculous pancreatic diabetes in subjects from the Indian subcontinent. Am. J. Hum. Genet. *71*: 964–968.
23. Hofmann, C., Obermeier, F., Artinger, M., et al. (2007). Cell-cell contacts prevent anoikis in primary human colonic epithelial cells. Gastroenterology *132*: 587–600.
24. Idris, M.M., Bhaskar, S., Reddy, D.N., Mani, K.R., Rao, G.V., Singh, L. et al. (2005). Mutations in anionic trypsinogen gene are not associated with tropical calcific pancreatitis. Gut *54*: 728–729.
25. Ip, S.P., Kwan, P.C., Williams, C.H., Pang, S., Hooper, N.M., Leung, P.S. (2003). Changes of angiotensin-converting enzyme activity in the pancreas of chronic hypoxia and acute pancreatitis. Int. J. Biochem. Cell. Biol. *35*: 944–954.
26. Jha, R. (2003/4) The Spatial Distribution of Protein Deficiency in Rural India in the Last Three Quinquennial Rounds of NSS. ASARC Working Paper.
27. Kaneko, K., Nagasaki, Y., Furukawa, T., Mizutamari, H., Sato, A., Masamune, A., et al. (2001). Analysis of the human pancreatic secretory trypsin inhibitor (PSTI) gene mutations in Japanese patients with chronic pancreatitis. J. Hum. Genet. *46*: 293–297.
28. Kang, H.G., Jenabi, J.M., Zhang, J., et al. (2007). E-cadherin cell-cell adhesion in ewing tumor cells mediates suppression of anoikis through activation of the ErbB4 tyrosine kinase. Cancer Res. *67*: 3094–3105.
29. Király, O., Wartmann, T., Sahin-Tóth, M. (2007). Missense mutations in pancreatic secretory trypsin inhibitor (*SPINK1*) cause intracellular retention and degradation. Gut *56*: 1433–1438.
30. Kloppel, G. and Maillet, B. (1991). Chronic pancreatitis: evolution of the disease. Hepatogastroenterology *38*: 408–412.
31. Kukor, Z., Toth, M. and Sahin-Toth, M. (2003). Eur. J. Biochem. *270*: 2047–2058. Kume, K., Masamune, A., Mizutamari, H., Kaneko, K., Kikuta, K., Satoh. M. et al. (2005). Mutations in the serine protease inhibitor Kazal Type 1 (*SPINK1*) gene in Japanese patients with pancreatitis. Pancreatology *5*: 354–360.
32. Kume, K., Masamune, A., Kikuta, K., Shimosegawa, T. (2006). [-215G>A; IVS3+2T>C] mutation in the *SPINK1* gene causes exon 3 skipping and loss of the trypsin binding site. Gut *55*: 1214.
33. Kuno, A., Yamada, T., Masuda, K., Ogawa, K., Sogawa, M., Nakamura, S. et al. (2003). Angiotensin-converting enzyme inhibitor attenuates pancreatic inflammation and fibrosis in male Wistar Bonn/Kobori rats. Gastroenterology *124*: 1010–1019.
34. Kuwata, K., Hirota, M., Shimizu, H., Nakae, M., Nishihara, S., Takimoto, A. et al. (2002). Functional analysis of recombinant pancreatic secretory trypsin inhibitor protein with amino-acid substitution. J. Gastroenterol. *37*: 928–934.
35. Laskowski, M., Wu, F. C. (1953). Temporary inhibition of trypsin. J. Biol. Chem. *204*: 797–805.
36. Layer, P., Yamamoto, H., Kalthoff, L., Clain, J.E., Bakken, L.J., Dimagno, E.P. (1994). The different courses of early- and late-onset idiopathic and alcoholic pancreatitis. Gastroenterology. *107*: 1481–1487.
37. Le Bodic, L., Bignon, J.D., Raguenes, O. et al. (1996). The hereditary pancreatitis gene maps to the long arm of chromosome 7. Hum. Mol. Genet. *5*: 549–554.
38. Le Maréchal, C., Masson, E., Chen, J.M., Morel, F., Ruszniewski, P., Levy, P., et al.. (2006). Hereditary pancreatitis caused by triplication of the trypsinogen locus. Nat. Genet. *38*: 1372–1374.
39. Lerch, M.M. and Halangk, W. (2006). Human Pancreatitis and the role of cathepsin B. Gut *55*: 1228–1230.

40. Leung, P.S. and Chappell, M.C. (2003). A local pancreatic renin-angiotensin system: endocrine and exocrine roles. Int. J. Biochem. Cell. Biol. *35*: 834–846.
41. Mahurkar, S., Idris, M.M., Reddy, D.N. et al. (2006). Association of cathepsin B gene polymorphisms with tropical calcific pancreatitis. Gut *55*: 1270–1275.
42. Mahurkar, S., Bhaskar, S., Reddy, D.N. et al. (2008). *TCF7L2* gene polymorphisms do not predict susceptibility to diabetes in tropical calcific pancreatitis but may interact with *SPINK1* and CTSB mutations in predicting diabetes. BMC Med. Genet. *9*: 80–87.
43. Masson, E., Le Maréchal, C., Chandak, G.R., Lamoril, J., Bezieau, S., Mahurkar, S. et al. (2008). Trypsinogen copy number mutations in patients with idiopathic chronic pancreatitis. Clin. Gastroenterol. Hepatol. *6*: 82–88.
44. Mathangi, D. C., Deepa, R, Mohan, V. et al. (2000). Long-term ingestion of Cassava (Tapioka) does not produce diabetes or pancreatitis in the rat model. Int. J. Pancreatol. *27*: 203–208.
45. McMillan, D., Geevarghese, P.J. (1980). Dietary cyanide and tropical malnutrition diabetes. In: Podolsky, S., Viswanathan, M. eds. Secondary diabetes: The Spectrum of the Diabetic Syndrome. New York: Raven Press, 239–253.
46. Mehrotra, R.N., Bhatia, E., Choudhury, G. (1997). Beta cell function and insulin sensitivity in tropical calcific pancreatitis from North India. Metabolism *46*: 441–444.
47. Mittal, N., Mehrotra, R., Agarwal, G., Rajeshwari, J., Choudhuri, G., Sikora, S. et al. (2002). The clinical spectrum of fibrocalculous pancreatic diabetes in north India. Natl. Med. J. India. *15*: 327–331.
48. Mohan, V., Snehalatha, C., Ramachandran, A., Jayashree, R., Viswanathan, M. (1983). Pancreatic beta cell function in tropical pancreatic diabetes. Metabolism *32*: 1091–1092.
49. Mohan, V., Ramachandran, A., Vijay Kumar, G., Snehalatha, C., Viswanathan, M. (1988). Insulin resistance in fibrocalculous (tropical) pancreatic diabetes. Horm. Metab. Res. *20*: 746–748.
50. Mohan, V., Chari, S.T., Hitman, G.A., Suresh, S., Madanagopalan, N., Ramachandran, A. et al. (1989). Familial aggregation in tropical fibrocalculous pancreatic diabetes. Pancreas. *4*: 690–693.
51. Mohan, V., Pitchumoni, C.S. (1998). Tropical chronic pancreatitis. In: Berger, H.G., Warshaw, A.L., Buchler, M.W. et al. eds. The Pancreas, vol. **1**. London: Blackwell Science: 688–697.
52. Noone, P.G., Zhou, Z., Silverman, L.M., Jowell, P.S., Knowles, M.R., Cohn, J.A. (2001). Cystic fibrosis gene mutations and pancreatitis risk: relation to epithelial ion transport and trypsin inhibitor gene mutations. Gastroenterology *121*: 1310–1319.
53. Ohmuraya, M., Hirota, M., Araki, M., Mizushima, N., Matsui, M., Mizumoto, T. et al. (2005). Autophagic cell death of pancreatic acinar cells in serine protease inhibitor Kazal type 3-deficient mice. Gastroenterology. *129*: 696–705.
54. Pitchumoni, C.S. (1985). Special problems of tropical pancreatitis. J. Clin. Gastroenterol. *13*: 941–959.
55. Rao, R.H. (1993). Is tropical diabetes malnutrition related? Diabet. Care. *16*: 941–945.
56. Rao, R.H., Yajnik, C.S. (1996). Malnutrition related diabetes: time to think? Diabet. Care. *19*: 1014–1017.
57. Rasmussen, J.E. (1988). Calcium and the skin. Arch. Dermatol. *124*: 443–444.
58. Reddy, D. N. (1998). Tropical pancreatitis. The Indian Experience. In Proceedings of the 8th World Congress of the International Gastro-Surgical Club, Strasbourg, France, pp. 249–253.

59. Rigat, B., Hubert, C., Alhenc-Gelas, F., Cambien, F., Corvol, P., Soubier, F. (1990). An insertion/deletion polymorphism in the angiotensin I-converting enzyme gene accounting for half of variance of serum enzyme levels. J. Clin. Invest. *86*: 1343–1346.
60. Rinderknecht, H., Renner, I.G., Douglas, A.P., Adham, N.F. (1978). Profiles of pure pancreatic secretions obtained by direct pancreatic duct cannulation in normal healthy human subjects. Gastroenterology *75*: 1083–1089.
61. Sahin-Tóth, M., Tóth, M. (2000). Gain-of-function mutations associated with hereditary pancreatitis enhance autoactivation of human cationic trypsinogen. Biochem. Biophys. Res. Commun. *278*: 286–289.
62. Shi, C., Zhao, X., Wang, X., et al. (2007). Potential effects of PKC or protease inhibitors on acute pancreatitis induced tissue injury in rats. Vasc. Pharmacol. *46*: 406–411.
63. Steer, M.L. and Waxman, I. (1995). Chronic pancreatitis. N. Engl. J. Med. *332*: 1482–1490.
64. Szilagyi, L., Kenesi, E., Katona, G. et al. (2001). Comparative *in vitro* studies on native and recombinant human cationic trypsins. Cathepsin B is a possible pathological activator of trypsinogen in pancreatitis. J. Biol. Chem. *276*: 24574–24580.
65. Szmola, R., Kukor, Z., Sahin-Toth, M. (2003). Human mesotrypsin is a unique digestive protease specialized for the degradation of trypsin inhibitors. J. Biol. Chem. *278*: 48580–48589.
66. Technical report series no. 28. (2004). Cyanogenic glycosides in cassava and bamboo shoots. A human health risk assessment. Food Standards Australia New Zealand July 2004.
67. Teich, N., Rosendahl, J., Toth, M. et al. (2006). Mutations of human cationic trypsinogen (*PRSS1*) and chronic pancreatitis. Hum. Mutat. *27*: 721–730.
68. Thomas, P.G., Augustine, P., Ramesh, H., Rangabashyam, N. (1990). Observations and surgical management of tropical pancreatitis in Kerala and southern India. World J. Surg. *14*: 32–42.
69. Tuescher, T., Rosman, J.B., Baillod, R. et al. (1987). Absence of diabetes in rural west African population with high carbohydrate/cassava diet. Lancet. *1*: 765–768.
70. Viswanathan, M. (1980). Pancreatic diabetes in India: an overview. In: Podolsky S, Viswanathan M, editors. Secondary diabetes: the spectrum of diabetic syndromes. New York: Raven; 1980. pp 105–116.
71. Wade, N. (2006). Before the Dawn: Recovering the Lost History of Our Ancestors. (Penguin Group, New York, USA, ISBN 1-59420-079-3.
72. Werner, J., Hackert, T., Hartwig, W., et al. (2005a). Alcoholic pancreatitis: Induction of Microcirculatory disturbances and inflammatory cascade by chronic alcohol intake. Pancreas *31*: 479.
73. Werner, J., Hartwig, W., Hackert, T., et al. (2005b). Alcoholic pancreatitis: Detailed characterization of microcirculatory disturbances and leukocyte adhesion. Pancreas *31*: 487.
74. Whitcomb, D.C. et al. (1996a) A gene for hereditary pancreatitis maps to chromosome 7q35. Gastroenterology *110*: 1975–1980.
75. Whitcomb, D.C. et al. (1996b) Hereditary pancreatitis is caused by a mutation in cationic trypsinogen gene. Nat. Genet. *14*: 141–145.
76. Whitcomb, D.C. (2004) Mechanisms of disease: Advances in understanding the mechanisms leading to chronic pancreatitis. Nat. Clin. Pract. Gastroenterol. Hepatol. *1*: 46–52.

77. Witt, H., Luck, W., Hennies, H.C., Claßen, M., Kage, A., Laß, U. et al. (2000) Mutations in the gene encoding mutations in the gene encoding the serine protease inhibitor, Kazal type 1 are associated with chronic pancreatitis. Nat. Genet. *25*: 213–216.

78. Witt, H., Sahin-Toth, M., Landt, O., et al. (2006) A degradation sensitive anionic trypsinogen (*PRSS2*) variant protects against chronic pancreatitis. Nat. Genet. *38*: 668–673.

79. Witt, H., Apte, M.V., Keim, V., Wilson, J.S. (2007) Chronic pancreatitis: challenges and advances in pathogenesis, genetics, diagnosis, and therapy. Gastroenterology *132*: 1557–1573.

80. Yajnik, C.S., Shelgikar, K.M. (1993) Fibrocalculous pancreatic diabetes in Pune, India. Clinical features and follow-up for 7 years. Diabetes Care. *16*: 916–921.

81. Yoko, N., Mitsuo, T., Taizo, Y. (2007) High glucose activates rat pancreatic stellate cells through protein kinase c and p38 mitogen-activated protein kinase pathway. Pancreas *34*: 364–372.

Human Genetic Variations in India: Origin and Disease Perspectives

11

K. Thangaraj

Centre for Cellular and Molecular Biology, Uppal Road, Hyderabad-500 007, India

Introduction

The most important question that has preoccupied evolutionary biologists/population geneticist for most of the last century is about the origin of man and his colonization in different parts of the world. Peopling of the Indian subcontinent is the result of several historical migrations, and this had a significant bearing to the extent vast diversity of Indian population. In India, there are 4635 anthropologically defined communities (Singh, 1992) that include endogamous caste and tribal populations, and they show striking diversity in terms of language, socio-cultural attributes and appearance. Understanding the enormous human biological diversity in India is important for several reasons. The high level of heterogeneity in Indian population, governed by high level of endogamy created by numerous social boundaries, along with the influence of general evolutionary forces such as genetic drift due to long-term isolation, has kept the Indian gene pool distinct from other continental populations. Major demographic events like migration, population bottleneck and population expansion leave genetic imprints where gene frequencies of the genome are altered. These imprints are passed onto successive generations, thus preserving the population's genetic history which determines health and diseases of the populations. The molecular genetic studies are now unraveling these unique patterns.

There are increasing number of findings identifying genetic polymorphisms that have major effect on a range of disorders. Genetic differences between people of populations are responsible for their variable susceptibility to disease, resistance to infection, and response to drug treatment. Most of the modifications in susceptibility to highly prevalent diseases in human populations are caused by variations that occur with a high allele frequency of >1% and tend to be evolutionarily older affecting gene regulatory function. On the other hand, rare genetic variations occur with an allele frequency of <1% especially as a result of new mutations and are of relatively recent origin in human evolution. The study of human genetic variation, therefore, is likely to enrich our understanding the genetic basis of diseases and the practice of clinical medicine.

Our study of ~15,000 individuals belonging to several tribal and caste populations, from all parts of India shows high level of genetic diversity compared to the populations of other parts of the world. Since India represents one of the largest sources of human biodiversity, study on Indian populations not only provide insight into their complex origin, history and relatedness, but also help in understanding molecular pathology of genetic diseases. Therefore, our interest has been to study both population history and molecular mechanism of diseases among Indian populations.

Is the Origin of Indian Populations Unique?

Indian populations can broadly be divided into four different linguistic families, namely, Austro-Asiatic, Dravidian, Tibeto-Burman, and Indo-European. Each population, affiliated to one of these linguistic families, has its own social and cultural identity, and practices endogamy, which helps to maintain its genetic architecture. Several studies have been carried out with maternally inherited mitochondrial DNA (mtDNA) markers to understand the population structure and genetic origins of Indian populations. These studies have shown that Indians are readily distinguishable from Europeans and East Asians, and interestingly, despite the linguistic (Austro-Asiatic languages) and cultural (the spread of rice-based economies) links, share few lineages in common with neighboring Southeast Asian populations.

Origin of Andaman and Nicobar Islanders

The Andaman and Nicobar Islands, located Southeast of the Indian subcontinent, form part of range of submarine mountains, 1130 km long, running from Cape Negrais in the Arakan Yoma range of Burma to Achin Head in Sumatra. This range separates the Bay of Bengal from Andaman Sea. The older rocks are early tertiary or late cretaceous. The newer rocks in Sentinel Islands contain fossils of radiolarians and foraminifers with superstructure of corals. The Andaman may have been linked to Myanmar by a land bridge during the ice age. There are two distinct tribal groups in Andaman and Nicobar Islands:(i) Negrito-who share physical features such as short stature, dark skin, peppercorn hair, scant body hair with African pygmies and other Asian Negrito people (Fig. 11.1), and (ii) Mongoloid – who share physical features with Chinese, Malays and Burmese. They speak dialects related to Mon-Khmer and the languages spoken in Vietnam, Malaysia, and part of northeast India. The origin of the Andaman "Negrito" and Nicobar "Mongoloid" populations has stimulated a wide range of speculation. Their origins are really a mystery. We do not know where they came from? And how long they have been there? Could these Islanders hold the key to the mystery of our own origins and the early history of human migration out of Africa!

Our previous study (Thangaraj et al. 2003), based on very small (400 base pair) hypervariable sequence (HVS) of mitochondrial DNA (mtDNA), revealed that the Andamanese "Negritos" have

Fig. 11.1 Picture of Andamanese: left: an Onge female; right: and Onge male.

closer affinities with Asians than Africans, while the Nicobarese showed close genetic affinities to Southeast Asians. In a parallel study (by other group from Oxford), confirming the Asian affinities of the mtDNA lineages of the Andamanese, analyses of museum specimens related two major lineage groups within mitochondrial haplogroups M2 and M4 that are found commonly throughout India. We recently provided molecular genetics evidence that migration of a first modern human "out-of-Africa" have settled within Andaman Nicobar Islands.

Since the Andaman and Nicobar Islanders are morphologically very different from Indian populations (Fig. 11.1), we have further analyzed the complete mtDNA sequence (16,569 base pairs) of five *Onges*, five *Great Andamanese* and five *Nicobarese* to reconstruct their origin. Analysis of the Onge and Great Andamanese's complete mtDNA sequences revealed the existence of two novel clades, which we named as to call M31 and M32. Importantly, none of the unique variation (mutation) distinguishing these two new lineages overlaps with the known Indian or East Asian mtDNA lineage. In our survey of 6500 samples from mainland India, none of the mtDNA sequences show similarity to Andamanese mtDNA sequences (M31 and M32). Therefore, we suggest that these two clades are likely to have evolved '*in situ*' on these islands and the founding mtDNA lineages N* and R* must have apparently disappeared completely from the existing Andaman Islanders (Onge and Great Andamanese). Our study suggest that two ancient maternal lineages have evolved in the Andaman Islands in genetic isolation independently from other South and Southeast Asian populations, probably since the initial penetration of the Northern coastal areas of Indian Ocean by the anatomically modern humans in their out-of-Africa migration about 50-70 thousand years ago. Analysis of the complete mitochondrial DNA sequence revealed that the Nicobarese belong to two lineages (B and F) commonly found in China, Malaysia, Myanmar and Thailand, suggesting their recent arrival from the east during the last 12 thousand years (Thangaraj et al., 2005).

Origin of Tribal and Caste Populations of India

The tribal groups constitute about 8% of the total Indian population (census 2001) and it has been suggested that they "may represent relic populations of shadowy origin but potentially of great genetic interest" (Cavalli-Sforza et al., 1994). The geographic position of the Indian populations provides good opportunity to study and address the question of early dispersal of modern human. The caste and tribal groups of India are considered socially and culturally the most stratified of all known societies in human history. The ancient Paleolithic migration by the modern humans was during their initial colonization of Eurasia. The early migration probably proto-Dravidian speakers, from the eastern horn of the Fertile Crescent, followed this. The Indo-European speakers, who might have arrived ~3,500 years ago, are the third potential source of Indian gene pool. The Austro-Asiatic (Khasi and Nicobarese) and Tibeto-Burman speakers with ties to East/Southeast Asia form the fourth major contributors. The most recent conquerors from Central Asia and the colonizers from Europe might also have added to this ethnic multiplicity.

Indo-European speaking populations are the most widely spread, which covers North, Central and Western parts of India. Dravidian populations are mainly confined to south India with some exceptions like Kurukh and Gondi speaking people, living in central and eastern parts of India. Austro-Asiatic speakers are dispersed mostly in the central and eastern parts of the country while the Tibeto-Burman speakers are concentrated in the northeastern part of the subcontinent. The origin and dispersal of these language families in South Asia remains still enigmatic.

During the last few years, genetic studies of mtDNA and Y chromosomal variation have helped in providing a substantial understanding of the human origins and dispersal patterns in South Asia. (Palanichamy et al., 2004; Metspalu et al., 2004; Thangaraj et al., 2005, 2006; Sengupta et al., 2006; Chaubey et al., 2007). Mitochondrial DNA analysis, in particular among the worldwide populations, has revealed continent-specific distributions of certain mtDNA lineage groups. Various studies on South Asian populations do not show any caste or language-specific distribution of major ancient lineages. More than 60% of the present day Indian maternal lineages are affiliated to mtDNA haplogroup 'M'.

We have shown that populations from Nicobar Islands, who speak Mon-Khmer language of the Austro-Asiatic language family, carry the highest frequency of the 9-bp deletion (Fig. 11.2), in the mtDNA between COII and tRNA Lysine gene, thus far reported among South Asian populations (Thangaraj et al., 2005). In contrast, we did not detect the 9-bp deletion among 302 Mundari samples, another linguistic branch of Austro-Asiatic speaking group from mainland India. Detailed molecular characterization of the Nicobarese samples using coding region SNPs classified them under haplogroup B5a1a, which is also common in China and other South East Asian populations. These findings suggest that the Austro-Asiatic (Mundari) populations of the Indian sub-continent and Nicobar Islands may have distinct genetic sources. Given the limited number of Mundari populations covered in our previous survey, we have recently analyzed mtDNA diversity in 3844 additional individuals from India, including 243 Mundari speakers from six populations not investigated previously, and, for the first time, populations from Nepal, to determine the genetic affinities of the population of mainland India with South East Asian populations, and to infer the origins of the 9bp deletion/insertion variants in the Indian subcontinent (Thangaraj et al., 2008).

We have also found a novel autochthonous deep-rooting M43 lineage, redefine haplogroup M6 and identify a rare west Eurasia-specific haplogroup H14 in Indian population. Based on the co-presence of haplogroup F1 in North Indian and Nepali populations, but the lack of B5a in the later, it seems likely that the migration of Southeast Asian maternal lineages to North India has occurred not only through the corridor of Nepal but also has involved gene flow along the southern slopes of the Himalayas within the Indian sub-continent. The occurrence of R6 in the Nepalese suggests

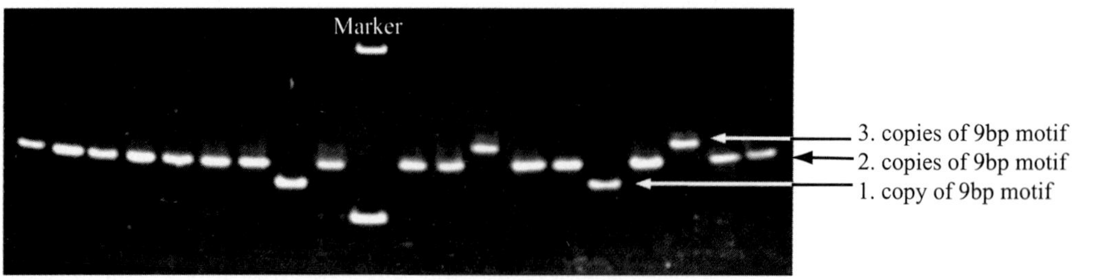

Fig. 11.2 The 9bp deletion /duplication: Most humans have inherited two tandem copies of the 9-bp motif (*CCCCCTCTA*), located in the noncoding region V, between the cytochrome oxidase II (*COII*) and tRNA lysine (*tRNAlys*) genes (*designated as 2 copies of 9bp motif*). Some individuals show the deletion of one copy of 9bp (*designated as 1 copy of 9bp motif*), while very few individuals show addition of an extra copy of 9bp (*designated as 3 copies of 9bp motif*).

gene flow from India to Nepal. Further, analyses of the B5a and F1 lineages in larger samples from Nepal and Eastern Uttar Pradesh should shed further light on the origin and spread of the Tharu and other populations of North India. Studies of mtDNA and Y-chromosomal markers in Nepalese populations would be useful for the investigation of this putative gene flow into India. Therefore, further evidence is needed to determine the mtDNA and Y-chromosome haplogroups/lineages of the Austro-Asiatic tribes of the northeastern region and their comparison with appropriate set of South and Southeast Asian populations (Thangaraj et al., 2008).

The Northeastern part of India is sandwiched by the marked presence of the young fold mountains of Eastern Himalayas on the northern side and the Indian Ocean on its southern side. The archaeological evidences, based on stone tools from the Garo hills of the Meghalaya region of Northeast India, suggest that this region might have been inhabited as early as in the Paleolithic period. Although the Northeastern region played a major role in the population dispersal as transect between India and East Asia, the extent and nature of mtDNA and Y-chromosome diversity of this region is not adequately studied. We have analysed the mtDNA and Y-chromosome SNP and STR data of the Austro-Asiatic populations from Northeast India, and found that the Austro-Asiatic Khasi tribes of Northeast India represent a genetic continuity between the populations of South and Southeast Asia.

Recently Migrated Populations

Siddis are the recent migrant population from Africa. Anthropological studies suggest that the Portuguese brought them to India as slaves during 15th to 18th century. There were very few who are also believed to have migrated during Ethiopia - India trade. Earlier, we have shown the presence of Y chromosome *Alu* insertion, an African-specific marker, in 40% of Siddis. The mitochondrial DNA hypervariable sequence also confirmed their close affinity with African populations (Thangaraj et al., 1998). Very recently, we have made high resolution analysis of the Y chromosome lineages of large number of Siddis, along with maternally inherited mitochondrial DNA markers. Our study showed that the Siddis have great genetic affinity with Bantu Speaking populations from South Africa. This suggests that they probably migrated through sea route and settled in India rather than migrating through land or coastal route. Interestingly, we also found high level of genetic admixture among the ancestral African population and the neighboring (Medar, Gram Vokkal, Kare Vokkal and Korova) Indian populations (unpublished information).

Is the Genetic Basis for the Diseases of Indians Unique?

From the above studies it is obvious that the genetic composition of each Indian population is unique and genetically diverse from the rest of the world. Therefore, one can presume that the genetic basis for various disorders among Indian populations may be different from rest of the world. We have analysed the DNA of individuals with various disorders and found that most of the disease causing mutations in Indian populations are unique and not reported in any other world populations. Therefore, it is a big challenge for us to find out genetic variation(s) that are associated with disease(s) among Indian populations.

Heterogeneity in Autosomal Gene

DAZL Gene

The autosomal *DAZL* (Deleted-in-Azoospermia-Like) gene, mapped on to the short arm of the human chromosome 3, is the precursor for the Y-chromosomal *DAZ* cluster, which encodes for putative RNA-binding proteins. Mutations in the *DAZL* have been reported to be associated with spermatogenic failure in Taiwanese population (Teng et al., 2002). As there was hardly any study among Indian populations, we have analyzed the entire coding sequences of exons 2 and 3 of *DAZL* in a total of 1010 men from Indian subcontinent, including 660 infertile men *598 non-obstructive azoospermia, 62 severe oligozoospermia* and 350 normozoospermic fertile control men, to investigate whether mutation(s) in the *DAZL* is associated with male infertility. Interestingly, none of our samples (1010) showed A386G (T54A) mutation, which was found to be associated with spermatogenic failure in Taiwanese population. In contrast, A260G (T12A) mutation was observed in both infertile and normozoospermic fertile control men, without any significant association with infertile groups (χ^2 = 0.342; p = 0.556). Similarly, we have found a novel A437G (I71V) mutation, which is also present in both infertile and normozoospermic fertile control men without any significant difference (χ^2 = 0.476; p = 0.490). Our study clearly demonstrates the complete absence of the A386G (T54A) mutation in Indian subcontinent and the other two mutations – A260G (T12A) and A437G (I71V) – observed were polymorpic. Therefore, we conclude that these mutations in the *DAZL* gene are not associated with male infertility in Indian men (Thangaraj et al., 2006).

Oestrogens Receptor Gene

Historically, oestrogens are considered female hormones but recent findings have shown that they are essential regulators of fertility in both men and women (Saunders, 2005). In males, oestrogen level is extraordinarily high in semen (Hess et al., 1997). They are synthesized from testosterone via the action of aromatase cytochrome P450 (Simpson et al., 1994). Oestrogens seem to play an important role in male infertility in adult mice (Robertson et al., 1999) and lead to reduced sperm production and motility in humans (Carani et al., 1997; Herrmann et al., 2002). Oestrogen signalling is mediated by oestrogen receptors (ERs), of which two subtypes have been reported – ERα and ERβ (Aschim et al., 2005). The ERβ gene has been mapped on the long arm of chromosome 14. It has 96% conserved amino acid sequence in the DNA binding domain and 58% conserved residues in the ligand binding domain, compared to ERβ (Tsukamoto et al., 1998). Recently, many sequence variants of ERβ gene have been described and shown to be associated with several diseases. *Rsa*I polymorphism (rs1256049) in exon 5 of ERβ has been shown to be associated with male infertility in Caucasian patients (Aschim et al., 2005). The CA repeat polymorphism in ERβ has not been studied so far in male infertility, however it has been shown to be associated with high Bone Mineral Density (BMD) in post-menopausal Japanese (Ogawa et al., 2000b) and New Mexican women (Scariano et al., 2004); and systemic blood pressure in post-menopausal Japanese women (Ogawa et al., 2000a). This repeat polymorphism has also been shown to be associated with variations in the levels of androgen and sex hormone binding globulin in blood of post-menopausal Swedish women (Westberg et al., 2001). As CA repeat polymorphism in male infertility has not been analyzed and *Rsa*I polymorphism in ERβ gene was found to be associated with male infertility, we have investigated both the polymorphisms in Indian men to find out their association with infertility. We have analysed 695 Indian men, including 443 infertile and 252 fertile men to evaluate the association of CA repeat length and *RsaI* polymorphisms in male infertility. Our results revealed no significant difference in the distribution of CA repeat length between infertile (mean ± SD 23.24 ± 2.06, median 24) and fertile men (mean ± SD 23.16 ± 2.27, median 24) (Fig. 11.3). The analysis of dosage effect by classifying samples into SS (short/short), SL (short/long) and LL (long/long) groups also did not show any significant difference between infertile and fertile men. Similarly, *Rsa*I polymorphism also did not show any significant difference between infertile and fertile men. Furthermore, the combined analysis of CA repeat and *Rsa*I polymorphisms by haplotyping showed that the distribution of haplotypes was not significantly different between fertile and infertile men. Our results suggest that CA repeat length and *Rsa*I polymorphisms in ERβ are not associated with infertility in Indian men (Khattri et al., 2007).

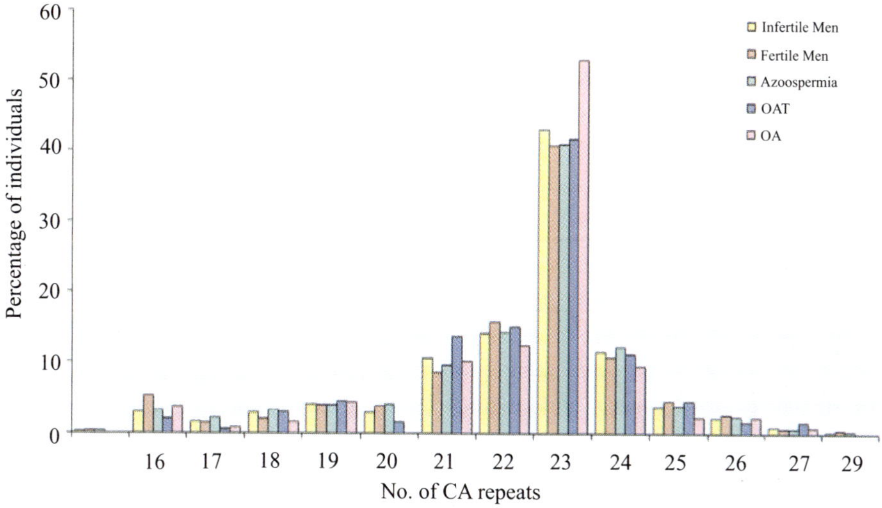

Fig. 11.3 Distribution of CA repeats in the estrogen receptor gene of various categories of infertile men.

DNA Polymerase γ

DNA polymerase γ (POLG) was found to be involved in the replication of mtDNA (Lestienne, 1987). Variations in the CAG repeat number of the catalytic sub-unit of the mitochondrial DNA polymerase gamma (POLG) have been speculated to be associated with male infertility (Rovio et al., 1999). The ten CAG repeats (10/10) was found to be the most common allele (88%), absence of which was found to be associated with male infertility. As no study on Indian population was conducted, so far, to support this view, we investigated the distribution of the POLG-CAG repeats in 509 oligoasthenozoospermic and 241 normozoospermic control Indian men from the same ethnic background. Our study suggested that the distribution of common allele (10/10) was almost similar in both infertile (75%) and normozoospermic (76%) men. Further, we had analyzed the CAG repeat number in as many as 1306 Indian men belonging to different ethnic, geographical and linguistic backgrounds and found the common allele (10 / 10) at a frequency of 78.4% (Fig. 11.4). Our study, therefore, suggests that the 10-CAG repeat is the most common allele present in Indian populations, but its absence and the occurrence of the other mutant homozygous (non 10 / non 10) genotype should not be understood as being specific to infertility. It, thus, suggests that the POLG-CAG repeat variation is not associated with male infertility in Indian populations, and hence is not a useful marker for screening infertile men (Rani et al., 2008).

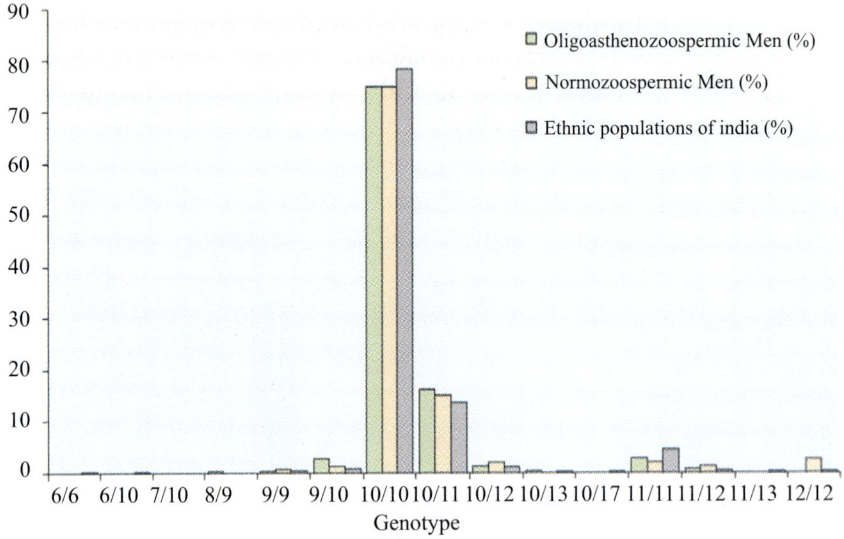

Fig. 11.4 Distribution of CAG genotype of the mitochondrial polymerase Gamma (*POLG*) in the oligoasthenozoo spermia individuals, fertile men and the men belonging to various ethnic populations of India.

Heterogeneity in X Chromosome

Androgens are important not only for secondary sexual differentiation in males but also have numerous other functions in both males and females. Androgens promote the enlargement of the skeletal muscles and affect human behavior, aggression and libido. Recent studies have also indicated that androgens inhibit the ability of some fat cells to store lipids by blocking a signal transduction pathway that normally supports adipocyte function. All these studies emphasize multiple roles of androgens in the human body right from the embryonic stage to adulthood. The variations in triplet CAG repeats have been found to be associated with large number of disorders. CAG repeat expansion in exon 1 of the androgen receptor (AR) gene has been reported to be associated with male infertility in some populations. We have analyzed the CAG repeat motif in the AR gene in 280 men with azoospermia and in 201 men with normal fertility. The mean number of CAG repeats in the AR gene of men with azoospermia was 21.7 ± 0.18, with a high incidence of repeat number 22. Among fertile-control men, the mean number of CAG repeats was 22.4 ± 0.19, with a predominance of repeat number 23. Comparison of fertile men and those with azoospermia on the basis of CAG repeats revealed that the number of CAG repeats in both groups were similar (Fig. 11.5), as revealed with a paired '*t*' test ($t = 0.04$; $P = 0.967$). Our study therefore suggests that the variation in the CAG repeat number of AR gene is not associated with male infertility in Indian populations (Thangaraj et al., 2002).

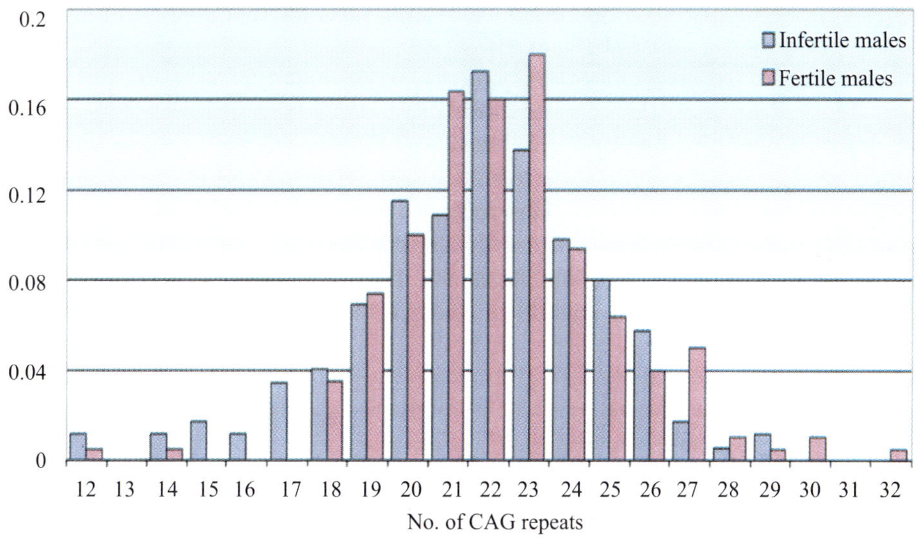

Fig. 11.5 Distribution of CAG repeats in the Androgen receptor gene on infertile and fertile Indian men.

We know, since 1990, that *SRY* (Sex-determining Region on Y chromosome) is the critical switch to initiate the cascade for testes development. Following SRY identification, many genes have been identified to play role in sex-determination. *SRY* and its successor *SOX9* (SRY like-homeobox 9) play major roles in this process, with support from *SF1* (steroidogenic factor 1) and opposition from *DAX1* (Adrenal hypoplasia congenita gene on the X chromosome, gene 1). *WT1* (Wilms' tumor 1) and other genes play minor roles. Mutations in major sex determining genes lead to the development of XY individuals as females (male to female sex reversals) with partial or complete gonadal dysgenesis. Upon testicular differentiation in genetically male individuals, androgens (testosterone and dihydrotestosterone) initiate the process of sexual secondary differentiation during prenatal stage. Androgen receptor (AR) is crucial for this action. The receptor-testosterone complex signals differentiation of Wolffian duct during embryonic life, regulation of secretion of leutinizing hormone by hypothalamic-pituitary axis, and spermatogenesis. The receptor-dihydrotestosterone complex promotes the development of external genitalia and prostate during embryogenesis, and is also responsible for the changes, which occur at puberty in males. The *AR* gene has been mapped on to the long arm (Xq11-12) of the X-chromosome. Our analysis of a large number of sex reversed (XY female; (Fig. 11.6) individuals revealed mutations only in the androgen receptor gene of a few individuals. Therefore, we have undertaken a genome-wide scan on nine familial cases of 46, XY sex

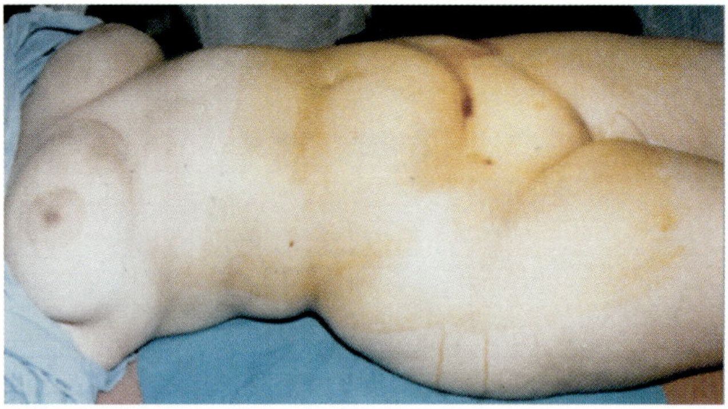

Fig. 11.6 Picture of male to female sex reversed individual (*XY, female*).

reversal, which revealed a LOD score of 5.70 at Xp11.21 (Rajender et al., 2006), suggesting that Indian sex reversals have different genetic basis for sex determination and differentiation.

Heterogeneity in Y Chromosome

Genes involved in the complex process of spermatogenesis are located on the long arm of the Y chromosome (Yq11). Most of these genes exist in multiple copies (Delbridge et al., 1997; Dechend et al., 2000), of which only some are functional (Prosser et al., 1996; Chai et al., 1997). Although 4 azoospermic factors (*AZFa, AZFb, AZFc,* and *AZFd*) have been mapped on the long arm of the Y chromosome, *AZFa, AZFb,* and *AZFc* are considered to be the most important for sperm production (Affara, 2001). Major candidate genes for *AZFa, AZFb,* and *AZFc* are *DFFRY/ USP9Y* (Mazeyrat et al., 1998; Sun et al., 1999), *RBMY* (Ma et al., 1993), and *DAZ* (Reijo et al., 1995), respectively. Large numbers of sequence-tagged sites (STS) markers have been mapped to *AZFa, AZFb,* and *AZFc* regions. Incidence of Y chromosome deletion is about 1 in 1000, and these deletions are often associated with male infertility. Deletion of the growth control gene on the Y chromosome (*GCY*) interval in the *AZFa* region was found to be associated with short stature (Kirsch et al., 2000). However, some of the Y chromosome deletions are neutral, which do not affect the fertility status and phenotype of the individual (Thangaraj et al., 2002b). Several studies have demonstrated that microdeletions in *AZF* regions cause male infertility. Deletion of each *AZF* region has been found to have a different phenotypic effect. Deletion of *AZF*a found to be associated with Sertoli cell–only syndrome type 1 (SCOS) phenotype, whereas the patient, who retained *DBY,* seemed a mild oligozoospermia. Deletions in the *AZFb* region have been found to be associated with azoospermia, oligozoospermia, and normozoospermia. Deletion of the *AZFc* region has been found to be associated with azoospermia and severe to mild oligozoospermia (Affara, 2001). Most of the STS-based studies on male infertility have been carried out with a few markers for each *AZF* region (Martinez et al., 2000). Hence, they fail to detect the Y chromosome deletion in many cases. Therefore, there is no collective opinion about the marker to be used for Y chromosome microdeletion analysis. For the first time, we have made an attempt to evaluate the association of Y chromosome microdeletion among Indian men with azoospermia using several STS markers from each *AZF* region. We have analysed the infertile men with a total of 6 STS marker (2 from each

AZF region), which are found to be commonly deleted in European populations. In contrary, we did not find any deletion in the *AZF*a region (sY127 and sY134) of Indian men (Thangaraj et al., 2003). This prompted us to select additional markers from the *AZFa* region. Interestingly, we did find the deletion of the newly used markers, suggesting that we need to identify Indian specific marker(s) that are associated with male infertility.

Mitochondrial DNA Variations and its Implications in Disease

Mitochondrian is an organelle that plays an important role in energy production by oxidative phosphorylation and mediation of cell death by apoptosis (Schapira, 2002). They are unique in the dual genetic origin and has its haploid genome inherited only through the maternal lineage. The mtDNA is a double stranded, compact and closed circle of 16569 bp. It encodes for 37 genes including 13 polypeptides, all of which are components of respiratory chain and a complement of rRNAs (2) and tRNAs (22) necessary for intra-organelle protein synthesis (Taanman, 1999). In addition, it also consists of a non-coding hypervariable region known as control region or displacement loop (D-loop).

Mitochondrial DNA is prone to mutations about ten times greater than nuclear genome due to lack of protective histones and its proximity to electron transport chain, exposing to high concentration of free radicals and also due to absence of DNA repair mechanisms (Cohen and Gold, 2001). The mitochondrial variations may vary from point mutations to the larger deletions (upto 14 kb) resulting in pathogenesis (Rotig et al., 1995). These variations cause impairment of oxidative phosphorylation machinery affecting the major organs such as heart, brain, eye and muscle that are in high-energy demand (Chinnery and Turnbull, 1997).

Epidemiology of Mitochondrial Disorders

Although the first report of a mitochondrial disease appeared in 1962, the genetics of mitochondrial disease came to the limelight only after the report of mitochondrial mutation in LHON (Leber's Hereditary Optic Neuropathy) in 1988. Since then, there has been an exponential increase in the number of mtDNA abnormalities linked to a wide variety of human diseases. More than 150 different types of mitochondrial defects have been reported (Wittenhagen and Kelley, 2003). Moreover these diseases are not rare as commonly believed and a conservative estimate of the prevalence of mtDNA disease indicate that at least one in 7000 individuals is suffering from mitochondrial disorder (DiMauro and Schon, 2003). This figure is likely to be a gross underestimate because many patients with mitochondrial mutation do not have any specific symptom and it is quite unfortunate that many of these individuals remain undetected unless and until the histochemical or a detailed genetic study is done. There is no such comprehensive estimate available for the Indian populations.

Clinical and Genetic Heterogeneity of Mitochondrial Diseases

Generally mitochondrial mutations would not cause diseases of a single organ rather manifest as syndromes, such as MELAS (Mitochondrial Encephalomyopathy, Lactic Acidosis and Stroke-like episodes), NARP (Neuropathy, Ataxia and Retinitis pigmentosa), MNGIE (Myoneurogastrointestinal disorder and encephalopathy) etc., affecting major organs. Severe neurodegenerative diseases like Parkinson's and Alzheimer's disease can also be caused because of the mitochondrial mutations.

Several mutations are also found to be associated with the Type II Diabetes Mellitus. MELAS is a progressive neurodegenerative disorder caused by the mutations either in tRNA genes (Goto et al., 1990) or in polypeptide coding region (Hanna et al., 1998). Recently, we have analysed several cases of MELAS. In an interesting case, we have found genetic heterogeneity of mitochondrial disease due to mutations, for the first time, in rRNA and polypeptide coding genes in that particular patient. Ours is the first study to establish the genetic heterogeneity of MELAS with mutations in both protein synthesis as well as protein coding regions in the same individual (Vanniarajan et al., 2006).

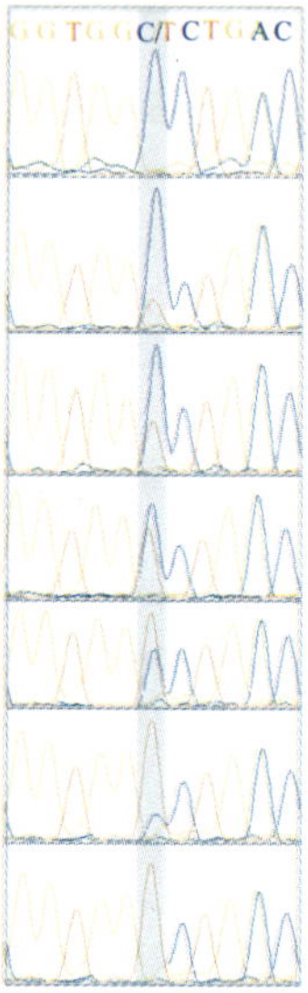

Fig. 11.7 Various proportions of heteroplasmy. Top panel shows pure wild type allele; lower panel shows pure mutant allele; while the remaining panels show different proportions of wildtype and mutant alleles.

Heteroplasmy

Mitochondrial mutations may not be homogeneous in all the cells. In cells that are highly proliferating like blood, these mutations are selectively eliminated, whereas their levels will increase in some tissues due to the negative selection and the increased mutant load will lead to progression of the disease (Chinnery et al., 2000). Heteroplasmy is the phenomenon of coexistence of both the normal and mutant nucleotide (Fig. 11.7). The disease is manifested when the level of the mutant mtDNA increases that can be very well explained by NARP and MILS (Maternally Inherited Leigh Syndrome), caused by a single pathogenic point mutation T8993G in ATPase 6 (Manfredi et al., 2002).

Mitochondrial diseases may also arise from processes other than germline mutations in mtDNA. Recent reports show somatic and homoplasmic mutations in mtDNA causing multiple neonatal deaths (McFarland et al., 2002). Generally, mtDNA deletions involved in pathogenesis are not inherited from the mother. The disease onset may be either early or late depending on the load of mutant mtDNA. Some mtDNA mutations may cause disease only when the bearer is exposed to an environmental toxin (Cohen and Gold, 2001).

Nuclear Genes Involved in Mitochondrial Diseases

Mitochondrial diseases may be caused not only due to mutations in mitochondrial genome but also due to the mutations in several nuclear genes. Maintenance and expression of mtDNA depend on many nuclear-encoded gene products. DNA polymerase γ (POLG) is the only DNA polymerase that has been identified to be involved in mammalian mtDNA replication. Mutations in POLG gene have been reported to associate with the Progressive External Ophthalmoplegia (PEO) characterized by mtDNA deletions (Goethem et al., 2001). POLG mutations were also found to have association with male infertility (Rovio et al., 2001).

Mitochondrial Transcription factor (TFAM), Adenine Nucleotide Translocator (ANT) and Twinkle are some of the proteins that colocalize with mtDNA in nucleiods. They are having important roles in the integrity of the mtDNA. Any alterations in these proteins will lead to the loss of integrity of mitochondrial genome. Mutation in gene coding for Twinkle has been found to be associated with the mtDNA deletion in case of autosomal dominant-PEO. (Spelbrink et al., 2001). Hence, the genes encoding these proteins can be screened to know the cause of the mtDNA rearrangements, especially deletions. Mutations in the structural components of the respiratory chain complexes I and II have been associated with severe neurological disorders of childhood such as Leigh's syndrome and leukodystrophy (DiMauro and Schon, 2003). Hence, the genes encoding the subunits of complexes of the respiratory chain are worth examining.

Genetic Heterogeneity of Mitochondrial Disease in India

As we have seen in the nuclear genes, variations in the mitochondrial genes and their implications in causing disease also unique among Indian populations. We have demonstrated that the Indian

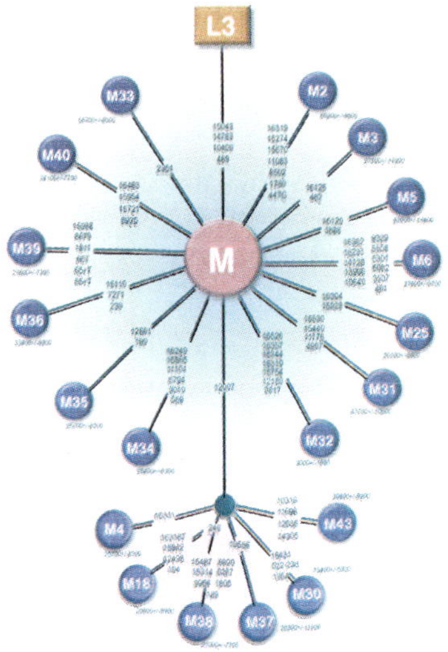

Fig. 11.8 Shows the *in situ* origin of several deep-rooted M lineages observed in our studies.

mtDNA pool consists of several deep rooting lineages of macrohaplogroup 'M' suggesting *in-situ* origin of these haplogroups in South Asia, most likely in India (Fig. 11.8). Since several evolutionarily associated polymorphisms are found in India, one can speculate that the pathogenic mutations which cause various diseases also can be unique to Indian populations. In order to address this question, we have analysed large number of individuals with various mitochondrial disorders and found several novel mutations associated with various disorders among Indians.

Unique Mitochondrial Variations Associated with Low Sperm Motility

Mitochondria are important components of sperm and supply energy for its motility. Mutations in the mtDNA of sperm result in either functionless or malfunctioning proteins, hence affecting sperm motility. A sequence level difference between gametic and somatic cells is a very rare phenomenon. A few studies have demonstrated the existence of differences between somatic and germ cells (Chong et al., 1995). We have analysed a unique case of 28-year-old man who, in spite of having a varicocele and a sperm concentration of 5 million/mL, of which 10% were motile and 20% had normal forms (oligoasthenoteratozoospermia [OAT]), was fertile. Analysis of mitochondrial genes revealed, for the first time, 9 missense and 27 silent mutations in the sperm's mitochondrial DNA (mtDNA) but not in the DNA from the blood cells. There was a 2-nucleotide deletion in the mitochondrial COII genes, introducing a stop codon, which might be responsible for low sperm motility (Thangaraj et al., 2003).

In another study, we have analyzed a total of 34 oligoasthenospermic men from our collection of 650 infertile men originating from eight different states. Our initial analyses of the complete mDNA of 10 oligoasthenozoospermic men revealed the presence of several sequence variants, but one variant at nucleotide position 11994 (C→T) was found in all the 10 samples analyzed. Interestingly, the T allele never before has been reported, either as a polymorphism or in association with a disease, anywhere in the world. Therefore, we screened for the presence of this mutation in the remaining 24 samples. Amazingly, we also found this mutation in all the remaining 24 samples, although none of the 150 proven-fertile (normozoospermic) men carried it. The C11994T mutation is located in the second base of a codon, resulting in an amino-acid change from threonine to isoleucine at amino-acid position 412 of the ND4 gene (Selvi et al., 2006). The ND4 gene is an essential subunit of complex I of the mitochondrial respiratory chain, and ND4 mutations have been implicated in mitochondrial diseases such as Leigh's syndrome and Leber hereditary optic neuropathy (Taylor and Turnbull, 2005). Because none of the fertile men analyzed in this study or in other studies around the world have been found to contain this C11994T mutation, the possibility of it being a neutral polymorphism was ruled out. However, this mutation was not found to be associated with low sperm motility in Portugal population (Pereira et al., 2008).

Novel mutations in Cardiovascular Diseases

A 65-year-old male had episodic atypical chest discomfort and bouts of dizziness unrelated to effort. Physical examination revealed an ejection systolic murmur without thrill, best heard at the parasternal region of fourth left intercostal space. Electrocardiography (ECG) showed first degree atrioventricular (AV) block with a PR interval of 0.36s and 24 h ambulatory ECG revealed episodic non-sustained ventricular tachycardia (NSVT) with intermittent second and third degree AV blocks. Echocardiography and angiocardiography showed asymmetrical mid ventricular septal hypertrophy with dynamic obstruction, with a maximum gradient of 47 mm Hg across the midcavity of the left ventricle (Fig. 11.9). He underwent dual chamber permanent pacemaker implantation for his episodic complete heart block and mid ventricular obstruction. Molecular genetic analyses revealed three novel mutations in mitochondrial 12s rRNA (ribosomal ribonucleic acid) at nucleotide

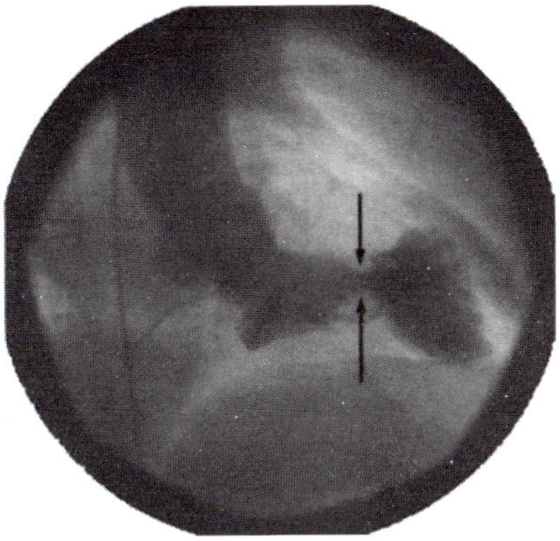

Fig. 11.9 Left ventricular angio RAO 30- view showing mid ventricular dynamic obstruction.

positions 869 C→T, 1391 T→C and 1556 C→T, as well as in ND1, ND4 and ND5 protein coding sequence genes of the mitochondria, revealed 3 mutations at nucleotide positions 3407 G→A, 11365 T→C and 12477 T→C, respectively (Prasad et al., 2006).

Similarly, we have observed novel mutations in a six-year-old male patient with hypertelorism, downward-slanting palpebral fissures, pectus exacavatum, micrognathia, low-set and posteriorly angulated ears, and webbed neck. Echo and ECG showed signs of severe hypertrophic obstructive cardiomyopathy (HOCM), right atrial hypertrophy, a wide QRS complex, left axis deviation, giant Q-waves, and negative left precordial complexes. Cytogenetic analysis of the patient revealed a normal male karyotype (46, XY). Neonatally there were feeding problems, hypotonia, and failure to thrive. We have analyzed the candidate genes for hypertrophic cardiomyopathy (HCM), including PTPN11 (Protein-tyrosine phosphatase, non-receptor type 11), beta-MHC (beta myosin heavy chain), MYBPC3 (Cardiac myosin binding protein) and Troponin T; and the complete mitochondrial genome (mtDNA) of this patient. Analysis of complete mtDNA revealed seven novel mutations in the mitochondrial genes (Dhandapany et al., 2007)

Novel Mutations in Neuromuscular Diseases

One of our studies of a patient with typical features of MELAS has given a new dimension to the diagnosis of MELAS. We did not find any mutations that has been known or associated with MELAS but we got a novel heteroplasmic mutation. This mutation in ND5 (C13565A) was found only in the tissue sample of the patient but not in the blood sample. This mutation is missense causing a change of serine to tyrosine at amino acid position 410. This mutation was found neither in controls nor in world populations (Vanniarajan et al., 2006a). Similarly we have sequenced the complete mitochondrial genome of a patient with Leigh syndrome (Fig. 11.10) and identified a novel mutation in ND4 gene (Vanniarajan et al., 2006b).

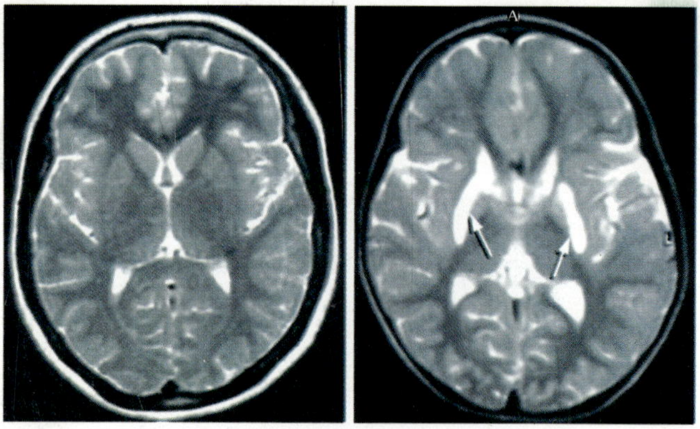

Fig. 11.10 MR images of a Leigh syndrome patient (*right panel*) showing bilateral basal ganglionic lesions, highly characteristic of Leigh disease (*shown by arrows*). Left: MRI of an age-matched control.

Conclusion

Our extensive genetic studies on Indian populations revealed that genetically distinct ethnic populations are inhabiting India. Our study suggests that India experienced the first wave of migration of early humans "out-of-Africa". Further, we observed *in situ* origin of several deep rooting mitochondrial lineages among Indians; suggesting that they are genetically different from rest of the world. Hence, one can speculate that the disease causing mutations could also be different in Indian populations. Analysis of autosomal, X chromosome, Y chromosome and mitochondrial genes revealed several novel and pathogenic mutations in Indian populations. From our study, it is obvious that the genetic architecture of the Indian populations are unique, therefore, it is absolutely essential to undertake extensive genetic study on Indian populations for understanding their origin and the genetic basis of the complex diseases.

Acknowledgements

I sincerely thank Dr. Lalji Singh, whose constant support and guidance made me possible to carry out the work presented in this article. I also thank all my colleagues and collaborators for their contributions in conducting experiments, evaluating patients and preparation of this manuscript.

References

1. Affara N.A, (2001). The role of the Y chromosome in male infertility. Exp. Rev. Mol. Med. 2001: 1–16.
2. Aschim E.L, Giwercman A, Stahl O, Eberhard J, Cwikiel M., Nordenskjold A, Haugen TB, Grotmol T, Giwercman YL, (2005). The RsaI polymorphism in the estrogen receptor-beta gene is associated with male infertility. J. Clin. Endocrinol. Metab. *90*: 5343–5348.
3. Borthwick G.M, Taylor R.W, Walls T.J, Tonska K, Taylor G.A, Shaw P.J, Ince P.G, Turnbull D.M, (2006). Motor neuron disease in a patient with a mitochondrial tRNAIle mutation. Ann. Neurol. *59*: 570–574.

4. Bowcock A.M, Ruiz- Linares A, Tomfohrde J, Minch E, Kidd JR, Cavalli-Sfroza. LL, (1994). High resolution of human evolutionary trees with polymorphic microsatellites. Nature. *268*: 455–457.
5. Carani C, Qin K, Simoni M, Faustini-Fustini M, Serpente S, Boyd J, Korach K.S, Simpson E.R, (1997). Effect of testosterone and estradiol in a man with aromatase deficiency. N. Engl. J. Med. *337*: 91–95.
6. Census of India (2001). website:http://indstat.com/india/demographics/scheduled_tribe_in_india Chai N.N, Salido E.C, Yen P.H, (1997) Multiple functional copies of the RBM gene family, a spermatogenesis candidate on the human Y chromosome. Genomics. *45*: 355–361.
7. Chaubey G, Metspalu M, Kivisild T, Villems R, (2007). Peopling of South Asia: investigating the caste–tribe continuum in India. BioEssays. *29*: 91–100.
8. Chinnery P.F, Turnbull D.M, (1997). Mitochondrial Medicine. QJ Med. *90*: 657–667
9. Chinnery P.F, Thorburn D.R, Samuels D.C, White S.L, Dahl H.M, Turnbull D.M, Lightowlers R.N, Howell N, (2000). The inheritance of mitochondrial DNA heteroplasmy: random drift, selection or both? Trends. Genet. *16*: 500–505.
10. Chong S.S, McCall A.E, Cota J, Subramony S.H, Orr HT, Hughes M.R, Zoghbi HY, (1995). Gametic and somatic tissue-specific heterogeneity of the expanded SCA1 CAG repeat in spinocere bellar ataxia type1. Nat. Genet. *10*: 344–350.
11. Dhandapany P.S, Sadayappan S, Vanniarajan A, Karthikeyan B, Nagaraj C, Gowrishankar K, Selvam G.S, Singh L, Thangaraj K, (2007). Novel mitochondrial DNA mutations implicated in Noonan syndrome. Int. J. Cardiol. *120*: 284–285.
12. Gold D.R, Cohen B.H, (2001). Treatment of mitochondrial cytopathies. Semin.Neurol. *21*: 309–325.
13. Danielson S.R, Wong A, Carelli V, Martinuzzi A, Schapira A.H, GinoA, (2002). Cortopassi Cells Bearing Mutations Causing Leber's Hereditary Optic Neuropathy Are Sensitized to Fas-induced Apoptosis. J. Biol. Chem. *277*: 5810–5815.
14. Deborah R, Gold M.D, Bruce H, Cohen M.D, (2001). Treatment of Mitochondrial Cytopathies. Semin. Neurol. *21*: 309–325.
15. Dechend F, Williams G, Skawran B, Schubert S, Krawczak M, Tyler- Smith C Schmidtke J, (2000). TSPY variants in six loci on the human Y chromosome. Cytogenet. Cell Genet. *91*: 67–71.
16. Delbridge M.L, Harry J.L, Toder R, O'Neill R.J, Ma K, Chandley A.C, Graves J.A, (1997). A human candidate spermatogenesis gene, RBM1, is conserved and amplified on the marsupial Y chromosome. Nature Genetics. *15*: 131–136.
17. DiMauro S, Schon E.A, (2003). Mitochondrial respiratory-chain diseases. N. Engl. J. Med. *348*: 2656–2668.
18. Goto Y, Nonaka I, Horai S, (1990). A mutation in the tRNA(Leu)(UUR) gene associated with the MELAS subgroup of mitochondrial encephalomyopathies. Nature. *348*: 651–653.
19. Hanna M.G, Nelson I.P, Morgan-Hughes J.A, Wood N.W, (1998). MELAS: a new disease associated mitochondrial DNA mutation and evidence for further genetic heterogeneity. J Neurol. Neurosurg. Psychiatry. *65*: 512–517.
20. Herrmann B.L, Saller B, Janssen O.E, Gocke P, Bockisch A, Sperling H, Mann K, Broecker M, (2002). Impact of estrogen replacement therapy in a male with congenital aromatase deficiency caused by a novel mutation in the CYP19 gene. J. Clin. Endocrinol. Metab. *87*: 5476–5484.

21. Hess R.A., Bunick D, Lee K.H, Bahr J, Taylor J.A, Korach K.S, Lubahn DB, (1997). A role for oestrogens in the male reproductive system. Nature *390*: 509–512
22. Khattri A, Singh L, Thangaraj K, (2006). Male Infertility: Genetic causes and their implications. Embryo Talk. *1*: 3–20.
23. Khattri A, Pandey R.K, Gupta N.J, Chakravarty B.N, Deendayal M, Singh L, Thangaraj K, (2007). CA repeat and *Rsa*I polymorphisms in ERβ gene are not associated with infertility in Indian men. Intl. J. Androl. *30*: 1–7
24. Kirsch S, Weiss B, De Rosa M, Ogata T, Lombardi G, Rappold GA, (2000). FISH deletion mapping defines a single location for the Y chromosome stature gene, GCY. J Med Genet. *37*: 593–599.
25. Lestienne P, (1987). Evidence for a direct role of the DNA polymerase gamma in the replication of the human mitochondrial DNA *in vitro*. Biochem Biophys Res Commun. *146*: 1146–1153.
26. Ma K, Inglis J.D, Sharkey A, (1993). A chromosome gene family with RNAbinding protein homology: candidates for the azoospermia factor AZF controlling human spermatogenesis. Cell *75*: 1287–1295.
27. McFarland R, Clark K.M, Morris A.A, Taylor R.W, Macphail S, Lightowlers R.N, Turnbull D.M, (2002). Multiple neonatal deaths due to a homoplasmic mitochondrial DNA mutation. Nat Genet. *30*: 145–146.
28. Manfredi G, Fu J, Ojaimi J, Sandlock J.E, Kwong J.Q, Guy J, Schon E.A, (2002). Rescue of a deficiency in ATP synthesis by transfer of MTATP, a mitochondrial DNA-encoded gene, to the nucleus. Nat Genet. *30*: 394–399.
29. Martinez M.C, Bernabe M.J, Gomez E, (2000). Screening for AZF deletion in a large series of severely impaired spermatogenesis patients. J. Androl. *21*: 651–655.
30. Mazeyrat S, Saut N, Sargent C.A, (1998). The mouse Y chromosome interval necessary for spermatogonial proliferation is gene dense with syntenic homology to the human AZFa region. Hum. Mol. Genet. *7*: 1713–1724.
31. Metspalu M, Kivisild T, Metspalu E, Parik J, Hudjashov G, Kaldma K, Serk P, Karmin M, Behar D.M, Gilbert MT, Endicott P, Mastana S, Papiha SS, Skorecki K, Torroni A, Villems R, (2004). Most of the extant mtDNA boundaries in south and southwest Asia were likely shaped during the initial settlement of Eurasia by anatomically modern humans. BMC Genet. *31*: 5–26
32. Ogawa S, Emi M, Shiraki M, Hosoi T, Ouchi Y, Inoue S, (2000a). Association of estrogen receptor beta (ESR2) gene polymorphism with blood pressure. J. Hum. Gene. *45*: 327–330.
33. Ogawa S, Hosoi T, Shiraki M, Orimo H, Emi M, Muramatsu M, Ouchi Y, Inoue S (2000b). Association of estrogen receptor beta gene polymorphism with bone mineral density. Biochem Biophys Res Commun. *269*: 537–541.
34. Palanichamy M.G, Sun C, Agrawal S, Bandelt HJ, Kong P.Q, Khan F, Wang C.Y, Chaudhuri T.K, Palla V, Zhang Y.P, (2004). Phylogeny of Mitochondrial DNA Macrohaplogroup N in India, Based on Complete Sequencing: Implications for the Peopling of South Asia. Am. J.Hum. Genet. *75*: 966–978.
35. Pandya A, King T.E, Santos F.R, Taylor P.G, Thangaraj K, Singh L, Jobling M, Tyler-Smith C, (1998). A polymorphic human Y chromosomal G to A transition found in India. Ind. J. Hum. Genet. *4*: 52–61
36. Prosser J, Inglis JD, Condie A, (1996). Degeneracy in human multicopy RBM (YRRM), a candidate spermatogenesis gene. Mamm Genome *7*: 835–842.

37. Rajender S, Rajani V, Gupta N.J, Chakravarty B, Singh L, Thangaraj K, (2006). *SRY* negative 46, XX male with normal genitals, complete masculinization and infertility. Mol. Hum. Rep. *12*: 341–346.
38. Rani D.S, Carlus S.J, Poongothai J, Jyothi A, Pavani K, Gupta N.J, Reddy A.G, Rajan M.M, Rao K,Chakravarty B, Singh L, Thangaraj K, (2008). CAG repeat variation in the mtDNA polymerase gamma is not associated with oligoasthenozoospermia. Int. J. of Androl. *(In press)*
39. Reijo R, Lee TY, Salo P, (1995). Diverse spermatogenic defects in humans caused by Y chromosome deletions encompassing a novel RNA-binding protein gene. Nat. Genet. *10*: 383–393.
40. Robertson K.M, O'Donnell L, Jones M.E, Meachem S.J, Boon W.C, Fisher C.R, Graves K.H, McLachlan R.I, Simpson E.R, (1999). Impairment of spermatogenesis in mice lacking a functional aromatase (cyp 19) gene. Proc Natl Acad Sci (USA) *96*: 7986–7991.
41. Rocher C, Taanman J-W, Pierron D, Faustin, Benard G, Rossignol R, Malgat M, Pedespan L, Letellier L, (1999). Influence of mitochondrial DNA level on cellular energy metabolism: implications for mitochondrial diseases. J. Bioenerg. Biomembr. *40*: 59–67.
42. Rötig A, Munnich A, (1995). Genetic Features of Mitochondrial Respiratory Chain Disorders. J. Am. Soc. Nephrol. *14*: 2995–3007.
43. Rovio A, Tiranti V, Bednarz A.L, Suomalainen A, Spelbrink J.N, Lecrenier N, Melberg A, Zeviani M, Poulton J, Foury, Jacobs H.T, (1999). Analysis of the trinucleotide CAG repeat from the human mitochondrial DNA polymerase gene in healthy and diseased individuals. European J. of Hum. Gene. *7*: 140–146.
44. Rovio, A. T., Marchington, D. R., Donat, S., Schuppe, H. C., Abel, J., Fritsche, E, (2001). Mutations at the mitochondrial DNA polymerase (POLG) locus associated with male infertility. Nature Genet. *29*: 261–262.
45. Saunders, P.T, (2005). Does estrogen receptor beta play a significant role in human reproduction? J. Clin. Endocrinol. Metab.*16*: 222–227.
46. Scariano J.K, Simplicio S.G, Montoya G.D, Garry P.J, Baumgartner R.N, (2004). Estrogen receptor beta dinucleotide (CA) repeat polymorphism is significantly associated with bone mineral density in postmenopausal women. Calcif Tissue Int.*74*: 501–508.
47. Schon E.A, DiMauro S, (2003). Medicinal and genetic approaches to the treatment of mitochondrial disease. Curr. Med.Chem. *10*: 2523–2533.
48. Sengupta S, Zhivotovsky LA, Roy K, Mehdi S. Q, Christopher A.E, Chow C-E, Lin AA, Mitra M, Sil S.K, Ramesh A, Rani M.V, Thakur C.M, Cavalli-Sforza L.L, Majumder P.P, Underhill P.A, (2006). Polarity and Temporality of High-Resolution Y-Chromosome Distributions in India Identify Both Indigenous and Exogenous Expansions and Reveal Minor Genetic Influence of Central Asian Pastoralists. Am. J.Hum. Genet. *78*: 202–221.
49. Simpson E.R, Mahendroo M.S, Means G.D, Kilgore M.W, Hinshelwood M.M, Graham-Lorence S, Amarneh B, Ito Y, Fisher C.R, Michael M.D, (1994). Aromatase cytochrome P450, the enzyme responsible for estrogen biosynthesis. Endocr. Rev. *15*: 342–355.
50. Singh K.S, (1992). People of India: An introduction. National Series. 1: 234.
51. Sun C, Skaletsky H, Birren B, Devon K, Tang Z, Silber S, Oates R, David C. Page D.C, (1999). An azoospermic man with a *de novo* point mutation in the Y-chromosomal gene USP9Y. Nature Genet. *23*: 429–432.
52. Thangaraj K, Reddy A.G, Singh L, (2002b). Is amelogenic gene reliable for gender identification in forensic casework and prenatal diagnosis? Intl. J. Legal. Med. *116*: 121–123.

53. Thangaraj K, Joshi M.B, Reddy A.G, Gupta N.J, Chakravarty B, Singh L, (2002). CAG repeat expansion in the androgen receptor gene is not associated with male infertility in Indian populations. J. Androl. *23*: 813–816.
54. Thangaraj K, Joshi M.B, Reddy A.G, Rasalkar A.A, Singh L, (2003). Sperm mitochondrial mutations as a cause of low sperm motility. J Androl. *24*: 388–392.
55. Thangaraj K, Subramanian S, Reddy A.G, Singh L, (2003). A unique case of deletion and duplication in the long arm of the Y chromosome in an individual with ambiguous genitalia. Am. J. Med. Genet. *116*A: 205–207.
56. Thangaraj K, Singh L, Reddy A.G, Rao V.R, Sehgal S.C, Underhill P.A, Pierson M, Frame I.G, Hagelberg E, (2003). Genetic affinities of the Andaman Islanders, a vanishing human population. Curr. Biol. *13*: 86–93.
57. Thangaraj K, Chaubey G, Kivisild T, Reddy A.G, Singh V.K, Rasalkar A.A, Singh L, (2005). Reconstructing the origin of the Andaman Islanders. Science. *308*: 996.
58. Thangaraj K, Sridhar V, Kivisild T, Reddy A.G, Chaubey G, Singh V.K, Kaur S, Agarawal P, Rai A, Gupta J, Mallick C.B, Kumar N, Velavan T.P, Suganthan R, Udaykumar D, Rashmi V, Mishra R, Khan A, Annapurna A, Singh L, (2005). Different population histories of the Mundari and Mon-Khmer speaking Austro-Asiatic tribes inferred from the mt.DNA 9-bp deletion/insertion polymorphism in Indian populations. Hum. Genet. *116*: 507–517.
59. Thangaraj K, Chaubey G, Reddy A.G, Singh V.K, Singh L, (2006). Unique origin of Andaman islanders: insight from autosomal loci. J. Hum. Genet. *51*: 800–804.
60. Thangaraj K, Chaubey G, Kivisild T, Rani D.S, Singh V.K, Ismail T, Carvalho-Silva D, Metspalu M, Bhaskar LVKS, Reddy A.G, Chandra S, Pande V, Prathap-Naidu B, Adarsh N, Verma A, Jyothi I.A, Mallick C.B, Shrivastava N, Devasena R, Kumari B, Singh A.K, Diwedi S.K.D, Singh S, Rao G, Gupta P, Sonvane V, Kumari K, Basha A, Bhargavi K.R, Lalremruata A, Gupta A.K, Kaur G, Reddy K.K, Rao A.P, Villems R, Tyler-Smith C, Singh L, (2008). Maternal footprints of southeast Asians in north India. Hum. Heredity. *28*: 1–9
61. Tsukamoto K, Inoue S, Hosoi T, Orimo H, Emi M, (1998). Isolation and radiation hybrid mapping of dinucleotide repeat polymorphism at the human estrogen receptor beta locus. J. Hum. Gene. *43*: 73–74.
62. Van Goethem G, Dermaut B, Löfgren A, Martin J.J, Van Broeckhoven C, (2001). Mutation of POLG is associated with progressive external ophthalmoplegia characterized by mtDNA deletions. Nat Genet. *28*: 211–212.
63. Vanniarajan A, Rajshekher G.P, Joshi M.B, Reddy A.G, Singh L, Thangaraj K. (2006a). Novel mitochondrial mutation in the ND4 gene associated with Leigh syndrome. Acta Neurol Scand. *114*: 350–353.
64. Vanniarajan A, Nayak D, Reddy A.G, Singh L, Thangaraj K, (2006b). Clinical and genetic uniqueness in an individual with MELAS. Am. J. Med. Genet .*141*B: 440–444.
65. Westberg L, Baghaei F, Rosmond R, Hellstrand M, Landen M, Jansson M, Holm G, Bjorntorp P, Eriksson E, (2001). Polymorphisms of the androgen receptor gene and the estrogen receptor beta gene are associated with androgen levels in women. J. Clin. Endocrinol. Metab. *86*: 2562–2568.
66. Wittenhagen L.M, Kelley SO, (2003). Impact of disease-related mitochondrial mutations on tRNA structure and function. Trends Biochem Sci. *28*: 605–611.

Heterochromatin on Human Y-Chromosome Reveals New Paradigms in Regulation of Gene Expression in Eukaryotes

12

Rachel A Jesudasan*[1] and Vidya Jonnalagadda[2]

[1]*Centre for Cellular and Molecular Biology, Hyderabad, India., *Corresponding Author;*
[2]*Bhavan's Vivekananda College, Secunderabad, India*

Regulatory Mechanisms for Gene Expression

The complexity of mechanisms involved in regulation of gene expression in eukaryotes is much greater than in prokaryotes. In a simple unicellular organism lacking a nuclear membrane partition between the DNA and cytoplasm, there are only two major modes of regulation of non-housekeeping genes. The first mode is blocking transcription by (a) disallowing RNA polymerase to bind to a promoter in the absence of a specific sigma factor, or (b) blocking the path of an RNA polymerase bound to a promoter. This can be achieved by either an inhibitor bound to an operator sequence or by the catabolite repression protein. The second mode is the premature termination of transcription by attenuation. These modes of regulation appear to suffice in modulating the expression of the simple genome of prokaryotic cells.

Eukaryotic genomes differ from prokaryotic genomes in two major aspects – (a) sequence content, and (b) packaging into chromatin. The wide range of regulatory mechanisms in eukaryotic cells reflects the complexity of a typical eukaryotic genome. In multicellular organisms with distinct tissues and organs, some genes are (a) always expressed in all cells, (b) never expressed in a certain cell type, (c) expressed only in certain developmental stages in a certain cell type, and (d) expressed rapidly only in response to specific environmental stimuli in some cell types. Gene expression is regulated at the levels of pre-transcription (to prevent transcription of "inappropriate" genes in a cell), initiation of transcription, post-transcriptional processing (such as alternative splicing), transport of mRNA to cytoplasm, and post-translational processing and modification of proteins.

Role of Repeat Sequences

In addition to 'classical genes', which conform to the 'one gene – one polypeptide' hypothesis, eukaryotic genomes may contain a substantial amount of non-genic, or so-called junk, DNA. In fact, less than 2% of the human genome is believed to code for proteins, while more than 50% appears to comprise various repeats (International Human Genome Sequencing Consortium, 2001). Of the repeated DNA, tandem repeats account more than 10% of the human genome and interspersed repeats comprise approximately 45% of the genome.

The evolutionary origins as well as the present day functions of various repeat elements in the human genome remain unclear, though they have been a topic of much intellectual speculation. Most repeat elements do not appear to be expressed; they lack the promoter sequences typical of eukaryotic genes, and their sequences do not appear to encode any known polypeptides. Regardless of these features, it appears that disallowing transcription from the repeat elements is advantageous for the cell in terms of conserving its resources for the expression of "real" genes. One strategy to preclude expression of repeat elements is to package them in a state of chromatin that is refractory to transcription.

Role of Chromatin in Transcription

Packaging of eukaryotic DNA into chromatin serves not only to condense the voluminous genome into a small nucleus, it also enables exquisite control of transcription by differential density of nucleosomes over specific regions of the genome. The differential packing of chromatin is evident as regions (bands) of darkly and lightly staining material in chromosomes stained with basic dyes. The darkly stained regions represent dense packing of chromatin and are termed heterochromatin. Lightly staining material represents looser packing of chromatin and is termed euchromatin.

Packaging regions of the genome not programmed to be expressed in a cell into heterochromatin is an effective strategy of pre-transcriptional control of gene expression. This notion is supported by the fact that reports of transcription from heterochromatic regions are very few (Tulin et al., 2002; Diaz et al., 1981; Rudert et al., 1995). Regions not known to contain expressible genes, such as centromeres, pericentromeres and telomeres, are packed in large blocks of constitutive heterochromatin. In addition, most of the tandem repeats in a genome appear to occur mainly as constitutive heterochromatin. Regions containing expressible genes may be packaged into facultative heterochromatin, which is nothing but epigenetically silenced euchromatin. Such regions can easily be converted into euchromatin, as observed with the stimulus-inducible chromosomal puffs of the polytene chromosomes of *Drosophila*.

Mammalian Y Chromosome

The mammalian Y chromosome is particularly rich in repetitive sequences as well as heterochromatin. Only partial sequence of the 58Mb-long human Y chromosome has been determined; a 40Mb block at the end of the long arm (q) of the human Y chromosome has proved recalcitrant to standard methods of sequencing. The resistance to sequencing is believed to be due to repetitive elements. This region, termed *Yq12* is also the major heterochromatic region on the Y chromosome, as shown in Fig. 12.1. So far, there was no evidence of transcription from this region (Skaletsky et al., 2003).

In general, the Y chromosome is believed to be poor in genes; Bachtrog and Charlesworth (2001) has listed only 15 Y-specific genes and 13 X-Y homologous genes on the Y chromosome. Analysis of the known sequence of the Y chromosome reveals around 340 putative genes (includes over 100 pseudogenes; UniGene Cluster analysis). In contrast, chromosome 1 has around 4220 genes spanning 263Mb. The Y-specific genes are relevant in testis formation and male fertility; they do not code for any housekeeping functions. This observation is consistent with the idea that the Y chromosome is dispensable for the survival of an individual though it is essential for the propagation of the species.

On the human Y chromosome, genes with known functions, such as SRY (sex determining region on Y) and male fertility genes have all been localized to the euchromatic regions (Fig. 12.2). The terminal region of the q arm, locus *Yq12*, is not yet fully sequenced; around 8 putative genes have been predicted to occur in the available partial sequence of the pseudoautosomal stretch in this region (UniGene Cluster analysis). Considering the fact that some genes have been identified

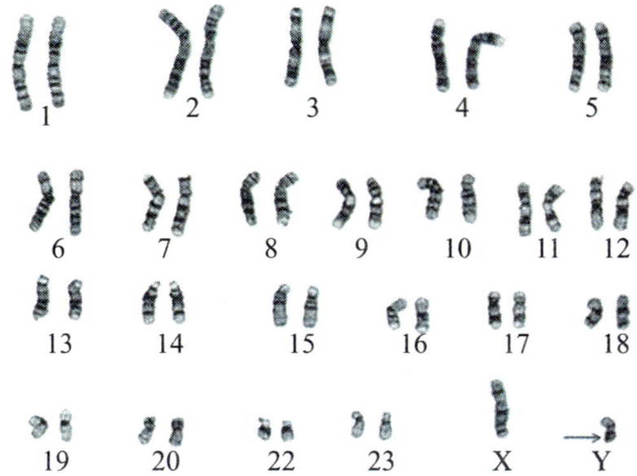

Fig. 12.1 A G-banded human male karyotype showing differential staining. Note the block of heterochromatin at the end of the q arm of the Y chromosome (indicated by arrow).

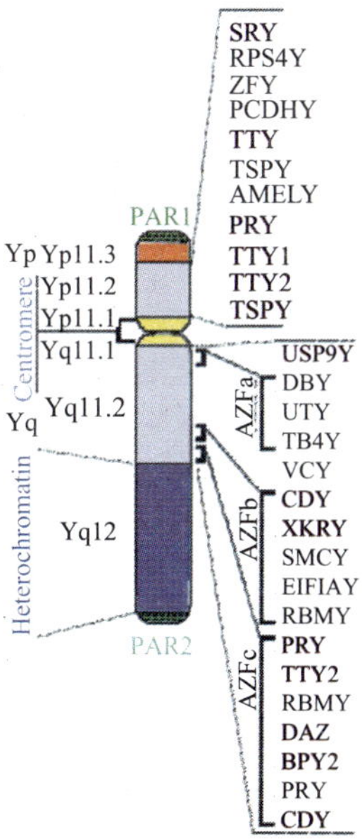

Fig. 12.2 Schematic representation of the human Y chromosome. No genes have been reported from the centromere (yellow) or *Yq12* heterochromatin (purple). All known genes are mapped to the euchromatin (blue) and PAR2 (Pseudo Autosomal Region 2; green). Genes labelled in maroon have no homologs on the X chromosome. AZFa, b, and c harbour the deletions seen in Azoospermia Factor a, b, and c, respectively. Note that there are no known genes in the *Yq12* heterochromatic region. Adapted from: The human Y chromosome, in the light of evolution: (Lahn et al., 2001).

in the heterochromatic Y chromosome of *Drosophila* (Carvalho et al., 2000), it is possible that the heterochromatic block of human Y chromosome too may harbor some genes.

A New Transcription Unit on the Human Y Chromosome

A study of evolution of snakes by Roychowdhary and coworkers revealed substantial differences in the heterochromatin content within the genomes of different species of snakes. A portion of the heterochromatinized sequences in the genome of the banded krait was resolved as a minor satellite peak by analytical ultracentrifugation. The DNA sequence in the minor satellite peak, termed Bkm (banded krait minor), was found to hybridize to homologous sequences in all eukaryotes, from yeast to man, in a sex-restricted fashion. Screening of a human Y-enriched genomic library with the Bkm DNA as a probe yielded a sex- and species-specific clone, which was called 102(d)2 (Singh et al., 1994).

The genomic clone 102(d)2 showed species-specific hybridization to the distal heterochromatic block on the q arm of the Y chromosome (*Yq12*) in multiple copies, when analyzed by Fluorescence *in situ* Hybridization (FISH) (Fig.12.3). This location corresponds to the 40Mb block of heterochromatin on the Y chromosome.

Analysis of hybridization of the 102(d)2 clone with tissues from various stages of developing human testes revealed an unexpected pattern. At 20 weeks, the hybridization signal from this probe

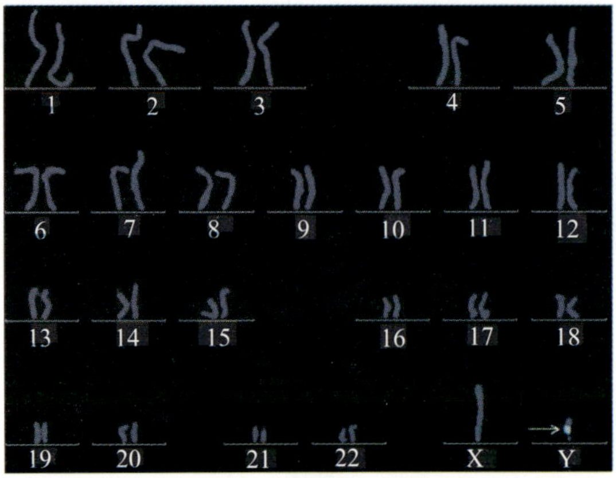

Fig.12.3 FISH analysis of human cells shows the genomic clone 102(d)2 hybridizes to the heterochromatic block on q arm of the Y chromosome. The region of hybridization appears as a bright cyan spot indicated by the arrow.

appeared as a single compact spot within the nucleus of each cell (Fig.12.4), indicating a condensed state of the cognate chromosomal region. At 32 weeks, the hybridization signal appeared as somewhat diffuse in around 30% of the nuclei, indicative of some decondensation of this chromosomal region in at least some cells. RNase treatment of the samples before hybridization to the probe did not alter the dimensions of the signals, indicating that these signals were due to decondensed DNA and not because of RNA. At 36 weeks, most nuclei in the testes sections showed a widely spread intense fluorescence signal; however, treatment with RNase prior to hybridization with the probe

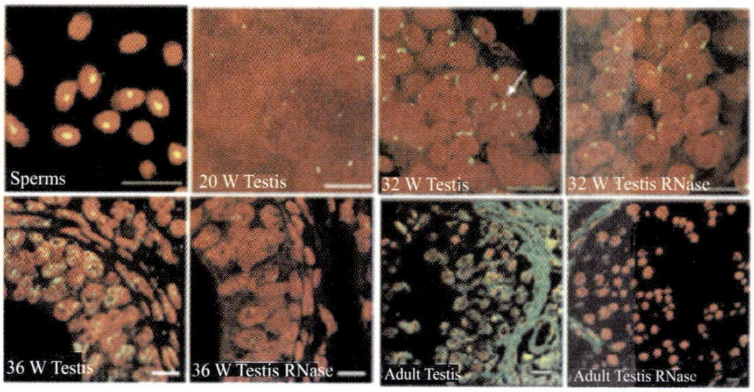

Fig. 12.4 FISH analysis of developing testis shows transcription from *Yq12* locus. Sections of human testis at 20 weeks, 32 weeks, 36 weeks, adult testis, and sperm samples were processed for FISH analysis using 102(d)2 as probe. Duplicate samples were analyzed for each time point; one with pretreatment with RNase and one without RNase pretreatment. Hybridizing material (green) appears yellow when against a red background (the nucleus). The Yq12 region studied here appears as a condensed signal in 20 weeks testis and in sperms. This region decondenses in 32 weeks testis and shows abundant expression from 36 weeks testis to the adult stage. Scale bar: 10 nm.

showed only a single dot-like signal in most cells. Thus, the diffuse signal observed at 36 weeks appears not to be due to chromosomal decondensation; instead it is due to transcription from this locus resulting in a substantial amount of nuclear RNA hybridizing to the FISH probe. The fluorescent signal was very bright in sections of adult testis; almost all the fluorescence could be ascribed to RNA, since treatment with RNase led to substantial reduction of the signal. No RNA hybridizing to the 102(d)2 probe was evident in mature sperm (Jehan et al., 2007).

To test if indeed the hybridization was due to RNA, purified polyadenylated RNA from testis at 20 weeks, 32 weeks and 36 weeks was analyzed by Northern blots (Fig. 12.5). Large heterogeneous transcripts were observed only in 36 weeks testis. Some signal was seen in 32 weeks testis after prolonged exposure (data not shown). Earlier developmental stages and other tissues did not show any signal. This confirmed the developmental stage-specific and tissue-specific expression from the human *Yq12* genomic locus.

Two Testis-Specific cDNA Clones Localizing to Yq12

A human testicular cDNA library was screened with 102(d)2 in order to isolate cDNA clones corresponding to the transcripts from the *Yq12* genomic locus. Two clones were obtained from the cDNA library. The 3′ poly (A) ends of the two cDNA clones were obtained by 3′ RACE reactions. The complete DNA sequences of genomic clone 102(d)2 (2189bp) and the two cDNA clones (816bp and 1441bp, respectively) were determined and deposited in the NCBI databank (Accession numbers AY598345, AY598346 and AY598347, respectively).

The two cDNA clones represent two distinct novel polyadenylated RNAs. Each hybridized to the *Yq12* region in normal metaphase spreads analyzed by FISH (Fig. 12.6). As observed with the 102(d)2 genomic clone, each of these two cDNAs also hybridized in multiple copies on the *Yq12* heterochromatic block (Jehan et al., 2007).

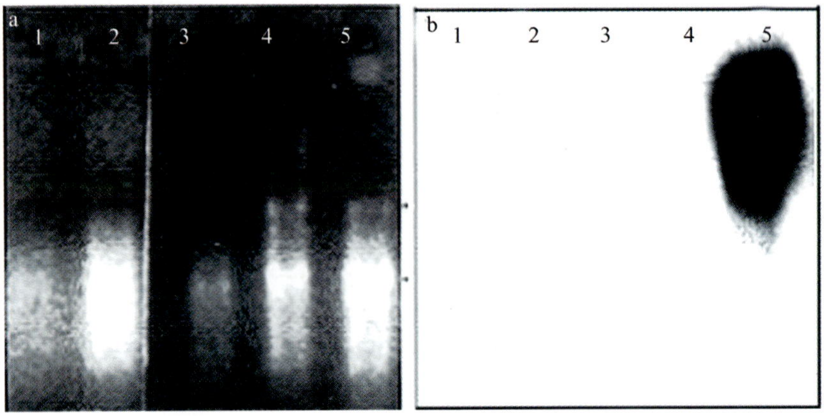

Fig. 12.5 Northern analysis reveals presence of a testis-specific polyadenylated RNA hybridizing to a probe for *Yq12*. Left panel (**a**) shows an ethidium bromide-stained agarose gel. Right panel (**b**) shows the corresponding Northern blot probed with the *Yq12* genomic clone 102(d)2. Lane 1 - 20 weeks testis; lane 2 - 36 weeks kidney; lane 3 - 36 weeks spleen; lane 4 - 32 weeks testis; lane 5 - 36 weeks testis. Small arrows indicate the positions of 18S and 28S rRNA. A heterogeneously sized signal is obtained in 36 weeks testis, confirming developmental stage-specific and testis-specific expression.

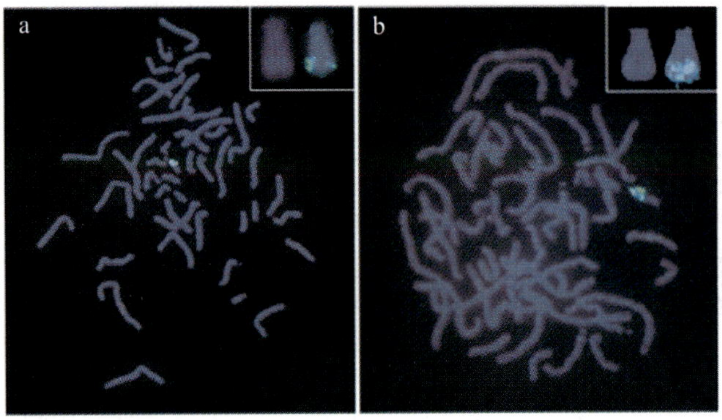

Fig. 12.6 FISH analysis of the two novel cDNAs shows multi-copy hybridization to the *Yq12* heterochromatin block. Human metaphase chromosomes were hybridized to fluorescently labelled cDNA clone AY598346 (panel **a**) and AY598347 (panel **b**). Hybridized probe appears as bright green signal. Insets show an expanded view of the Y chromosome: left panel shows staining for heterochromatin (note the bright fluorescence in the *Yq12* region); right panel shows hybridization of the cDNA probe (note the multi-copy hybridization in the *Yq12* heterochromatin region).

When used as probes for Northern analysis of multi-tissue polyadenylated RNA, each cDNA clone hybridized to heterogeneously sized transcripts only in testis tissue (Fig. 12.7). The size of the transcripts hybridizing to the cDNA clone AY598346 was smaller than those hybridizing to the cDNA clone AY598347, indicating that these cDNA clones come from two different genes.

The detection of at least two distinct transcripts that appear to arise from a heterochromatic block on the Y chromosome is significant on three counts. First, ever since the coining of the term 'heterochromatin' by Emil Heitz in 1928 (Heitz, 1928), heterochromatic regions have been

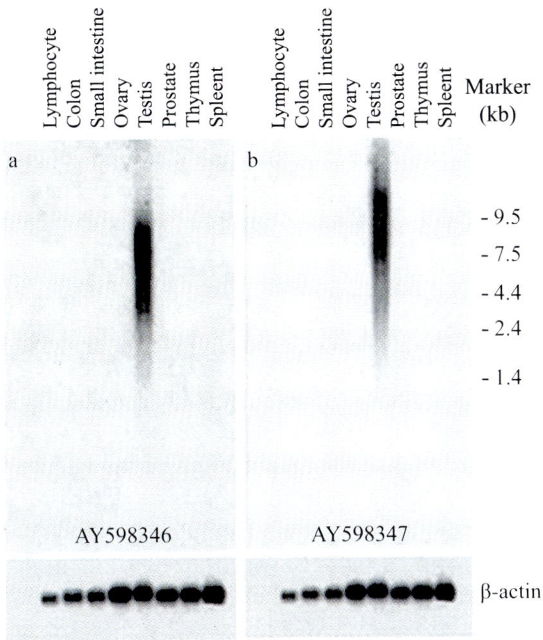

Fig. 12.7 Northern blot analysis of a multi-tissue panel of human polyadenylated RNA preparation probed with cDNA clones AY598346 (panel **a**) and AY598347 (panel **b**) showing testis-specific expression of the corresponding RNAs. Note the difference in size of the transcripts hybridizing to the two probes. Lower panels show hybridization of the same blots to beta-actin (positive control).

accepted as highly condensed loci, which are functionally inert. Constitutive heterochromatin, in fact, is generally considered to be bereft of 'genes'. Facultative heterochromatin is believed to alternate between heterochromatic and euchromatic states, depending on the requirement for transcription of DNA. Thus, the extent of compaction or decondensation of chromatin structure is considered a manifestation of gene activity from the corresponding region (Weintraub and Groudine, 1976; Guttenbach et al., 1989). Secondly, the Y chromosome is often considered a genic wasteland, with only ~20 proteins known to be expressed from this 60Mb chromosome. Finally, the role, if any, of repeat sequences in the normal development and functioning of cells is not yet well established. There are some reports of transcription of repeated sequences of the eukaryotic genome (Diaz et al., 1981; Fornace and Mitchell, 1986; Kimura et al., 2001; Tulin et al., 2002; Jolly et al., 2004, Toure et al., 2005), but no significant roles have been attributed to these elements in terms of regulation of gene expression. It is possible that repeat sequences express functional non-coding RNAs (Mattick and Makunin, 2006).

Thus, the decondensation and subsequent expression from the distal heterochromatic block of the long arm of human Y chromosome comprising repeat elements opens up new questions regarding the significance of heterochromatin and repeat sequences in general, and the Y chromosome in particular. Decondensation and condensation of satellite sequences was previously reported from within the testicular tissues of human and mouse embryos, but transcription was observed only in mouse (Rudert et al., 1995), and not in human (Speed et al., 1995). The observation of development-stage specific expression of RNA, as evident from FISH, Northern blots, and isolation of cDNA clones, indicates that the heterochromatic block composed of repeat sequences in the human Y chromosome is not transcriptionally inert as previously believed.

Genomic Loci for the cDNA Clones

One question that arises is whether the two cDNA clones are parts of much larger repeats present within the genome. To answer this question, the human genome nucleotide database was searched to identify genomic loci showing homology to the two clones isolated from the testis cDNA library. Homology searching identified two sequences, both on the Y chromosome – (a) a 98kb BAC clone, AC068123, and (b) sequence X06228, described as *DYZ1*. Comparison of sequences of the AC068123 BAC clone and the X06228 *DYZ1* sequence showed that the 98kb BAC clone contained within it 27 tandem repeats of the 3.5kb *DYZ1* clone.

A homology search of a human transcript plus genome database (http://www.ncbi.nlm.nih.gov/) using the two cDNA clones yielded a human testicular cDNA named AK128024. The sequence of AK128024 (1986bp) could be aligned with the cDNA clone AY598347 (1441bp) end to end with 92% identity over a stretch of 600bp. Thus these two transcripts appear to be transcribed from a contiguous region. The sequences of AK128024 and AY598347 were merged to yield a long continuous sequence of 2151bp, which we named AK47. The expression of AK128024 and AK47 RNA in human testis tissue was confirmed by RT-PCR (Fig. 12.8).

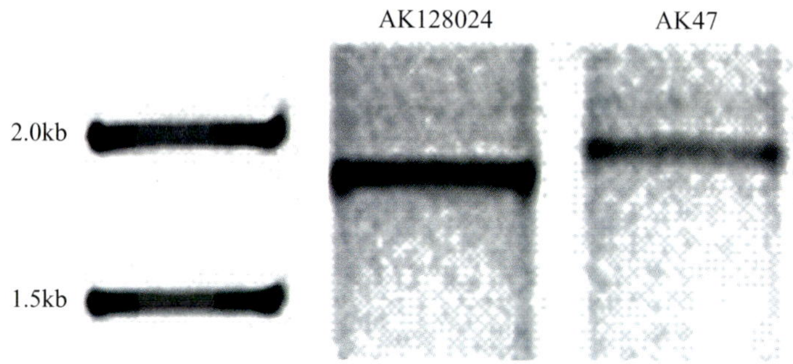

Fig. 12.8 RT-PCR analysis of human testis RNA using primers specific for AK128024 and AK47 RNA.

Comparison of the genomic sequence in the BAC clone AC068123 with the sequences of cDNAs AY598346 and AY598347 showed that the two cDNA sequences spanned the complete BAC clone, alternating with each other repeatedly (represented in Fig. 12.9). Since the BAC clone comprises entirely DYZ1 sequences, it appears that the two cDNAs, AY598346 and AY598347, arise from transcription of the DYZ1 repeats (Jehan et al., 2007).

The only sequence information previously available regarding the *Yq12* region was the presence of repeated copies of *DYZ1* (~ 4000 repeats) and *DYZ2* (~ 2000 repeats) (Manz et al., 1992).

Transcription of the *DYZ1* sequences appears to begin during embryonic stages. The transcripts are confined to the nuclei at embryonic stages but localize mainly in the cytoplasm in adult testicular tissue. Some species of testicular RNAs are transcribed earlier in development and sequestered as translationally repressed messenger ribonucleoprotein particles (mRNPs) until required (Kleene, 2003 and references therein, Schafer et al., 1995). The DYZ1 and DYZ2 transcripts appear to fall in the same category since chromosomal decondensation is observed only in 32 day testis but the RNAs persist through maturation (see Fig. 12.4).

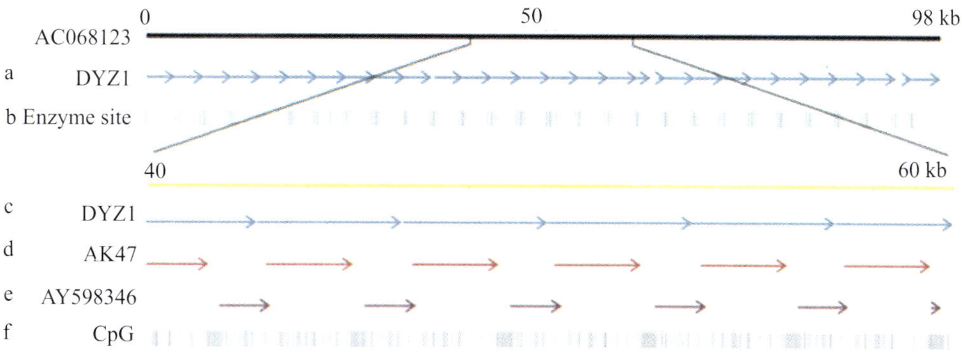

Fig. 12.9 Mapping of DYZ1 repeats and the cDNA clones on to the 98kb BAC clone (AC068123). **a**- the BAC sequence comprises tandem repeats of DYZ1; **b**- position of restriction enzyme sites on the BAC clone; **c**- an expanded view of a 20kb region of the BAC sequence illustrates the tandem repetition of the DYZ1 repeats; **d & e** the two cDNA clones (AK47 and AY598346) alternate with one another on the BAC clone in the same orientation, indicating that they are transcribed from the same strand; **f**- position of CpG dinucleotides within the 20kb region.

The two cDNA clones arising from the *Yq12* region are polyadenylated, as confirmed by sequencing and 3′RACE PCR. It is not clear at the moment whether the heterogeneity in the sizes of the two transcripts is due to abnormally long poly(A) tails or due to multiple start/stop sites or both. Some testicular transcripts are known to have very long poly(A) tails, which are later reduced to the size of other potentially translated poly(A) RNAs found in cytoplasmic fractions (Kleene, 1989).

What do the Two cDNAs Encode?

No coding sequences or long ORFs could be detected in the sequences of the two cDNA clones (AY598346 and AY598347) by computer analysis (Fickett's TestCode (Fickett, 1982); Genescan (Tiwari et al., 1997); GenScan (http://genes.mit.edu/GENSCAN.html)). A conceptual translation of these three sequences did not show homology to any proteins in the sequence databases. In fact, the GenScan program identified some short open reading frames within the 98kb BAC sequence, although none of these putative gene sequences were present on either of the two cDNAs. By these criteria, no single copy gene appears to be buried within the sequence of the 98kb region. Therefore, these two cDNAs appear to be mRNA-like non-coding RNAs (ncRNAs). They are likely to belong to the category of class 2 genes of human Y as per the classification scheme of Lahn et al. (2001), since these are also Y ampliconic sequences with no homology to the X chromosome. Similar transcripts have been described from the euchromatic *Yq* (Foote et al., 1992).

Sequence analysis of the two cDNA clones by the Tandem Repeats Finder (TRF) program (Benson, 1999) revealed a consensus sequence GGAAT. This highly conserved pentameric motif is common to repeat sequences on all human chromosomes, including the human satellite III on pericentromeric regions of chromosomes 9 and 15. Thus, the fact that the two cDNA clones hybridise only with the *Yq12* region upon FISH analysis suggests that the organization of the GGAAT pentamer is different at different chromosome regions. This micro-heterogeneity in repeat elements between various chromosomes may reflect diverse functions they may have within a cell.

Trans-Spliced Transcript with Sequences from Yq12 and 1p36.3

A homology search of the UTR database (http://bighost.area.ba.cnr.it/BIG/UTRHome/) with the 816-nucleotide long sequence of the AY598346 cDNA yielded an unexpected finding. A 67nt region of AY598346 showed 98.5% identity to the entire 67 nucleotide-long 5′UTR of testicular *β SV13* isoform of *CDC2L2* mRNA (AF067525), which is transcribed from chromosome 1 (locus *1p36.3*). The only difference between the two sequences was that a T residue in the AY598346 cDNA was replaced with a C residue at position 57 in the 5′UTR of the *CDC2L2* mRNA (Jehan et al., 2007).

Surprisingly, a homology search of the genomic DNA database with just the 67 nucleotide long sequence from the AY598346 cDNA did not show any identity to the *β SV13 - CDC2L2* gene: this means that this sequence is not present in the cognate genomic locus at chromosome *1p36.3*. In other words, sequence corresponding to the 67nt UTR appears to be present only on the Y chromosome and on the *β SV13* isoform mRNA of *CDC2L2*. This indicates that the *β SV13* isoform mRNA is a chimeric RNA produced by *trans*-splicing involving sequences from the Y chromosome and chromosome 1. Since the AY598346 ncRNA is expressed only in the testis, the putative *trans*-splicing reaction can take place only in testis.

The actual presence of the putative *trans*-spliced transcript was confirmed by RT-PCR of RNA from testis and brain (Fig. 12.10). Primers specific for internal exons of the *CDC2L2 β SV13* isoform yielded a PCR product from RNA from both tissues. However, PCR done with primers spanning the 67 nucleotide 5′UTR and a portion of the *CDC2L2 β SV13* isoform yielded a PCR product only from the testis RNA (Jehan et al., 2007).

The possibility that a cryptic copy of the *CDC2L2 β SV13* isoform gene was present on the Y chromosome also was ruled out by Southern blots of genomic DNA from normal male and female humans. Identical bands were observed in male and female DNA samples digested with two different restriction enzymes using *CDC2L2* as a probe (not shown here; refer to Jehan et al., 2007), indicating that the Y chromosome does not carry a copy of the *CDC2L2 β SV13* isoform gene. In addition, quantitative real time PCR using primers to *CDC2L2* with DNA from individuals with genotypes XO, XX, XY and XYY showed that the strength of the signal was not enhanced by the presence of increasing number of Y chromosomes in the sample. Thus the only explanation for the presence of the testis-specific chimeric RNA is *trans*-splicing. This process is represented in Fig. 12.11.

A testis-specific isoform of *CDC2L2* RNA characterized by its unique 5′UTR had been reported previously (Gururajan et al., 1998). The 67nt 5′UTR reported in their study shows 98.5%

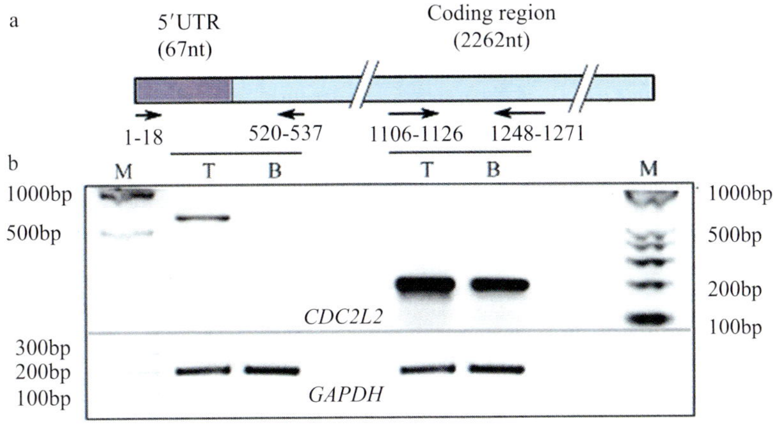

Fig. 12.10 RT PCR shows the chimeric transcript is testis-specific. Panel **a** shows a diagrammatic representation of the *CDC2L2 β SV13* isoform mRNA. Arrows indicate primers used for PCR. Panel **b** shows resolution of PCR products for *CDC2L2* (upper panel) and *GAPDH* (positive control; lower panel) on an agarose gel. M – size markers; T – testis; B – brain. A chimeric product of the expected size was observed only in the testis and not in the brain. Primers to exons common to both brain and testis isoforms of the *CDC2L2* mRNA show amplification in both the tissues.

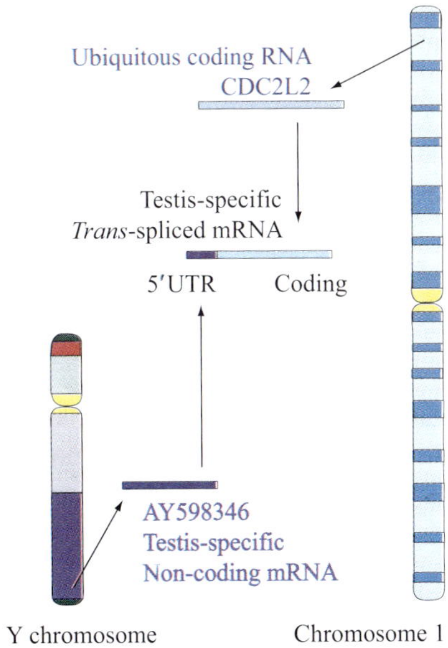

Fig. 12.11 Diagrammatic representation of the testis-specific chimeric RNA generated from the ubiquitously expressed *CDC2L2* mRNA (blue) from chromosome number 1 and the non-coding RNA (purple) from *Yq12* region on the Y chromosome. Note that the two chromosomes, their banding patterns, as well as all the RNA species are not drawn to scale.

identity to the *Yq12* derived 67nt sequence. Splicing of eukaryotic precursor RNAs to produce mRNA is a well-characterized process. Compared to the numerous examples of *cis*-splicing, cases of *trans*-splicing between two distinct mRNAs that result in the production of chimeric RNAs are few (Kikumori et al., 2002; Peatson et al., 2004; Takahara et al., 2002). The generation of mRNA for a testis-specific *CDC2L2* cell division kinase by *trans*-splicing to a non-coding RNA from *Yq12*, which provides the 5′UTR for this mRNA, appears to be a novel regulatory role for the Y-heterochromatin in testis.

Future Directions

This, incidentally, is the first report of putative control of a transcript from an autosome by a product of the Y chromosome. The Y chromosome is generally considered genetically isolated, as it does not recombine with the rest of the genome during the process of spermatogenesis. However, the human Y chromosome apparently interacts with the rest of the genome at least at the RNA level, if not at the DNA level. As this is the first report of such an interaction, it is not known whether this is an isolated case or one of many such interfaces waiting to be discovered. Characterization of *CDC2L2* transcripts in testes of other organisms like mouse, zebra fish and *Drosophila* will reveal whether this is a universal phenomenon. It will also be interesting to determine if any other transcripts are produced from the heterochromatin region of the Y chromosome in other model systems, and determine if these transcripts interact with autosomal RNAs. In addition, it is important to determine what, if any, role the *trans*-spliced UTR plays in the regulation of expression of the *CDC2L2* transcript in the human testis.

The acquisition and retention of sequences that specifically benefit male fecundity and promote spermatogenesis in particular seem to be a global theme in the evolution of the Y chromosome. The present study demonstrates that the heterochromatic block of the human Y chromosome, thus far described as transcriptionally inactive, is transcribed specifically in the testis. One species of non-coding RNA probably modulates translation of the *β SV13* isoform of *CDC2L2* within testis by *trans*-splicing. As some *CDC2L2* isoforms, besides being substrates for caspases, are associated with cyclins, a fine tuning of the cell division related and apoptotic activities of *CDC2L2* in testis is perhaps achieved in-house via the non-coding transcripts from *Yq12* heterochromatin. It is worth recalling that McClintock (1951) had predicted that the study of heterochromatin would lead to the understanding of regulation of euchromatic gene expression. *Trans*-splicing of a part of non-coding transcript from a Y chromosome with an autosome-derived coding mRNA to generate a testis-specific isoform may be one such example of regulation of euchromatic gene expression by heterochromatin.

Acknowledgements

We gratefully acknowledge the contributions by Zeenath Jehan, S. Vallinayagam, Suman Pradhan, Shrish Tiwari, Hemakumar Reddy, and Amritha Suresh. This research was funded by a grant from the Department of Biotechnology, India.

References

1. Bachtrog, D., and Charlesworth, B. (2001). Towards a complete sequence of the human Y-chromosome. Genome Biol. (Reviews) *2,* 1016.1–1016.5.

2. Benson, G. (1999). Tandem repeats finder: a program to analyze DNA sequences. Nuc. Acids Res. *27,* 573–580.
3. Carvalho, A.B., Lazzaro, B.P., and Clark, A.G. (2000) Y chromosomal fertility factors kl-2 and kl-3 of *Drosophila melanogaster* encode dynein heavy chain polypeptides. PNAS (USA) *97,* 13239–44.
4. Diaz, M.O., Barsacchi-Pilone, G., Mahon, K.A., and Gall, J.G. (1981). Transcripts from both strands of a satellite DNA occur on lampbrush chromosome loops of the newt Notophthalmus. Cell *24,* 649–59.
5. Fickett, J.W. (1982). Recognition of protein coding regions in DNA sequences. Nucl. Acids Res. *10,* 5303–5318.
6. Foote S., Vollrath, D., Hilton A., and Page, D.C. (1992). The human Y chromosome: overlapping DNA clones spanning the euchromatic region. Science *258,* 60–66.
7. Fornace, A.J., Jr., and Mitchell, J.B. (1986). Induction of B2 RNA polymerase III transcription by heat shock: enrichment for heat shock induced sequences in rodent cells by hybridization subtraction. Nucleic Acids Res. *14,* 5793–5811.
8. Gururajan, R., Lahti, M., Grenet, J., Easton, J., Gruber, I., Ambros, P.F., and Kidd, V.J. (1998). Duplication of genomic region containing the *Cdc2L1-2* and *MMP21-22* genes on human chromosome *1p36.3* and their linkage to *D1Z2.* Genome Res. *8,* 929–939.
9. Guttenbach, M., Schmid, M., Jausch A., and Vogt, P. (1989). The Y chromosome of the mouse is decondensed in Sertoli cells. Chromosoma *97,* 429–433.
10. Heitz, E. (1928). Das Heterochromatin der Moose. Jahrb. Wiss. Botanik *69,* 762–818.
11. International Human Genome Sequencing Consortium. (2001) Initial sequencing and analysis of human genome. Nature *409*: 860.
12. Jehan, Z., Vallinayagam, S., Tiwari, S., Pradhan, S., Singh, L, Suresh, A., Reddy, H.M., Ahuja Y. R., and Jesudasan, R.A. (2007) Novel noncoding RNA from human Y dital heterochromatic block (Yq12) generates testis-specific chimeric *CDC2L2.* Genome Res. *17,* 433–440.
13. Jolly, C., Metz, A., Govin, J., Vigneron, M., Turner, B.M., Khochbin, S., and Vourih, C. (2004). Stress induced transcription of satellite III repeats. J. Cell Biology *164,* 25–33.
14. Kikumori, T., Gilbert, C.J., and Gagel, R.F. (2002). Naturally occurring heterologous *trans*-splicing cellular transcripts during infection. FEBS Lett. *522,* 41–46.
15. Kimura, R.H., Choudary, P.V., Stone, K.K., and Schmid, C.W. (2001). Stress induction of Bm1 RNA in silkworm larvae: SINEs, an unusual class of stress genes. Cell Stress Chaperones. *6,* 263–272.
16. Kleene, K.C. (1989). Poly (A) shortening accompanies the activation of translation of five mRNAs during spermatogenesis in the mouse. Development *106,* 367–73.
17. Kleene, K.C. (2003). A possible meiotic function of the peculiar patterns of gene expression in mammalian spermatogenic cells. Mechanisms of Development *106,* 3–23.
18. Lahn, B.T., Pearson, N.M., and Jegalian, K. (2001). The human Y chromosome, in the light of evolution. Nature Reviews Genetics *2,* 207–215.
19. Manz, E., Alkan, M., Buhler, E., and Schmidtke, J. (1992). Arrangement of *DYZ1* and *DYZ2* repeats on the human Y-chromosome: a case with presence of *DYZ1* and absence of *DYZ2.* Molecular and Cellular Probes *6,* 257–259.
20. Mattick, J.S., and Makunin I.V. (2006). Non-coding RNA. Human Molecular Genetics *15,* R17–R29.

21. McClintock, B. (1951). Chromosome organization and genic expression. Cold Spring Harb Symp Quant Biol. *16,* 13–47.

22. Peatson, A.E., Eviskov, A.V., Graber, J.H., de Vries, W.N., Hollbrook, A.E., Solter, D., and Knowles, B.B. (2004). Retrotransposons regulate host genes in mouse oocyte and pre-implantation embryos. Dev. Cell *7,* 597–606.

23. Rudert, F., Bronner, S., Garnier, J.M., and Dolle, P. (1995). Transcripts from opposite strands of gamma satellite DNA are differentially expressed during mouse development. Mamm. Genome *6,* 76–83.

24. Schafer, M., Nayernia, K., Engel, W., and Schafer U. (1995). Translational control in spermatogenesis. Developmental Biology *172,* 344–52.

25. Singh, L., Panicker, S.G., Nagaraj, R., and Majumdar, K.C. (1994). Functional significance of molecular organization of sex chromosomes. Proc. Indian Natn. Sci. Acad. *B60,* 455–470.

26. Skaletsky, H., Kuroda-Kawaguchi, T., Minx, P.J., Cordum, H.S., Hillier, L., Brown, L.G., Repping, S., Pyntikova, T., Ali, J., Bieri, T., et al., (2003). The male-specific region of the human Y chromosome is a mosaic of discrete sequence classes. Nature *423,* 825–837.

27. Speed, R.M., Vogt, P., Kohler, M.R., Hargreave, T.B., and Chandley, A.C. (1995). Chromatin condensation behaviour of the Y chromosome in the human testis.1. Evidence for decondensation of distal *Yq* in germ cells prior to puberty with a switch to Sertoli cells in adults. Chromosoma *102,* 421–427.

28. Takahara, T., Kasahara, D., Mori, D., Yanagisawa, S., and Akanuma, H. (2002). The *trans*-spliced variants of Sp1 mRNA in rat. Biochem. Biophys. Res. Comm. *298,* 156–162.

29. Tiwari, S., Ramachandran, S., Bhattacharya, A., Bhattacharya, S., and Ramaswamy (1997). Prediction of probable genes by Fourier analysis of genomic sequences. CABIOS *13,* 263–270.

30. Toure A., Clemente, E.J., Ellis, P., Mahadevaiah, S.K., Ojarikre, O.J., Ball, P.A.F., Reynard, L., Loveland, K.L., Burgoyne, P.S., and Affara, N.A. (2005). Identification of novel Y chromosome encoded transcripts by testis transcriptome analysis of mice with deletions of the Y chromosome long arm. Genome Biology *6,* R 102.

31. Tulin, A., Stewart, D., and Spradling, A.C. (2002). The *Drosophila* heterochromatic gene encoding poly (ADP-ribose) polymerase (*PARP*) is required to modulate chromatin structure during development. Genes & Dev. *16,* 2108–2119.

32. Weintraub, H., and Groudine, M. (1976). Chromosomal subunits in active genes have an altered conformation. Science *193,* 848–856.

Index